Successful Grant Proposals in Science, Technology, and Medicine
A Guide to Writing the Narrative

There are many resources on grant writing in science, technology, and medicine, but most do not provide the practical advice needed to write the narratives of grant proposals. Designed to help novice and experienced investigators write compelling narratives and acquire research funding, this book is a detailed guide to the content, organization, layout, phrasing, and scientific argumentation of narratives.

The authors draw on more than 25 years of research and analysis of grant proposals, having worked extensively with investigators at different levels, from pre-doctoral students to senior scientists. They have used this experience to design a framework for scientific writing that you can apply directly to narratives. The guidelines and advice that they offer are applicable across many funding agencies, including the NIH and NSF. Featuring many real-life examples, the book covers a range of topics, from organizational alternatives to best practices in grammar and editing, overview visuals, and working with contributors.

Sandra Oster is a discourse linguist and a freelance scientific writer and editor through Oster-Edits, specializing in grant proposals and research papers. She has worked for over 25 years in grant writing, 11 years of which were at Oregon Health & Science University. She continues to give seminars on grant writing, research paper writing, and scientific English within and outside the USA. Sandra is also an attorney.

Paul Cordo is a Professor of Biomedical Engineering at Oregon Health & Science University and Chief Technology Officer for AMES Technology, Inc., a medical device start-up company that he founded. He has been funded by NIH grants for over 30 years as a basic and clinical neuroscientist, and brings practical expertise – as a writer, reviewer, and manager of grants – to this book.

Successful Grant Proposals in Science, Technology, and Medicine

A Guide to Writing the Narrative

Sandra Oster
Oster-Edits

and

Paul Cordo
Oregon Health & Science University and AMES Technology, Inc.

CAMBRIDGE
UNIVERSITY PRESS

CAMBRIDGE
UNIVERSITY PRESS

University Printing House, Cambridge CB2 8BS, United Kingdom

Cambridge University Press is part of the University of Cambridge.

It furthers the University's mission by disseminating knowledge in the pursuit of education, learning, and research at the highest international levels of excellence.

www.cambridge.org
Information on this title: www.cambridge.org/9781107038097

First published 2015

Printed in the United States of America by Sheridan Books, Inc.

A catalog record for this publication is available from the British Library

Library of Congress Cataloging in Publication Data
Oster, Sandra, 1949– author.
Successful grant proposals in science, technology, and medicine : a guide to writing the narrative / Sandra Oster, Oster-Edits and Paul Cordo, Oregon Health & Science University and AMES Technology, Inc.
    pages   cm
Includes index.
ISBN 978-1-107-03809-7 (hardback) – ISBN 978-1-107-65930-8 (paperback)
1. Technical writing.   2. Proposal writing for grants.   I. Cordo, Paul, author.   II. Title.
T11.O88   2014
506.8'1–dc23

                                                                2014002478

ISBN 978-1-107-03809-7 Hardback
ISBN 978-1-107-65930-8 Paperback

In memory of:
Werner and Ann Oster
Harriette Oster

Contents

Preface and Acknowledgments

History of Our Collaboration

We met for the first time when Paul, a neuroscientist and biomedical engineer, was a new assistant professor seeking his first NIH R01 grant and trying to get his first sole-author manuscript published. First submissions of both failed, and the critiques focused on unclear writing. Paul contacted a local university, looking for help with scientific writing, and he was directed to Sandra, a discourse linguist with expertise in technical writing and scientific English. After a 2-month, intensive one-on-one training program, he resubmitted both documents. The grant was funded and the manuscript was accepted. Since then, Paul and Sandra have maintained a close relationship, both professionally and as friends. They began a study of narratives to grant proposals that continues to this day.

Fifteen years later, when Paul became the head of his research institute, he hired Sandra as a full-time scientific writer and editor. In 3 years, 21 of the 22 faculty members had at least one NIH R01, and the institute became the second best-funded department in the university.

Sandra and Paul decided to put what they had learned from studying narratives and helping investigators write and revise narratives into book form in order to share this experience with other researchers. This book is the result of that effort.

Our Approach to Writing the Narrative

There are many resources on grant writing on the market, but many focus on grantsmanship, or they review existing information provided in publications from NIH, such as PHS 398, or from NSF. Our focus is on the prose sections of grant applications – the narrative and the abstract.

The purpose of this book is to describe the *genre* of the narrative for investigators. To our knowledge, there has been no concentrated effort to characterize the narrative of the scientific grant proposal as a genre, and we hope our work will motivate others to contribute to this characterization. This work is not definitive, but it takes an important step forward in pinpointing features of content, organization, and scientific argumentation in a narrative, which can help investigators better understand a funding agency's requirements for its narrative and can help investigators compose narratives that hold together logically while meeting those requirements, yet doing justice to the novelty and significance of the proposed research.

To provide a resource that goes beyond existing advice and instruction, we have included in this book many features of narratives and of scientific English that have not been previously identified and described, such as scientific argumentation, organizational alternatives to major sections of narratives, and different types of classic definitions.

By working closely with investigators, we were able to evaluate the descriptive and prescriptive adequacy of the book's content, its clarity and readability, and its usefulness to both novice and experienced investigators. Our grant-writing seminars for investigators provided us with many opportunities to experiment with different informational designs, allowing us to settle on explanations and visuals that present basic and advanced information about narratives and scientific writing, without talking down to investigators and without requiring exceptional reading concentration. With this said, however, this book is dense, is not a fast read, and cannot be digested quickly.

We draw on information from discourse linguistics, rhetoric, and scientific methodology. We seldom provide hard and fast rules on how to organize a narrative or on how to explain proposed research. We provide guidelines and alternatives that investigators can manipulate and apply while they plan, organize, draft, and revise narratives of grant proposals for submission to a wide variety of funding agencies, not just to NIH or NSF.

To a large extent, we focus our descriptions of content and organizational alternatives on scientific argumentation because investigators not only need to describe their proposed research, but they also need to convince reviewers that their proposed research is worthy of a fundable score. Through scientific argumentation – especially through the relationship between the review of previous research and the analysis of it – investigators can help reviewers understand the novelty and significance of their proposed research.

Our targeted readers are investigators who write grant proposals, whether they be scientists, clinician-scientists (e.g., in medicine, dentistry, and nursing), engineers, post-doctoral fellows, or graduate students. We expect professional, scientific, medical, and technical writers and editors will also find this book useful in its breadth of content and its focus on scientific argumentation. Many features of narratives and of scientific English that we identify are also relevant to other types of documents, such as research papers and review articles.

This book was written for both native and non-native speakers of English. From decades of working with both, we have come to realize that both groups need to work on clarity of language and on vocabulary, phrasing, syntax, and the grammar of scientific English. As a result, we include some information on scientific English throughout the chapters, and Chapter 8 focuses on manipulating features of scientific English in order to shorten the text.

Investigators will need to evaluate our guidelines and organizational alternatives in order to decide which to apply when describing their proposed research and when seeking to convince reviewers that their proposed research is significant and novel. In deciding how to apply our guidelines and alternatives, investigators should be guided by 3 types of constraints: (1) the genre of the narrative in scientific grant proposals, (2) the submission requirements from their targeted funding agency, and (3) the unique character of their proposed research.

This book focuses primarily on the first factor – the genre of the narrative. Each chapter provides information on necessary content and organizational alternatives to include in the narrative in order to persuade reviewers of the novelty and significance of the proposed research, and of the abilities of the investigator to execute the proposed research successfully in the proposed time frame and budget.

One of the features of this book that distinguishes it from others is that we provide many examples throughout the chapters. These examples have been excerpted from many grant applications that passed through our hands over the years and others that were funded without our assistance. At times we have lightly edited the examples to reinforce points that we are making.

We are grateful to the contributing investigators who allowed us to use excerpts from their narratives, and we identify them in the following list. We truly appreciate their generosity and kindness, and their patience in responding to our questions. We are also indebted to many other investigators with whom we have worked in seminars and who have provided us with information about their research areas and particular scientific methods.

Albert Agro, PhD	Peter Larsson, PhD
Antonio Baptista, PhD	Catherine Leclair, MD
Bruce Bebo, PhD	David Levitz, PhD
Curtis Bell, PhD	Nannan Liu, PhD
Nathan Bryson, PhD	Karen Lomond, PhD
Paul Cordo, PhD	Owen McCarty, PhD
Paula Deming, PhD	Claudio Mello, MD, PhD
Steven Dow, PhD	Catherine Morgans, PhD
Robert Duvoisin, PhD	Colin Parrish, PhD
Alison Edelman, MD	Robert Peterka, PhD
Alvin Eisner, PhD	Patricia Prelock, PhD
Victor Han, PhD	Hemachandra Reddy, PhD
Steven Hanson, PhD	David Rossi, PhD
Sharon Henry, PT, PhD	Sandra Rugonyi, PhD
Monica Hinds, PhD	Mark Smeltzer, PhD
Fay Horak, PT, PhD	Paul Tratnyek, PhD

Jesse Jacobs, PhD   Tania Vu, PhD

Jeffrey Jensen, MD   Ruikang (Ricky) Wang, PhD

Christopher Keator, PhD   John Weiss, PhD

Esther Klabbers, PhD   Vijayshree Yadav, MD

Special thanks go to Katrina Halliday, PhD, and Megan Waddington, our editors; and Christina Sarigiannidou, our production editor, all of whom have been incredibly patient with us. Even as this book approached maturity, it was still hard for us to let it go without one more review.

We cannot forget our families and friends who graciously gave us the time and assistance we needed to pursue this project – with special thanks to Linda, Emily, and Luke Cordo, and to Rose and Arthur Bloom.

Each chapter ends with figures and tables, and then cases (extended examples).

The Scientific Grant Proposal and Its Narrative

A *scientific grant proposal* is a request for funds to conduct original research in a discipline or a subject area typically associated with science, technology, or medicine. A *principal investigator*[1] (PI) is the researcher who is responsible for the content of a scientific grant proposal and for submitting it to a funding agency on time, and who will be responsible for directing, developing, and executing the research on time. The PI is usually the person writing most of the narrative, if not the entire scientific grant proposal. The narrative comprises the major prose sections of the scientific grant proposal, including the abstract.

There may also be *co-investigators* (Co-Is) who help the PI design and develop the proposed research and write the narrative, and who will help the PI execute the research. Sometimes there are *collaborators* who will perform a significant part of the research and are, to a certain degree, responsible for designing their limited portion of the proposed research, subject to the PI's approval. Finally, there may be *contractors* hired to perform a relatively small portion of the research, but who have not helped in designing or developing the research project.

---

[1] Depending on submission requirements, close synonyms to *principal investigator* are *principal researcher, project director*, and *program director*.

There are many very good resources on the market dealing with strategies and other information to select funding mechanisms from public and private agencies that support scientific, technical, and medical research. However, there are far fewer resources dealing with how to write the narrative of the scientific grant proposal that describes and argues for the proposed research.

The focus of this book is on the narrative of a scientific, technical, or medical grant proposal that is typically submitted to a funding agency by an individual PI (and possibly involving Co-Is and collaborators) who is proposing novel research and who usually has preliminary data to support the grant proposal.[2] This book provides guidelines to help PIs describe the narrative and to argue for the scientific merit of the proposed research. You need to convince reviewers that you are credible and that if you are given the opportunity to execute your research as proposed, you will produce novel, valid, significant, and relevant findings.

The information in this book about writing the narrative can be applied to a variety of grant proposals from public and private funding agencies that support scientific, technical, and medical research. The first chapter introduces terms and concepts that will be used and expanded upon in later chapters.

1.1 Funding agencies

The PI submits a grant proposal to a funding agency, which applies criteria to decide which research projects to fund. These criteria encompass the scientific merit of the proposed research as evaluated by reviewers, research priorities within the funding agency, and the availability of funds.

Each funding agency has its own *submission requirements*, sometimes called *submission guidelines*, for narratives, and you must follow them for serious consideration. There are many similarities in submission requirements across funding agencies, but there are notable differences. Before writing the narrative, you need to read these guidelines, become familiar with them, and follow those that are required and seriously consider following those that are recommended, regardless whether you agree with them and whether guidelines in this book are consistent with them.

> Guideline 1.1   *If any of the information in this book conflicts with the submission requirements from your targeted funding agency, follow those from the funding agency.*

1.2 Your reviewers

Each funding agency has its own procedures and criteria for selecting reviewers to serve on review panels, and each gives reviewers evaluation criteria for determining the merit of scientific grant proposals. Some panels consist exclusively of technical

---

[2] Some grant proposals do not need preliminary studies, such as an R21 from National Institutes of Health.

reviewers who are your peers in your discipline; others consist of both technical and lay reviewers. When submitting a grant application, you need to keep in mind that the funding agency, not the reviewers, ultimately approves research projects for funding. As a consequence, when writing a grant proposal, you need to satisfy 2 types of readers: the reviewers and the funding agency.

Guideline 1.2 *Write scientific grant proposals for 2 readers: the reviewers and the funding agency.*

The less technical and less knowledgeable the reviewers are in your proposed area of investigation, the more you need to lower the technical level of your explanations, definitions, vocabulary, and visuals. However, even if your reviewers have your approximate education, skills, and experience in your area of investigation, often times you will need to explain information to them – *information that you know they already know* – in order to facilitate their evaluation of you and your proposed research.

Guideline 1.3 *Do not assume that the reviewers' background in science, technology, or medicine lessens your responsibilities as an investigator to explain experimental fundamentals in your proposed research.*

There will be times in this book when you will be encouraged to explain something to your reviewers that they may already know.

Basic, safe assumptions about your reviewers – whether technical or lay – are that they: (1) are educated beyond high school for lay reviewers and well beyond undergraduate studies for technical reviewers, (2) are fluent readers of English, whether English is their native language or not, and (3) may likely be reading your narrative under less than ideal circumstances, when their reading comprehension might not be in top form, such as when they are tired or stressed. In brief, narratives to scientific, technical, and medical grant proposals need to be written with tired and stressed reviewers in mind, as explained in Guideline 1.4:

Guideline 1.4 *Structure your narrative for tired or stressed reviewers, and write:*
(a) *As **clearly** as possible so that your reviewers can readily understand your prose without a lot of mental effort.*
(b) *As **concisely** as possible so that your reviewers do not need to read through extra words and sentences to access your key information.*
(c) *As **consistently** as possible, repeating key terms, so that your reviewers can follow your discussion.*
(d) *With **details** so that your reviewers can understand precisely what you are proposing and can evaluate its scientific merit.*
(e) *With **research objective**, **aims**, and **hypotheses** made obvious.*
(f) *With transparent **layout** and **organization** so that the arrangement of information reinforces your research design, contributes to your credibility, and helps reviewers anticipate, locate, and retrieve information from the narrative with little effort.*

Reviewers take their role of reviewer seriously. They *try* to understand your narrative. However, only you can be sure of what you really mean. Remember:

● **Reviewers cannot read clarity into an unclear text.** It is your responsibility to write clearly and to make your topics and themes clear.

● **Reviewers cannot read quickly through a narrative that is not concise.** It is your responsibility to write as concisely as possible while maintaining clarity and representing your proposed research.

● **Reviewers cannot supply details for general or vague passages.** It is your responsibility to write precisely.

● **Reviewers cannot perceive the structure of your content where there is no or little organization.** It is your responsibility to organize and to lay out your text in a way that reviewers easily understand the narrative and can readily access its information.

This book provides specifics on how to write clearly, precisely, concisely, and with transparent organization and high readability.

An important concept about the review process for narratives of scientific, technical, and medical grant proposals is that it is influenced by social and political realities: reviewers often have biases about how grant proposals *should* be written. You need to write in a style associated with a person of higher education, not only because this style will help you express your ideas clearly and precisely, but also because your reviewers expect such a style from a PI who has achieved a high level of education. In this book, this style is called *scientific English*.

1.3   Scientific judgments

Your proposed research will most likely not be considered meritorious if you fail to convince reviewers that you are credible; that your research will produce novel, valid, significant, and relevant results; and that you will follow accepted legal, administrative, and ethical standards in the execution of your methods. Tables 1-1 and 1-2 present review criteria for scientific grant proposals targeted for National Institutes of Health (NIH) and National Science Foundation (NSF). These review criteria are generally representative of criteria for many funding agencies. Each review criterion reflects at least one issue related to credibility, validity, novelty, significance, and relevance.

● **Credibility**. Credibility deals with how believable you are and how you impress your reviewers in terms of your qualifications to execute the proposed research successfully and on time. You need to convince your reviewers that you and your research team have the scientific and technical knowledge, skills, experience, and intelligence to: (a) timely and successfully execute the proposed methods, (b) analyze and interpret the results logically and objectively, and (c) disseminate results from your research.

● **Validity**. You need to convince reviewers that your research design and methods will yield accurate, reliable data that will likely support your hypothesis. Thus, in the narrative, you need to present a precise, reasonable hypothesis based on your discipline's current body of knowledge about your research topic. You also need to identify, justify, and explain your methodological approach. With this information, reviewers should be able to evaluate your proposal and determine whether your experimental approach involves confounding variables that will likely compromise the assessment of your results – results that should be based on sound reasoning and that should advance scientific and technical knowledge in novel and significant ways.

● **Novelty**. In the narrative, you need to convince your reviewers that your methodological approach will lead to the discovery of new information in your discipline. The perception of novelty emerges from an effective review and analysis of the previous research on which your proposed research is based.

● **Significance**. You also need to convince your reviewers that your research results will be *important*, as judged against current scientific and societal knowledge, norms, and values. You need to argue that your discoveries will ultimately have a substantial, positive impact on your discipline and/or on the larger domain of science, technology, or medicine; and on society in general.

● **Relevance**. Your proposed research needs to be appropriately targeted to the types of research that your funding agency funds and to the appropriate funding mechanism. There may be times when your proposed research may be appropriate for more than one funding unit within a funding agency, in which case you need to decide which funding unit should receive your grant proposal. Likewise, there may be times when your proposed research may be appropriate for more than one funding mechanism, with some tweaking of the research approach and design.

● **Legal, administrative, and ethical standards**. You need to provide reviewers with information for them to conclude that you are following current legal, administrative, and ethical standards expected of investigators. For example, if your experimental design calls for subjects, you must discuss at what point in the proposed procedures that you will inform them of potential risks and will seek informed consent. For subjects, you may also need to ensure that the ethnic, racial, and gender composition of your subject groups is representative of the local population. Equally important, you may need to describe how you will achieve and control the confidentiality of each individual subject's identity and data.

1.4 Persona of a scientist

In the narrative to a scientific grant proposal, you need to sound like a scientist. If you cannot sound like a scientist, your credibility will be questioned.

Regardless whether you are an investigator in basic, applied, clinical, or translational research, you need to evaluate previous research and describe your proposed research in ways to reflect your familiarity with scientific methodology. You also need to be accurate – that is, to be as factually correct as possible and not to give the appearance of subjectivity. You need to support conclusions, substantiate the likely accuracy of your claims, and indicate that you have not designed your research arbitrarily. In other words, you need to indicate that you carefully crafted a research design that controls or eliminates variables that may influence the data that you will collect, that may affect your analysis of the collected data, and that may ultimately impact your results and outcomes.

> **Guideline 1.5** *Through your narrative, you need to convey that:*
> *(a) You are a highly trained scientist who is in control of scientific methodology.*
> *(b) You have support for your conclusions.*
> *(c) You are aware that your claims need substantiation.*
> *(d) Your research design is not arbitrary.*

Throughout the following chapters, guidelines are offered to help you describe your proposed research, not only in a conservative writing style of an educated person but also in a style of a scientist.

1.5   Constraints on the narrative

When you prepare to write a narrative, you need to consider 3 major constraints that influence your decisions about its content, organization, phrasing, and layout: (1) the genre of the narrative in scientific grant proposals, (2) current submission requirements from the funding agency, and (3) the novelty of your proposed research.

1.5.1   Overview of constraints

Most chapters in this book address the 3 major constraints that will influence the design and composition of your narrative. Each constraint is introduced here.

● **Genre constraints.** By the submission date, your narrative should look like a typical narrative that other professionals in your field would submit – typical in terms of appearance, organization, content, and phrasing. Of course, the key question is, what are these typical features? This book provides information to help you better understand the genre of the narrative to a grant proposal in science, technology, and medicine; and to help you draft the narrative. However, if you follow only the information in this book, your narrative will likely be unsuccessful because your narrative also needs to meet constraints from the funding agency and those imposed by the novelty of your proposed research.

● **Constraints from a funding agency.** Constraints from a funding agency are the *submission requirements* that a funding agency publishes. By the time you submit your scientific grant proposal, the narrative needs to follow all submission requirements from your targeted funding agency – yet still reflect genre constraints and do justice to the novelty of your proposed research. Submission requirements range from content specifications to format, organization, and font specifications. Two major federal funding agencies with extensive submission requirements are NIH (www.nih.gov/ grants/funding/phs398/phs398.html [November 19, 2014]) and NSF (http://www.nsf. gov/pubs/policydocs/pappguide/nsf13001/gpgprint.pdf [November 19, 2014]). Many other funding agencies adopt features of current and previous NIH, and NSF submission requirements as their own.[3]

● **Constraints from your novel proposed research.** Constraints on the narrative are not only imposed from the genre and the funding agency, but also from the novelty of your proposed research. This constraint implies a document design that allows you to describe novel aspects of your proposed research while your description follows submission requirements and reflects the genre of the narrative.

The need for the narrative to reflect all 3 constraints is one reason why the narrative is hard to write. Guideline 1.6 offers suggestions for how to draft a narrative that respects all 3 constraints:

> **Guideline 1.6**   *Write a narrative that respects and reflects constraints from the genre, funding agency, and the novelty of your proposed research design by:*
> (a) *Including specific information that is phrased with terminology from the funding agency's submission requirements and your discipline.*
> (b) *Organizing and producing headings for the **major sections**, according to the genre of the narrative, unless submission requirements specify otherwise.*
> (c) *Organizing and producing headings for **subsections** to indicate the content and novelty of your proposed research and its historical context, unless submission requirements specify otherwise.*
> (d) *Drafting sentences that explicitly indicate the significance and novelty of your proposed research.*

1.5.2   Generic content and typical sections

Most of this book focuses on features of a generic narrative to a scientific grant proposal. When you write the narrative, you will need to manipulate these features in order to produce a narrative that is responsive to the genre of the narrative, submission requirements, and the novelty of your proposed research.

Table 1-3A presents generic information for the narrative. Different funding agencies may use different terms for this information in their submission requirements. For example, one funding agency may use the term *objective* for the purpose of the proposed research; another may use the term *goal*. Regardless of the terms used by funding agencies to describe their required content, most require similar information. In addition, most

---

[3] For example, the National Multiple Sclerosis Society notes that "(Our) template parallels NIH format." (http://www.nationalmssociety.org/for-professionals/researchers/get-funding/research-grants/index.aspx; Download instructions for online submission of research grant applications; accessed November 19, 2014)

funding agencies identify the major sections of a narrative that they want, although they sometimes name and organize these major sections differently. Table 1-3B identifies different names for major sections of a narrative. The order that funding agencies want these sections sequenced may also vary. For example, the purpose of the proposed research is often first presented in a major section that is variously entitled **Aims**, **Specific Aims**, **Introduction**, or **Research Objective**. Sometimes, however, a funding agency may want the narrative to open with a section that provides background information.

If the submission requirements of your funding agency do not specify the names it prefers for the major sections, you can select section names from Table 1-3B. The following list gives an overview of generic content in the narrative.

- **Topic**. The topic is the specific area, problem, or question that you intend to investigate in the proposed research. The topic is identified in the **Aims Section** and is reiterated throughout the narrative in the same key terms as initially used in the **Aims Section**.

- **Research purpose**. The research purpose is an explanation of what you intend to accomplish in your proposed research. Although funding agencies vary somewhat in the vocabulary they use to describe the research purpose, there are 3 levels of research purposes that can be associated with proposed research: (1) the long-term *goal*, (2) the proposed *research objective* of the current research, and (3) the proposed scientific and/or technical *aims*. The research purpose(s) is first identified in the **Aims Section** and is usually repeated in the **Methods Section**. Chapter 2.2.4 discusses the research purpose in detail.

- **Background**. Background information consists of your and other researchers' relevant previous research that directly relates to your proposed research. You explain this previous research in order to place your proposed research into an historical research context and to provide a backdrop against which to claim novelty and significance for your proposed research. Some funding agencies require background information to be located in one major section that is named, for example, **Background Section** or **Literature Review**. Other funding agencies may require 2 separate sections for background information: (1) a major section presenting primarily other research-ers' relevant research, which is traditionally termed **Background**, and (2) a major section describing only your relevant previous research, usually called **Preliminary Studies** or **Progress Report**. Background information is also briefly included in the **Aims Section** and in a rationale subsection(s) in the **Methods Section**. Chapters 2, 3, 4, and 5 describe how to explain background information.

- **Significance**. Funding agencies expect you to explain the significance – that is, the importance – of your proposed research. Significance is explained with, and emerges from, background information. Funding agencies, such as NSF, sometimes require significance in dedicated (separate) sections or subsections, called **Broader Impacts** and **Technical Merit**. Other funding agencies, such as NIH, require that significance be explained in a separate background section, entitled **Significance**. Chapter 2.2.5 further explains significance information.

- **Novelty**. Novelty is how your particular topic of research is innovative, unique, or distinct from previous research. Novelty is explicitly identified or emerges from your review of background research. Novelty is first identified in the **Aims Section**, is explained in the **Background Section** and the **Preliminary Studies Section**, and is again mentioned in the **Methods Section**. Some funding agencies require a separate section for novelty. (See Chapter 2.3 and the relationship between analysis in scientific argumentation and novelty.)

- **Proposed work/methods**. *How* you intend to accomplish your research objective and aims is your methods. *How* includes your research design; the focus of your research, such as subjects, objects of study, or specimen; materials, equipment, and tools that you will use in executing your methods; the procedures you will perform over the funding period to achieve your research objective and aims; and analytical procedures for the data that you will collect. Your proposed work is first very briefly characterized in the **Aims Section** and is later described in the **Methods Section**. The **Methods Section** is covered in Chapters 5 and 6 of this book.

- **Investigators' qualifications**. A part of most grant proposals is a curriculum vitae, a biosketch (biographical sketch), a bioparagraph, or a résumé that identifies the key personnel, including the PI; and their education, professional experience, and publication history. A career or training grant might also have a major section of the narrative in which the PI addresses these points. Reviewers use information to assess the investigators' credibility, qualifications, and skills in order to judge whether they will successfully perform their proposed methods and achieve their proposed research objective and aims in a timely manner.

1.6   The NIH model

The organization of sections and section names selected for the generic narrative of a scientific proposal in this book are closely similar to those of the pre-January 25, 2010, NIH narrative.

NIH is the major funding agency in the United States in terms of dollars that Congress allocates for academic research and development. In fiscal year 2013, the NIH had a budget of $29.15 billion, in contrast to the amount allocated to NSF, at about $6.9 billion.[4]

The sections to the NIH narrative were changed to a certain degree for most narratives of grant proposals submitted to NIH on or after January 25, 2010. Table 1-4 gives the previous and the current overall organization of the narrative to an NIH grant

---

[4] http://nexus.od.nih.gov/all/2013/05/08/funding-operations-for-fy2013 (accessed November 19, 2014) and http://www.nsf.gov/about/congress/113/highlights/cu13_0409.jsp (accessed November 19, 2014). However, NIH funding is primarily for human and health services, in contrast to NSF funding, which is primarily for research in the physical sciences and engineering.

proposal. The most distinctive changes (arguably) for most grant proposals submitted to NIH on or after January 25, 2010, follow in this list:

● **Specific Aims Section**. The previous NIH **Specific Aims Section**, which was a recommended length of 1–2 pages, must now be no longer than 1 page for the majority of NIH funding mechanisms.

● **Background Section**. The pre-January 25, 2010, NIH **Background Section** is now 2 background sections, termed the **Significance Section** and the **Innovation Section**.

● **Preliminary Studies/Progress Report Section**. The previous NIH **Preliminary Studies/Progress Report Section** is now a subsection – usually the first subsection – in the **Approach Section**.

● **Research Design and Methods Section**. The previous NIH **Research Design and Methods Section** is now in the **Approach Section**.

● **Introduction to the Competitive Renewal**. The previous NIH **Introduction for Competitive Renewal** has been reduced from 3 pages to 1 page for most resubmitted NIH funding mechanisms. (See Chapter 7.2.)

● **Page length**. Most NIH narratives (in particular, the R01) have been reduced from the former 25 pages to 12 pages, plus the one-page **Specific Aims Section** and an **Introduction** for a resubmission.

The pre-January 25, 2010, NIH model captures basic content of the narrative with organizational simplicity (e.g., Table 1-4A) and allows PIs to address typical content found in narratives to scientific grant proposals (including the current NIH model's focus on significance). Perhaps due to these features, many funding agencies in the United States, such as the American Heart Association[5] (see Table 1-5C), still basically follow the pre-January 25, 2010, NIH model.

If your targeted funding agency does not specify the organization and the names of major sections to its narrative, you can choose among those in Table 1-5A–1-5D. You can also create your own unique arrangement for this high-level organization, but your end result needs to be a document that reflects the 3 constraints on the narrative discussed in Chapter 1.5.1 – constraints imposed by the genre, the funding agency, and the novelty of your own proposed research.

The generic content and organizational features of the narrative to scientific grant proposals identified in Table 1-5 are based on narratives submitted to funding agencies in the United States. These generic features may need modification before submitting the grant proposal to funding agencies outside the United States.

---

[5] Note, however, that in addition to the pre-January 25, 2010, NIH sections (**Specific Aims, Background and Significance, Preliminary Studies**, and **Research Design and Methods**), the American Heart Association also requires a section entitled **Ethical Aspects of the Proposed Research**. See http://my.americanheart.org/professional/Research/FundingOpportunities/SupportingInformation/Creating-the-Research-Plan-for-Strategically-Focused-Research-Networks–17-Pages_UCM_450711_Article.jsp (accessed November 28, 2014).

1.7   Selecting an organization for the narrative

As earlier mentioned, this book adopts the basic organization and major section names for a narrative submitted to an NIH narrative before January 25, 2010 (Table 1-4A), except that this book adds an ending section, **Future Plans**, as shown in Table 1-5A.

Chapters 2 through 6 of this book describe the major sections and different types of generic content of the narrative. Chapters 7 and 8 describe other prose sections, including the abstract, and other critical features of the narrative, such as citations and strategies to shorten the narrative in order to meet length requirements or to make room for additional information.

Chapter 2.   The Aims Section[6]
Chapter 3.   The Background Section[7]
Chapter 4.   The Preliminary Studies/Progress Report Section
Chapter 5.   The Methods Section, Part 1
Chapter 6.   The Methods Section, Part 2
Chapter 7.   Other Prose Considerations
Chapter 8.   Technical Features of Sentences

1.8   Readability

You can include all of the right kind of information in the narrative, yet reviewers might still have significant problems in understanding your proposed research. To describe and to argue for your proposed research, you need to write a highly readable text. The term *readability* refers to the mental and physical effort that readers – most notably your reviewers – need to exert in order to readily understand your text on a first reading, to follow the logic of your discussion on a first reading, and to locate and quickly retrieve particular information on a subsequent reading.

Text that has *high* readability is relatively easy to understand and to follow on the first reading. Text with *low* readability requires a lot of concentrated reading, mental effort, and possible guessing on the first or the first few readings in order to understand, follow, or locate information.

---

[6] In this book, the name **Aims Section** is used instead of the **Introduction Section** for the first major section of the narrative because funding agencies, such as NIH, use the term **Introduction** to refer to a preface to the narrative that is required upon resubmission of a grant proposal. See Chapter 7.2 for a discussion of an NIH **Resubmission Introduction**.

[7] NIH divides its former **Background Section** into the **Significance Section** and the **Innovation Section**.

Guideline 1.7 *Aim to write a narrative that promotes high readability.*

When reviewers find themselves struggling to understand content, they often assume that the problem is in the content, not in the writing. As a result, they might dismiss your ideas, or they might not take your ideas as seriously as they would if you had written the narrative differently. In addition, when reviewers find themselves understanding your content but noticing what they consider to be problems in how it is written, they might question your education, research skills, and credibility.

And when reviewers have to struggle in order to understand the content of your narrative, the added effort and time makes them less likely to be enthusiastic about the proposed research or less inclined to give you the benefit of the doubt when they are not certain about the accuracy of your assertions.

Guideline 1.8 *For your grant proposal to receive a fundable score, you need to excite your reviewers about your proposed research and to motivate them to champion your proposal throughout the evaluation process.*

Although readability might seem like a minor issue, it is not. As we discuss in this and subsequent chapters of this book, many factors affect the readability of the text.

1.8.1 Readability, layout, and font

Reviewers have many grants to evaluate in a relatively short time, and they need to move quickly to retrieve information from various sections when they discuss grant proposals with other reviewers or program directors. You can help them read, understand, and locate information in the narrative by laying out the text in ways to enhance readability.

Layout is a visual characteristic of a page involving the arrangement of text and images relative to the white space. Effective layout helps reviewers remember your organization and arguments.[8] Guideline 1.9 gives strategies for using white space to help reviewers quickly perceive the organization of the content, which can help them access the content:

Guideline 1.9 *Use white space to help reviewers concentrate, read, and understand the text quickly and to help them remember where information is located for later retrieval.*
(a) Skip lines between sections, subsections, and paragraphs (however, see Chapter 8.6.2).
(b) Single-space text within paragraphs.
(c) Indent the first line of a paragraph 4–5 spaces.

---

[8] www.nichd.nih.gov/about/meetings/2010/Documents/Hopmann_Scholars_Workshop.pdf (accessed November 28, 2014).

(*d*) *Locate key information in the focus positions of paragraphs and subsections.*
(*e*) *Keep paragraphs relatively short, such as less than 10 lines.*
Exception: *Use relatively longer paragraphs to bury information within the middle of the paragraph.*

Readability can be greatly affected by white space within and between paragraphs, subsections, and sections. However, a writing problem emerges: using white space effectively to enhance readability may extend the narrative beyond the prescribed maximum length. Because of this problem, Chapter 8 gives strategies to shorten the text. Guideline 1.9 is explained in the following list.

● **White space within and between paragraphs**. The text in the narrative is usually single-spaced, unless submission requirements specify otherwise. The little bit of white space provided by an indented first line of a paragraph[9] or a skipped line between paragraphs will help your reviewers read more quickly and more readily understand the organization and the hierarchical relationship across sections, subsections, and paragraphs. Also, indenting or skipping a line between paragraphs contributes to a more readable text. However, both indenting paragraphs and skipping a line between paragraphs may increase the total length of the narrative beyond the length limit specified in the submission requirements (see Chapter 8.6.2).

● **Paragraph length**. A relatively short paragraph – such as one less than 10 lines – can help your reviewers read the paragraph without slowing down or losing concentration in its middle sentences. In contrast, a relatively long paragraph, such as one 15 lines or longer, can slow readers, can make them work to maintain concentration, and can help to bury information located in middle sentences of the long paragraph.

You can divide a relatively long paragraph into 2 or more shorter paragraphs to improve readability. You should evaluate the content, argument structure, and phrasing of sentences and then divide it: (a) into passages of related information, where each chunk – that is, each related group of sentences – covers a separate topic; (b) before a sentence that shifts to a different grammatical subject or topic – especially if the grammatical subject is *we*; (c) after a sentence beginning with *therefore*, *thus*, or *consequently*, in order to more closely associate the effect, result, conclusion, or consequence with its preceding premise or cause; and (d) after a ***review-analysis*** pair or after an ***analysis-review*** pair in a scientific argument in order to keep the related information together (see Chapter 3.6).

Case 1-1 presents a 27-line **Preliminary Studies** subsection consisting of one, 16-sentence paragraph. This paragraph can be divided into shorter paragraphs to improve its readability and to make its organization and scientific argument clearer.[10] The symbol ¶

---

[9] Note, however, that the first paragraph in a (sub)section is often not indented.
[10] For improved readability, this new paragraph (sentences 2 to 14) can be revised in either of two ways: (1) it can be further divided into 2 review paragraphs, at sentence 12; thus, sentences 3–11 would comprise one review paragraph, and sentences 12–14 would comprise a second review paragraph; or (2) the figure can be included in the paragraph rather than after it (not shown).

indicates where the paragraph can be divided, to create chunks of related information and topics. Sentence 1 identifies the topic and the purpose of the preliminary research. Sentences 2 to 14 review preliminary research and results, and sentence 15 analyzes the results by offering a conclusion about the methods. Sentence 16 shows the connection between the preliminary methods and the methods proposed for aims 1 and 2.

Once you divide a relatively long paragraph into shorter paragraphs, you need to examine the first sentences of the resulting shorter paragraphs. If these first sentences use personal pronouns (such as *she, he, it, they,* or *them*) or demonstrative pronouns (such as *this* or *these*), and if their antecedents are now in the previous paragraph, you need to repeat the antecedent term rather than use the pronoun. An exception to the use of pronouns in the first sentence of a paragraph is *we*, which is acceptable because its antecedent is not in the preceding text; it refers to a member(s) of the research team who will be conducting the proposed research.

A problem with creating 2 relatively short paragraphs is that the text will use more textual space than will the paragraphs if combined into one longer paragraph. Thus, there is tension between creating short paragraphs for readability and combining the short paragraphs into a longer paragraph to conserve space.

There are ways to compensate for the lower readability of longer paragraphs that you do not subdivide as explained in Chapter 8.6.2.

● **Paragraph and subsection foci.** You can help draw the reviewers' attention to particular information in a paragraph or a subsection by positioning the information close to white space rather than in the middle of a paragraph. The focus positions of a paragraph are: (1) in the first couple sentences of a paragraph, (2) in the last one or 2 sentences of a paragraph, and (3) in a short paragraph of just one or 2 sentences. In this book, positions (1) and (2) – the beginning and ending of a paragraph – are termed the *focus positions* of a paragraph.

Since focus positions are adjacent to white space, if you want readers **not** to notice particular information, you place that information in the middle of a longer paragraph, such as one 15 lines long, in order to bury it. Such information might be problematic preliminary data that you need to provide but that you hope reviewers will not especially notice.

● **Emphatic font.** Another way to draw the reviewers' attention to information in the middle of a longer paragraph that you do not subdivide is to use emphatic font or to use underscoring in order to highlight particular information or organizational cues. Chapter 8.6.2 discusses emphatic font that is used to compensate for the lower readability of longer paragraphs.

1.8.2   Readability and font

You can decrease or improve the readability of the narrative, depending on your font selection. With the ease of changing fonts that word processing programs provide, you might be tempted to use a variety of font styles in a grant proposal. Readability can be affected by the number and types of font that you use and how they are used. A wide variety of font styles, or a couple of font styles used inconsistently, can distract reviewers from the content and can lower the readability of your text. Guideline 1.10 presents strategies for using font:

> **Guideline 1.10** *To avoid visual distractions, consider consistently using 4 or fewer styles of font throughout the narrative:*
> *(a) Use one standard, non-bold and non-italicized font for the majority of your text, such as standard Times Roman 12 point or Arial 11 point.*
> *(b) Use one type and size of bold font for headings, such as Arial bold 11 point.*
> *(c) Use one visually distinctive font for emphasis, such as **bold italics**.*
> *(d) Use conventional font that is discipline-specific, such as lower-case italics for names of genes.*

Funding agencies often have specifications about accepted styles and sizes of font. If a font size is not specified, you can use an 11- to 12-point font for most of the narrative. Text in a smaller font may be harder for reviewers to read quickly – even if they use reading glasses.

1.8.3 Readability and lists

Lists help organize information in sentences and paragraphs. Thus, they can improve the readability of the text by providing white space and structuring the content. A list can also provide an ***itemization scheme*** to help readers identify or retrieve discrete items in the list.

> **Guideline 1.11** *Use lists to help organize information in sentences and paragraphs.*

A list consists of 2 parts: (1) a **lead-in**, which announces what the list is about and sometimes identifies the number of items in the list, and (2) the **listed items**. The listed items need to be in parallel grammar (see Chapter 8.6.3).

1.8.4 Readability and vocabulary

While many features of vocabulary are addressed throughout this book, 2 key features are needed for a clear and highly readable narrative: (1) consistent vocabulary and (2) standard vocabulary.

> **Guideline 1.12** *Throughout all sections of a narrative, use vocabulary consistently and use standard vocabulary.*

● **Consistent vocabulary**. For clarity and readability, you should use consistent terms for key concepts throughout the entire narrative of the grant proposal. This means that you should repeat **key terms** rather than use synonyms or pronouns for the key terms. For example, if you refer to *aims* in the **Aims Section**, then throughout the narrative, you need to use the key term *aims*, not *specific aims*, not *SAs*, and not *objectives*. By repeating key terms, you make it easier for reviewers to recognize discrete concepts that you are discussing.

> **Guideline 1.13** *Repeat key terms; use synonyms and pronouns sparingly.*

Before writing the narrative, you need to decide on the key term that you will use for each key concept, and then you need to use each key term consistently.

- **Standard vocabulary.** To enhance your credibility and to help reviewers readily understand your narrative, you should use standard terms from your discipline.

> **Guideline 1.14** *Use standard terms; use uniquely defined terms sparingly.*

A **standard term** is a vocabulary item that has one precise meaning and that has been accepted by the professionals in your discipline for a particular concept. It is appropriate and expected that you use standard vocabulary in the narrative to reflect your familiarity with and your understanding of the discipline and to use the language of your peers. Using standard terms, however, does not give you license to use unnecessarily technical terms or to omit explanations or definitions.

1.9   Scientific argumentation, the scientific method, and the narrative

In the narrative to a scientific grant proposal, you need to describe your proposed research in such a way that you convince your reviewers and the funding agency that your proposed research has scientific merit, that you are capable of executing the research, that your proposed research reflects sound scientific methodology, and that your results will be significant, novel, valid, and relevant. NIH notes the need "for [reviewers] to be convinced that there is a reasonable probability that YOU, using your approach, your resources, and your collaborators, can do it."[11]

You can convince your reviewers through the use of *scientific argumentation*. As used in this book, the term *scientific argumentation* consists of particular content that describes your proposed research and clarifies its significance and novelty in order to convince reviewers that your proposed research reflects features of sound scientific methodology, is meritorious, and should receive a fundable score. A scientific argument is first briefly described in the **Aims Section**, and then it is further detailed and elaborated upon in subsequent sections of the narrative.

---

[11] Video at http://enhancing-peer-review.nih.gov/application_changes_video.html (accessed November 28, 2014).

A scientific argument consists of 5 types of content, which are represented by the abbreviation **TRACS**.

(1) Topic – The specific area, concept, or question that you are investigating or discussing.

(2) Review of previous research – Methods and scientific observations (collected data and results) from your and others' research that identify what is currently known in the scientific, technical, or medical literature about your research topic.

(3) Analysis – Your educated opinion about what is not yet known about the topic, given the previous research. Sometimes the analysis is followed by a review of additional research that serves to bolster the analysis.

(4) Connection to your proposed research – An explanation about how your proposed research is relevant to the reviewed research. Connection information is commonly expressed in terms of your proposed research purpose (goal, objective, and aims), your proposed methods, and your expected outcomes.

(5) Significance – A type of analysis that identifies your educated opinion about the importance of your proposed research. The scientific method does not necessarily entail the pursuit of research areas that society in general or your scientific peers in particular will consider important. However, realities of research funding require that you pursue research that society and/or professionals in your field do consider significant. Thus, significance usually needs to be explicitly addressed (however, see Chapter 2.2.6).

Subsequent chapters of this book explain how these 5 types of information are integrated into the major sections of the narrative.

1.10   Speculations, hypotheses, and predictions

Most basic, applied, clinical, and translational grant proposals in science, technology, and medicine present hypotheses that your proposed research is designed to confirm or to refute. Because the hypothesis plays a central role in scientific methodology, it is also central to your narrative. In this section, speculations, hypotheses, and predictions are briefly defined. Chapter 2.2.3 explains how to phrase them and how they relate to scientific argumentation.

1.10.1   Speculations

In the context of a narrative to a scientific, technical, or medical grant proposal, a speculation is a statement that gives the PI's educated guess about an aspect of scientific investigation. There are different types of speculations, 2 of which are the hypothesis and the prediction. Speculations are derived from analyses of others' research, your own research, or both. If you do not clearly relate your speculation or its relevance to your own or others' previous research, the speculation will be considered logically weak.

1.10.2   Hypothesis defined

According to *Stedman's Medical Dictionary*, a hypothesis is:

A conjecture advanced for heuristic purposes, cast in a form that is amenable to confirmation or refutation by conducting definable experiments and the critical assembly of empiric data; ***not*** to be confused with assumption, postulation, or ***unfocused*** speculation.[12] (Emphasis added)

A hypothesis, as a conjecture, is a kind of speculation. A hypothesis presents a speculation about a specific, tentative, ***unproven*** observation. The hypothesis, therefore, is a claim that needs to be substantiated before becoming a fact. A hypothesis usually identifies a predictor–outcome relationship[13] or an impactor–impactee relationship.

Not all speculations are hypotheses, but all hypotheses are speculations. You may have many speculations, but only those that you intend to pursue through the design and execution of your proposed methods are your actual research hypotheses. Thus, as soon as you label a speculation *our hypothesis*, you have promoted a mere *conjecture* to the status of an *observation* to be studied through the scientific method. Examples 1, 2, and 3 illustrate hypotheses:

Example 1
The hypothesis behind the proposed research is that the staphylococcal accessory regulator is a major regulatory switch controlling the expression of *S. aureus* virulence factors.

Example 2
We hypothesize that infected monocytes serve as the primary means of disseminating *Bp* from the gut during enteric infection.

Example 3
We hypothesize that particular physical therapy treatments are the most efficacious when applied to patients with lower back pain, who have similar clinical and neuromuscular measures.

1.10.3 Prediction defined

In the context of a narrative to a scientific, technical, or medical grant proposal, a prediction is an educated guess about the result or the outcome to a proposed experiment. The following examples illustrate predictions.

Example 4
We predict that cells lacking the Ifitm proteins will have trouble controlling cell activation, potentially leading to increased cell expansion, inappropriate selection (autoimmunity), and/or the inability to control cell differentiation (memory B cells versus plasma cell production).

---

[12] http://www.medilexicon.com/medicaldictionary.php?t=43074 (accessed November 28, 2014).
[13] Hulley, Stephen B. *et al. Designing Clinical Research: An Epidemiologic Approach*, 2nd Ed. Philadelphia, PA: Lippincott Williams & Wilkins; 2001, p. 52.

Example 5
We predict that contrast-enhanced ultrasound will accurately quantify intrauterine blood flow in women with fibroids and will identify differences in blood flow supplying each uterine fibroid.

A prediction is not a hypothesis. A hypothesis is the underlying theory to the prediction, whereas the prediction is bound to your particular proposed experimental methods. (See Chapter 2.2.3.)

1.11   Preparing the narrative

Writing the narrative of a grant proposal takes time, possibly weeks to months. This effort can take even more time if you are essentially designing and organizing your research while writing. To write as efficiently as possible, before drafting you need to clarify to yourself the title of your grant proposal and its key terms, the organization and timing of your methods, the basic format and headings that you will use throughout the narrative, and the relationship between each proposed aim and the methods for each proposed aim.

1.11.1   The title

The title is the headline for the grant proposal, and it carries a significant burden beyond informing reviewers about the topic of your proposed research and piquing their interest. The title functions as a repository of key terms that you will use throughout the narrative and abstract to establish themes of discussion. Because of these communication burdens, the title needs to be as clear as possible. Also, the title (and accompanying abstract) needs to use key terms since they will help online databases identify your grant proposal.

Guideline 1.15 *Phrase the title to a grant proposal with key terms that you intend to use throughout the narrative and abstract.*

The following pointers can help you compose an informative title:

● **Submission requirements.** You need to check whether your targeted funding agency has requirements for titles, and then you need to follow them closely to avoid a possible administrative rejection of your submission. For example, NIH requires "a descriptive title" no longer than 81 characters, including spaces and punctuation,[14] and

---

[14] NIH PHS398, Section 4.1. http://grants.nih.gov/grants/funding/phs398/phs398.html (accessed November 28, 2014).

notes that a resubmission should have the same title as the previous submission. However, NSF has no specific requirements for titles, but notes that they should "be brief, scientifically or technically valid, intelligible to a scientifically or technically literate reader, and suitable for use in the public press."[15]

● **Terminology, grammar, and punctuation.** The title to a grant proposal should be phrased with key terms from the narrative. Key terms in the title address:

(1) Main topic.
(2) Broad topic (optional).
(3) Key methodological feature or research design (optional).
(4) Research purpose (optional).
(5) Intended major outcome (optional).

Key terms that you select for a title will be those that you will likely first repeat in the **Aims Section**, in the sentences that give the research objective, aims, or hypothesis. Example 6a gives the title to a grant proposal that provides a broad topic (*Role of Ifitm/Fragilis proteins*), a more specific main topic (*intracellular shuttles*), and a key methodological feature (*cell activation*). Example 6b presents the first sentence in the **Aims Section** to the same grant proposal, with the key terms from the title repeated in bold italics.

Example 6
a.

Role of Ifitm/Fragilis proteins as	*intracellular shuttles during*	*cell activation*
Broad Topic	Main Topic	Methodological Feature

b.
In this R21 application we propose to test the hypothesis that the *Ifitm proteins* are molecular *shuttles* taking a set of *protein* modifiers (cargo) from an *intracellular* compartment to the membrane after *cellular activation*.

Since the title must bear the key terms for such critical information as the topic and key methodological features, you should not use synonyms. Also, you should not use acronyms, pronouns, or abbreviations for the key terms in the title unless you are desperate for space. And the title should give a good first impression by reflecting appropriate grammar and punctuation characteristic of scientific English to support your credibility.

● **Grammar of a title.** We offer several ways to phrase a title to a scientific grant proposal.

(1) **Noun with modification.** In Example 7, the noun *Genes* is pre-modified with *Song-related* and post-modified with the prepositional phrase *in the Songbird Brain*.

---

[15] http://www.nsf.gov/pubs/policydocs/pappguide/nsf14001/gpgprint.pdf (II-7; accessed November 28, 2014)

Example 7
Song-related Genes in the Songbird Brain

For clarity and readability, key nouns should not be pre-modified with a string of adjectives, such as in Example 8. An alternative is to revise the title to eliminate the least important or unneeded terms. In Example 8, the adjective phrase *unusual and novel* is omitted since *the role of Ifitm/Fragilis proteins* is not novel; it is the research about the *Ifitm/Fragilis proteins* that is novel. The adjective *unusual* can also be omitted since it can be interpreted as a conclusion based on the completed research. Since the research is at the proposal stage and has not yet been funded or completed, the use of *unusual* does not logically fit the title.

Example 8
~~Unusual and novel~~ *role of Ifitm/Fragilis proteins as intracellular shuttles* during cell activation

Nouns in titles should not be post-modified with a long series of prepositional phrases, such as 4 in a row, which can make the title difficult to readily understand. In Example 9, the noun *Combined oral contraceptives* is modified by 4 prepositional phrases: *as a potential adjunct, for weight loss, in obese individuals,* and *in an urban setting*. Not all of the information in this series is critical in order for reviewers to understand key elements of the proposed research. The last prepositional phrase, as shown by the cross-out in Example 9, can be eliminated.

Example 9
Combined oral contraceptives as a potential adjunct for weight loss in obese individuals ~~in an urban setting~~[16]

A title to a grant proposal cannot express all key terms, and the writer needs to make decisions about which terms to use in the title and which to use for the first time in the narrative. The phrasing in Example 9, with the last prepositional phrase *in an urban setting* omitted, reflects the decision to clarify the setting within the text.

(2) **Verb-related term with modification**. A title with a verb-related structure indicates the purpose of the proposed research. A title can begin with a verb, either in the infinitive (**to + verb**) form of the verb (e.g., *To quantify*) or present participle (**verb + ing**) (e.g., *Quantifying* in Example 10). A title can also begin with a noun form of a verb, called a nominalization. In Example 11, the title leads with a noun form of the verb *develop*, which is *Development*. Titles that begin with verb-related terms should not be general. Some general verbs, like *testing* or *researching*, do not give useful information and are considered not **space-worthy** in a title, in the sense that they state the obvious.

Example 10
Quantifying proprioceptive acuity in individuals with motor disability

---

[16] The original title did not include the information *in an urban setting*.

Example 11
Development of a Non-human Primate Model for Alzheimer's Disease to Conduct Translational Research

(3) Noun: Noun. In this title form, the information *before* the colon (:) is the broad topic phrased as a noun, and the information *after* the colon is the more specific topic, also phrased as a noun. In Example 12, the key noun *Lipoic acid* is identified as a type of *treatment*, and the disease is identified as *acute optic neuritis*. The key term following the colon characterizes the type of research study as *pilot*.

Example 12
Lipoic Acid as a Treatment for Acute Optic Neuritis: A Pilot Study

(4) Questions. A title to a scientific grant proposal can be phrased as a *wh* or a *yes/ no* technical question. A *wh* technical question usually begins with a *wh* question word: *what, when, where, who,* and *how.*[17] The *wh* question form implies that the proposed research will answer the question posed in the title. A *yes/no* technical question usually begins with the auxiliary *do* or *does*, or a modal, such as *can*, and its answer is either *yes* or *no*. Example 13a gives a title to a grant proposal phrased as a *wh* technical question and Example 13b, as a *yes/no* technical question. Example 13c shows a noun-with-modification structure for the title. Notice all 3 types of structures highlight all of the same key terms: *role, glutamate transporters, glia,* and *neurons*.

Example 13
a. *What **role** do **glutamate transporters** have in **glia and neurons?***
b. *Do **glutamate transporters** have a **role** in **glia and neurons?***
c. ***Role** of **glutamate transporters** in **glia** and **neurons***

There is an important distinction between a *wh* and a *yes/no* technical question, which the titles in Examples 13a and 13b illustrate. Example 13a is phrased in such a way as to imply that researchers have already determined that glutamate transporters do have a role in glia and neurons, and that now, in the proposed research, investigators will be elucidating *what* this role is. In contrast, the title in Example 13b implies that the proposed research is focusing on an earlier stage of research than is implied in Example 13a: determining whether glutamate transporters do or do not have a role in glia and neurons.

Regardless of the structure of the title that you choose, you need to phrase it with key terms that you will then repeat throughout the narrative of the grant proposal.

1.11.2   Overview visuals

An *overview visual* identifies the relationship among the proposed aims and the proposed methods, and the timing of your proposed methods. There are 2 primary

---

[17] *How* is considered a *wh* question word because the grammar of a *how* question matches the grammar of a *wh* question.

types of overview visuals (Chapter 5.2.2): (1) an overview table that identifies the relationship among the proposed aims and the proposed methods, and (2) an overview flow chart, which shows the sequence of the tasks comprising the methods. You should create these overview visuals *before* drafting a heading outline (Chapter 1.11.3) or the narrative, to help you make decisions about content, organization, time schedule, and key terms.

Overview visuals are also useful when integrated into the text (with very slight modification). Table 1-6 gives an example of an overview worksheet that you can fill out before drafting the methods. You can use this worksheet while drafting the methods, to keep you focused on the alignment of your proposed aims with the appropriate proposed methods. You can then slightly modify the overview worksheet into an overview table and locate it in the introduction to the **Methods Section**. (See Chapter 5.2.2.)

1.11.3   Heading outline

A *heading outline* is a vertical list of your intended sections and subsections, and their headings, in an itemization scheme (sometimes called the serialization) that you intend to use in the final draft of the narrative. By drafting a heading outline before writing the narrative, you can better focus on its content and organization throughout the drafting period, and you can usually write more efficiently. Also, you can use the overview table and overview flow chart to efficiently draft the heading outline.

● **Advantages of drafting a heading outline**. Drafting a heading outline before writing the narrative necessitates your mentally organizing and formatting the content of the narrative into a tangible, organizational hierarchy, which can help you make writing decisions about content and organizational alternatives before actually drafting the text. An *organizational hierarchy* is the relationship among categories of content, presented in sections, subsections, and sub-subsections (sometimes indicated in this book as (*sub*)*sections*). It is far easier – and much faster – to test organizational alternatives for a **Background Section**, for example, with a heading outline of its subsections rather than while composing the text, rearranging actual subsections, or rephrasing their headings. Also, after drafting the heading outline and then composing the text of the narrative, you can transform the heading outline into a **Table of Contents**. Table 1-7 identifies different itemization schemes from which to select when organizing the narrative.

Case 1-2 gives a heading outline to a narrative of a grant proposal. As is typical in heading outlines and as shown in this example, each subsection level after the first level (e.g., **B.1**) is indented to help the PI, with just a few eye movements, distinguish lower and higher levels in the organizational hierarchy, but such indentation of headings is not necessarily used in the actual narrative.

Guideline 1.16 gives characteristics of a heading outline and formatting strategies for use while you draft a clearly organized narrative:

> **Guideline 1.16** *In the heading outline:*
> (a) *Select an itemization scheme that you expect to use for the headings throughout all (sub)sections of the narrative.*
> (b) *Use font and capitalization in the headings that you expect to use in the final draft of the narrative.*
> (c) *Phrase and punctuate the headings as you expect them to be phrased and punctuated in the final draft. That is, use articles (a, an, the) appropriately, and use verb forms of concepts over noun forms (see Chapter 8.6.3).*

Writing a heading outline will force you to make decisions about the organization of the narrative and its format ***before*** drafting, thus freeing you to concentrate on the language and the specific content ***while*** drafting.

● **Signaling the organizational hierarchy through font**. You can distinguish levels in the organizational hierarchy by changing the font of headings at each different level.

> **Guideline 1.17** *Use a different style of font for headings at each level in the organizational hierarchy; the font needs to be distinct in at least one feature from the style used in the headings at other levels.*

For example, in Case 1-2, the second-level heading **D.2 Embryo Preparation** is distinct from the third-level headings Normal group:, OTB group:, and VVL group: in terms of style (standard and bold font) and underscoring.

● **Itemization schemes**. You should plan on using an itemization scheme for (sub) sections to help reviewers more easily follow the organizational hierarchy and to help them track or refer to your (sub)sections when discussing your narrative. An itemization scheme also allows you to cross-reference concisely to another (sub) section.

Notice that there are no fourth-level headings in Case 1-2, and this entire heading outline shows no more than 3 levels of subsections per major section.

> **Guideline 1.18** *Use no more than 3 levels of subsections in each major section of the narrative, and itemize each level.*

For readability, it is useful to limit the number of levels within each section to 3 or fewer levels since reviewers may have difficulty relating fourth-level content to higher-level content. An example of an alpha-numeric itemization scheme for a fourth-level heading is **A.1.a.(1)**. One strategy for avoiding problems associated with fourth-level headings is to use bulleted, vertical lists with headings instead of the fourth-level, itemized headings.

> **Guideline 1.19** *Instead of using fourth-level itemized headings for fourth-level subsections, consider presenting the information as a bulleted list in a third-level subsection.*

- **Layout, phrasing, capitalization, and punctuation of headings.** The *layout* of headings deals with the spatial arrangement of text in relation to the headings and white space. Three common layout alternatives for headings are: (1) centered stand-alone, (2) left stand-alone, and (3) run-in.

In a stand-alone heading, the heading is by itself on a line, and the text begins on a line under the heading. In a run-in heading, the text immediately follows on the same line. A stand-alone heading does not usually end in punctuation, but a run-in heading does. Guideline 1.20 offers a standard layout pattern for headings:

> **Guideline 1.20**
> (a) *For first-level headings – that is, headings to the main sections of the narrative – consider using left stand-alone or centered headings.*
> (b) *For second-level subsection headings, consider using left stand-alone headings of a different font than that used for first-level headings, or run-in headings.*
> (c) *For headings of third-level subsections, consider using run-in headings that are in a different font or capitalization from that used in any higher level, run-in headings.*

How you phrase each heading depends on the major section in which it occurs, as explained in the following guideline:

> **Guideline 1.21**
> (a) *In the **Aims Section**, phrase the lead words in the headings as verbs to indicate the purpose of the proposed research. (See Chapter 2.1.)*
> (b) *In the **Background Section**, phrase headings to subsections as nouns, with the terms reflecting the key terms used in the **Aims Section**. (See Chapter 3.1.2.)*
> (c) *In the **Preliminary Studies/Progress Report Section**, phrase the headings as nouns or as short sentences. Each heading should represent the most important content in the subsection – namely, the main result or conclusion. (See Chapters 4.4.2 and 4.12.)*
> (d) *In the **Methods Section**, repeat the aims from the **Aims Section** as the first-level headings. For second-level procedural subsections, consider using **verb + ing** headings or headings phrased as nominalizations. (Chapter 6.2.)*

Headings are usually bold and in a sans serif font style, such as **Arial**. Sometimes, headings to second- and third-level headings are underscored or italicized. The purpose of the bold, italics, and underscoring is to make the headings visually distinct from the text and from other headings at other levels in the organizational hierarchy.

The capitalization style for headings can vary, depending on the section level, as explained in Guideline 1.22:

> **Guideline 1.22** *Consistently use the same capitalization style for the headings at the same level in the organizational hierarchy:*
> (a) *For first-level headings, consider using a standard capitalization style or all upper-case letters.*
> (b) *For second-level headings, consider using a standard capitalization style.*
> (c) *For third-level headings, consider using a sentence-capitalization style.*

You might not be accustomed to writing a heading outline before drafting such a complex document as the narrative to a scientific grant proposal. However, given the need to write a strong narrative and given the tough competition for grant monies, you may need to change your writing habits. Heading outlines are important in helping you test organizational alternatives before writing, in helping you make content decisions, and in helping you write more efficiently.

1.12   Achieving unity across all sections of the narrative

Unity is the major outcome of a well-written narrative. The narrative needs to be singular in purpose (research objective and related aims), with all the (sub)sections providing consistent, harmonious information that will explain how the PI intends to achieve the research objective through methods designed to achieve the proposed aims.

How is unity created? This and subsequent chapters of the book include many features that, when executed, can produce a unified narrative. Some of these features are listed in this guideline:

> **Guideline 1.23** *To produce a unified narrative:*
> (a) *Repeat key terms from the title and the **Aims Section** throughout the narrative. (See Chapters 1.8.4 and 1.11.1.)*
> (b) *Consistently phrase the topic, research objective, aims, and hypothesis in the same key terms wherever mentioned in the narrative. (See Chapter 1.5.2.)*
> (c) *In the **Aims Section**, craft a research objective that covers the proposed aims and no others. (See Chapter 2.2.4.)*
> (d) *In the **Aims Section**, briefly mention relevant preliminary research, the proposed research design, and the key methods for each aim. (See Chapter 2.1.)*
> (e) *In **Background** (sub)sections, cross-reference to the relevant proposed aims that you identify in the **Aims Section**. (See Chapters 3.5 and 3.7.)*
> (f) *In the **Preliminary Studies** (sub)sections, cross-reference to the relevant proposed aims and to the relevant methods in the **Methods Section**. (See Chapters 5.6.2 and 8.6.4.)*

(g) In the **Methods Section**, use the proposed aims – as phrased and as sequenced in the **Aims Section** – as the organizational principle for the **Methods Section**. (See Chapter 5.2.1.)

(h) In the introduction to the **Methods Section**, include an overview visual and the research objective as you initially phrased it in the **Aims Section**. (See Chapter 5.4.1.)

(i) In the **Methods Section**, identify any shared methods across aims. (See Chapter 6.11.)

(j) Create an ending section to the narrative that clarifies how successful completion of the proposed research will produce contributions to scientific knowledge and will further the progression in the PI's line of research. (See Chapter 6.12.)

(k) Wherever relevant in any (sub)section of the narrative, relate significance and novelty to the proposed research. (See Chapter 3.9.)

(l) In all (sub)sections of the narrative that include scientific argumentation, review key studies so that the analysis that you offer seems reasonable to the reviewers. (See Chapter 3.5.)

(m) In all (sub)sections of the narrative that include scientific argumentation, include the connection in order to relate the previous research to the proposed research. (See Chapter 2.2.4.)

(n) When integrating contributions from collaborators into the narrative, edit the text so that the language and layout in the contributions match the language and layout used in the rest of the narrative. (See Chapter 7.8.2.)

1.13   Anonymity, symbols, abbreviations, and font

We have kept all examples in the book anonymous. Where an investigator is mentioned by name in the text, we use a generic name, such as *Dr. John Smith*. Also, to conserve textual space and to assist in presenting all examples anonymously, we have replaced all citations with the last names of authors with (*refs*).

In this book, we use certain symbols, abbreviations, and font types, which are explained in the following chart:

*	ungrammatical
‡	grammatical but not usually appropriate for scientific English
Co-I	co-investigator
CV	curriculum vitae
LOI	letter of intent
MET	materials, equipment, and tools
NIH	National Institutes of Health
NSF	National Science Foundation

PI	principal investigator
(refs)	citations by author's family name within the text, which have been omitted
SOS	subjects, objects of study, and specimens
TRACS	components of a scientific argument: topic, review, analysis, connection, and significance
bold italics in text	the term is listed in the glossary
bold italics in examples	the passage is discussed in the text
italics in text	an excerpt from an example included in the text
bold, initial upper case	names of major sections in the narrative, such as the **Methods Section**
superscripted numbers at the beginnings of sentences in examples and cases	numbers indicate the sequence of sentences in the examples and cases; included for reference purposes
superscripted numbers at the ends of sentences in examples and cases	footnote numbers from the original text

Table 1-1. National Institutes of Health, core review criteria. From: http://grants.nih. gov/grants/guide/notice-files/NOT-OD-09-025.html (accessed November 28, 2014).

Core review criteria. Reviewers will consider each of the five review criteria below in the determination of scientific and technical merit, and give a separate score for each. An application does not need to be strong in all categories to be judged likely to have major scientific impact. For example, a project that by its nature is not innovative may be essential to advance a field.

Significance. Does the project address an important problem or a critical barrier to progress in the field? If the aims of the project are achieved, how will scientific knowledge, technical capability, and/or clinical practice be improved? How will successful completion of the aims change the concepts, methods, technologies, treatments, services, or preventative interventions that drive this field?

Investigator(s). Are the PD/PIs, collaborators, and other researchers well suited to the project? If Early Stage Investigators or New Investigators, do they have appropriate experience and training? If established, have they demonstrated an ongoing record of accomplishments that have advanced their field(s)? If the project is collaborative or multi-PD/PI, do the investigators have complementary and integrated expertise; are their leadership approach, governance, and organizational structure appropriate for the project?

Innovation. Does the application challenge and seek to shift current research or clinical practice paradigms by utilizing novel theoretical concepts, approaches or methodologies, instrumentation, or interventions? Are the concepts, approaches or methodologies, instrumentation, or interventions novel to one field of research or novel in a broad sense? Is a refinement, improvement, or new application of theoretical concepts, approaches or methodologies, instrumentation, or interventions proposed?

Approach. Are the overall strategy, methodology, and analyses well-reasoned and appropriate to accomplish the specific aims of the project? Are potential problems, alternative strategies, and benchmarks for success presented? If the project is in the early stages of development, will the strategy establish feasibility and will particularly risky aspects be managed?

If the project involves clinical research, are the plans for (1) protection of human subjects from research risks, and (2) inclusion of minorities and members of both sexes/genders, as well as the inclusion of children, justified in terms of the scientific goals and research strategy proposed?

Environment. Will the scientific environment in which the work will be done contribute to the probability of success? Are the institutional support, equipment and other physical resources available to the investigators adequate for the project proposed? Will the project benefit from unique features of the scientific environment, subject populations, or collaborative arrangements?

Table 1-2. Merit Review Criteria from the National Science Foundation. From: http://www.nsf.gov/pubs/policydocs/pappguide/nsf13001/gpgprint.pdf (III-1; accessed November 28, 2014).

All NSF proposals are evaluated through use of two National Science Board approved merit review criteria. In some instances, however, NSF will employ additional criteria as required to highlight the specific objectives of certain programs and activities.

The two merit review criteria are listed below. Both criteria are to be given full consideration during the review and decision-making processes; each criterion is necessary but neither, by itself, is sufficient. Therefore, proposers must fully address both criteria. (GPG Chapter II.C.2.d.(i) contains additional information for use by proposers in development of the Project Description section of the proposal.) Reviewers are strongly encouraged to review the criteria, including GPG Chapter II.C.2.d.(i), prior to the review of a proposal.

When evaluating NSF proposals, reviewers will be asked to consider what the proposers want to do, why they want to do it, how they plan to do it, how they will know if they succeed, and what benefits could accrue if the project is successful. These issues apply both to the technical aspects of the proposal and the way in which the project may make broader contributions. To that end, reviewers will be asked to evaluate all proposals against two criteria:

- **Intellectual Merit:** The Intellectual Merit criterion encompasses the potential to advance knowledge; and

- **Broader Impacts:** The Broader Impacts criterion encompasses the potential to benefit society and contribute to the achievement of specific, desired societal outcomes.

The following elements should be considered in the review for both criteria:

1. What is the potential for the proposed activity to:
   a. Advance knowledge and understanding within its own field or across different fields (Intellectual Merit); and
   b. Benefit society or advance desired societal outcomes (Broader Impacts)?

2. To what extent do the proposed activities suggest and explore creative, original, or potentially transformative concepts?

3. Is the plan for carrying out the proposed activities well-reasoned, well-organized, and based on a sound rationale? Does the plan incorporate a mechanism to assess success?

4. How well qualified is the individual, team, or organization to conduct the proposed activities?

5. Are there adequate resources available to the PI (either at the home organization or through collaborations) to carry out the proposed activities?

Table 1-3. Generic content of the narrative and corresponding sections. Narratives of grant proposals to different funding agencies have similar content **(A)** that funding agencies call different names and **(B)** organize differently.

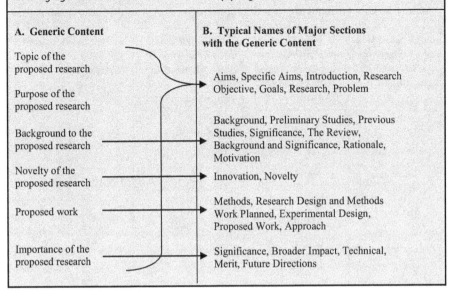

A. Generic Content

Topic of the proposed research

Purpose of the proposed research

Background to the proposed research

Novelty of the proposed research

Proposed work

Importance of the proposed research

B. Typical Names of Major Sections with the Generic Content

Aims, Specific Aims, Introduction, Research Objective, Goals, Research, Problem

Background, Preliminary Studies, Previous Studies, Significance, The Review, Background and Significance, Rationale, Motivation

Innovation, Novelty

Methods, Research Design and Methods Work Planned, Experimental Design, Proposed Work, Approach

Significance, Broader Impact, Technical, Merit, Future Directions

Table 1-4. The NIH Model. Correspondences between major sections of the NIH narrative **(A)** before January 25, 2010, and **(B)** on or after January 25, 2010.

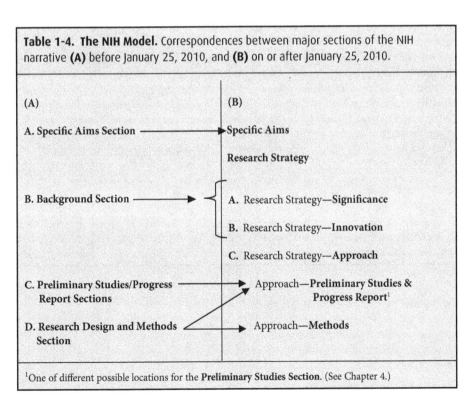

(A)

A. Specific Aims Section

B. Background Section

C. Preliminary Studies/Progress Report Sections

D. Research Design and Methods Section

(B)

Specific Aims

Research Strategy

A. Research Strategy—**Significance**

B. Research Strategy—**Innovation**

C. Research Strategy—**Approach**

Approach—**Preliminary Studies & Progress Report**[1]

Approach—**Methods**

[1]One of different possible locations for the **Preliminary Studies Section**. (See Chapter 4.)

Table 1-5. Conventional organization of narratives for scientific grant proposals.

A. Organization of a Narrative, as Indicated in This Book	B. NIH Model: NIH Submission Requirements on/after January 25, 2010	C. Organization from American Heart Association[2] Submission Requirements	D. Organization from NSF Submission Requirements[3]

A. Aims
 	
B. Background
 	
C. Preliminary Studies
 	
D. Methods
 	
E. Future Plans (Ending)
	Specific Aims
	
	Research Strategy
	
	A. Significance
	
	B. Innovation
	
	C. Approach
	
	  C.1 Preliminary Studies[1]
	  C.2 Methods
		1. Specific Aims
		
		2. Background and Significance
		
		3. Preliminary Studies
		
		4. Research Design and Methods
		
		5. Ethical issues
			1. Introduction
			  1.1 Objectives
			  1.2 Background
			
			2. Preliminary Studies
			  2.1 Results from prior NSF support
			  2.2 Preliminary research
			
			3. Experimental Methods & Procedures
			
			4. Technical Merit
			
			5. Broader Impacts

[1] NIH submission requirements have flexibility in terms of the location of the preliminary studies in the Approach Section. Their location before the **Methods Section** is useful when the preliminary research is closely related to all of the proposed aims. See http://grants.nih.gov/grants/funding/phs398/phs398.html (accessed November 28, 2014).

[2] The American Heart Association has different types of grants, the narratives of which vary in length; the organization in Table 1-5C reflects the submission requirements of a 12-page grant proposal, located at http://my.americanheart.org/professional/Research/FundingOpportunities/SupportingInformation/Creating-the-Research-Plan—12-Pages_UCM_320675_Article.jsp (accessed November 28, 2014).

[3] NSF identifies content for its narrative but gives few submission requirements for the organization. The arrangement in Table 1-5D is one of many different ways to organize a narrative submitted to NSF. Another possible arrangement, for example, is to reorganize **Section 2.1** as **Section 6: Prior NSF Support**. This alternative is useful when the prior NSF grant deals with a different line of research from that in the current grant proposal. See http://www.nsf.gov/pubs/gpg/nsf04_23 (pages 22-23; accessed November 28, 2014).

Table 1-6. An overview worksheet.

Funding Agency and Deadline:

Hypothesis:

Research Objective of Proposed Work:

Aims	Methods		
	Experiment/Task #	Task Description	Year/Time Line

Table 1-7. Three hierarchical organizations for (sub)sections of the narrative.

A. Standard	B. Alpha-numeric	C. Numeric
I. XXXXX	A. XXXXX	1. XXXXX
A. XXXXX	A.1 XXXXX	1.1 XXXXX
1. XXXXX	A.1.a XXXXX	1.1.1 XXXXX
2. XXXXX	A.1.b XXXXX	1.1.2 XXXXX
B. XXXXX	A.2 XXXXX	1.2 XXXXX
C. XXXXX	A.3 XXXXX	1.3 XXXXX
1. XXXXX	A.3.a XXXXX	1.3.1 XXXXX
2. XXXXX	A.3.b XXXXX	1.3.2 XXXXX
D. XXXXX	A.4 XXXXX	1.4 XXXXX
II. XXXXX	B. XXXXX	2. XXXXX

Case 1-1. Dividing a long paragraph into 3 paragraphs. The paragraph as printed here is 27 lines long. ¶ indicates where the paragraph can be divided into shorter paragraphs (see sentences 2 and 15).

Distribution of GABAergic cells in the NCM. [1]To determine whether the NCM contains GABAergic cells and to reveal the possible distribution of such cells, we optimized a published immunocytochemical assay for GABA (refs), based on our standard ICC protocol (refs), and we analyzed GABA immunostaining in serial brain sections throughout the NCM. ¶ [2]Our results revealed numerous GABAergic cells distributed throughout the NCM (Fig. 1) and adjacent auditory processing areas (field L and CMHV). [3]At least two kinds of GABAergic cells were readily distinguished: (1) [4]A smaller cell, with mostly triangular soma and no obvious processes (Fig. 1A). [5]This cell type is numerous and appears to be evenly distributed throughout the NCM. [6]The staining intensity ranged from very weak to moderately strong. [7]And (2) a larger cell, with round and strongly labeled soma and containing a few thick processes (Fig. 1B). [8]This cell type was much less numerous and scattered throughout the NCM. [9]We estimated that 25–35% of the neuronal cells in the NCM are GABAergic, based on the proportion of labeled cells relative to total neuronal counts (data not shown). [10]These findings were independently confirmed by in situ hybridization for GAD65, a glutamate decarboxylase isoform, and a specific GABAergic cell marker. [11]For that purpose, we PCR-amplified and cloned a cDNA fragment representing zebra finch GAD65. [12]After confirming the fragment identity by sequencing, we conducted in situ hybridization with a specific riboprobe, revealing a moderately abundant expression of GAD65 in the NCM, compared to other telencephalic areas (Fig. 1C); and numerous labeled cells in the NCM (Fig. 1D). [13]These findings clearly demonstrated the presence of GABAergic cells in the NCM. [14]Of particular significance here, we have determined that our basic ICC protocol can be applied to slices as prepared for electrophysiological recordings. ¶[15]Our GABA ICC method, therefore, will provide the basic methodology for identifying GABAergic neurons in the recorded slices (Specific Aim 1A). [16]Our data also provided the basis for postulating a prominent role of GABA in the physiology of the NCM circuitry, which will be tested further in Specific Aims 1 and 2.

Case 1-2. Example of a heading outline to a narrative.

Introduction to the Resubmission

A. Specific Aims
Aim 1: Determine in vivo changes in the growth of the OFT wall.
Aim 2: Determine dynamic variations of hemodynamic forces in the OFT under normal and altered hemodynamic conditions over the cardiac cycle.

B. Background and Significance
B.1 The Chick Heart at Early Developmental Stages
B.2 Influence of Mechanical Stimuli on Heart Development
B.3 Imaging the Developing Heart
B.4 Biomechanical Modeling of the Developing Heart
B.5 Significance

C. Preliminary Studies
C.1 FDOCT Images Allow Quantification of OFT Wall Motion
C.2 2D OCT Images Allow 4D Reconstruction of the OFT
C.3 OCT Images Reveal Differences in OFT Wall Motion under Normal and Altered Hemodynamic Conditions
C.5 Imaged-based FEM Model of OFT Indicates the Feasibility of the Study
C.6 FEM Model of OFT Reveals Effects of Cardiac Cushions on Flow
C.7 Microarray Analysis Shows Up-regulation of Collagen and Integrin after OTB

D. Research Design and Methods
D.1 The Research Team
D.2 Embryo Preparation
    a) Normal group:
    b) OTB group:
    c) VVL group:
D.3 Aim 1: Determine in vivo changes in the growth of the OFT wall.
    a) OCT Imaging of the Chick OFT
    b) Modification of the OCT System for *in vivo* OFT Data Acquisition
        ● Mounting stage:
        ● Determination of blood flow velocity:
    c) 4D Reconstruction of the Chick OFT
    d) Data Analysis
    e) Potential Difficulties and Solutions
D.4 Aim 2: Determine dynamic variations of hemodynamic forces in the OFT under normal and altered hemodynamic conditions over the cardiac cycle.
    a) Real-time Pressure Measurements
    b) FEM Models of the OFT
        ● Validation of FEM models:
        ● Analysis of FEM results:
    c) Gene Expression Analysis
        ● Real time-polymerase chain reaction (RT-PCR):
        ● 3D reconstruction of in situ hybridization (ISH) and immunohistochemistry:
        ● Collagen stain:
    d) Data Analysis and Statistics
    e) Potential Difficulties and Solutions

E. Summary

The Aims Section

The **Aims Section** is the first major prose section in the narrative of a scientific grant proposal. Some funding agencies call this first section the **Introduction**. The **Aims Section** provides the *conceptual framework* for the proposed research in the sense that it introduces the proposed research topic or research area, the significance and novelty of the proposed research, the purpose of the proposed research, the proposed methodological approach, and the proposed aims. What is called the *aims* in this book is variously called *objectives* or *specific aims* by different funding agencies.

Figure 2-1 shows the basic layout of the **Aims Section**. Figure 2-2 shows the relationship among the long-term goal, the research objective, and the aims in the **Aims Section**. Cases 2-1 and 2-2 show **Aims Sections** from the narratives of scientific grant proposals addressing basic research; Case 2.3, applied research; and Case 2.4, clinical research.

2.1   Overview of length, content, organization, and layout

The **Aims Section** ranges from one-half page to 2 or 3 pages, depending on submission requirements and on the overall length of the narrative. Different funding agencies have different requirements for the **Aims Section**. In most cases, NIH requires that its first section of the narrative, which it terms the **Specific Aims Section**, not be longer than one page.[1] NSF does not specify the name or length for its **Aims Section**.[2] However, the first section of an NSF narrative is typically called the **Introduction**, which corresponds to the **Aims Section** and **Background Section**, combined. In unsolicited NSF narratives, the **Introduction** is typically 3–5 pages long.

---

[1]   http://grants.nih.gov/grants/funding/phs398/phs398.html (page I-17; accessed November 28, 2014).

[2]   http://www.nsf.gov/pubs/policydocs/pappguide/nsf13001/nsf13_1.pdf (II-8; accessed November 28, 2014).

Regardless of the length of the **Aims Section**, it consists of 2 or 3 units of information, which we term *passages*. Figure 2-1 identifies the 3 passages of an **Aims Section**. Taken together, the 3 passages present information that comprises a scientific argument (see Chapter 1.9). The passages in the **Aims Sections** of Cases 2-1 to 2-4 have their content (i.e., scientific arguments) identified in the right-hand column and their sentences identified after the case title.

> **Guideline 2.1** *Organize the **Aims Section** into 2 or 3 passages, and include information to comprise at least one scientific argument.*

● **Passage 1 – Lead paragraph(s) before the proposed aims.** Passage 1 consists of the lead paragraph(s) or subsection(s) in the **Aims Section**, up to the proposed aims, not including any sentence that introduces the proposed aims. Passage 1 is about one-half to two-thirds of the entire length of the **Aims Section**. Passage 1 in NIH narratives is usually 1–3 paragraphs long, in contrast to Passage 1 in NSF narratives, which is sometimes 3–5 pages long because it covers information that typically corresponds to both the **Aims** and **Background Sections**.

Passage 1 is aimed at an educated reader, but not necessarily an educated reader who is highly specialized in the proposed area of research, and it presents content that comprises at least one scientific argument, including the research objective of the proposed research, the major hypothesis of the proposed research, and the methodological approach.

● **Passage 2 – Proposed aims.** Passage 2 consists of any sentence that introduces the proposed aims, and the paragraph(s) that describes the proposed aims. The proposed aims are presented in emphatic font, such as bold or underscored font, and in a vertical, enumerated list, as shown in Cases 2-1 to 2-4.

> **Guideline 2.2** *Present your proposed aims in emphatic font and in a vertical, enumerated list so that reviewers can quickly identify them.*

The following list gives a brief overview of Passage 2.

(1) Lead-in. A lead-in announces the upcoming aims and provides a transition between Passages 1 and 2. To conserve space, the lead-in is usually attached to the last paragraph of Passage 1. The lead-in can be as brief as *Our specific aims are:* or it can be a long sentence. The lead-ins to the proposed aims in Cases 2-1, 2-2, 2-3, and 2-4 are in sentences 5, 20, 14, and 22, respectively.

(2) Proposed aims. Each proposed aim, presented in emphatic font for readability, is phrased with an infinitive research purpose verb, such as the research purpose verbs *identify* and *determine*. Chapter 2.2.4 details how to phrase and organize the aims.

A good cache of research purpose verbs for one project's proposed aims is in a funding agency's request for proposal (RFP) for that project. A funding agency, when describing types of projects it is soliciting, frequently uses key terms, including key research purpose verbs, which you should consider using in your proposed aims.

> **Guideline 2.3** *Examine RFPs for key research purpose verbs to include in your proposed aims.*

(3) **Aims paragraph (optional).** Although technically optional, each aim is usually followed by a short paragraph, called the *aims paragraph*. An aims paragraph for a particular aim briefly characterizes the methods (or tasks) that you are proposing in order to achieve that aim. An aims paragraph can also include information comprising a scientific argument, such as: (a) a brief review of research related to the particular aim, including definitions of key terms relevant to that aim; (b) an analysis of what is not known about the topic or research area related to the particular aim, including a minor hypothesis (a hypothesis related to the individual aim; see Chapter 2.2.3), or technical questions related to the particular aim; and (c) expected outcomes for the methods designed to achieve the particular aim.

The **Aims Section** in Case 2-2 has no aims paragraphs, unlike the **Aims Sections** in Cases 2-1, 2-3, and 2-4. The exclusion of aims paragraphs sometimes occurs when data-collection methods for each aim are closely similar to each other and are broadly characterized in Passage 1, as shown in sentence 12 of Case 2-2.

● **Passage 3 – Ending paragraph (optional).** Passage 3, which follows the proposed aims, is an optional ending paragraph that discusses the novelty, significance, and/or outcome(s) of the proposed aims. Sometimes this information is mentioned only in Passage 3, and sometimes the information is already mentioned in Passage 1, in which case it can be repeated in Passage 3 for emphasis. Case 2-4 shows an **Aims Section** with an ending paragraph (sentence 32) that explains the importance of the proposed research in this line of research.

2.2   Scientific argumentation

The **Aims Section** presents information comprising a scientific argument for the proposed research (see also Chapter 1.9). As summarized by **TRACS**, this information is: (a) the proposed research topic or research area, (b) a brief review of what is known about the proposed research area, (c) an analysis of what is not yet known about the topic or research area, (d) the research purpose, proposed methods, and/or outcomes that connect the reviewed information and the analysis to the proposed research, and (e) significance. Items (b) and (c) comprise the *rationale* or *motivation* for the scientific grant proposal.

The information comprising a scientific argument can occur in various arrangements. However, in general, the topic and the review (e.g., sentences 1–4 in Case 2-1) lead Passage 1 of the **Aims Section**.

2.2.1 TRACS: The topic

The topic of your proposed research is the specific scientific, technical, or medical research area that you propose to investigate. The topic is usually located in Passage 1, in a focus position of a paragraph: in the first 2 or the last one or 2 sentences of a paragraph (see Chapter 1.8.1). The topic is phrased in key terms, some of which are mentioned in the title of the grant proposal (see Chapter 1.11.1). Most of the key terms that comprise the topic in the **Aims Section** are repeated in subsequent sections of the narrative. The following guideline indicates how to use key terms to reinforce the topic:

> **Guideline 2.4** *Focus attention on the topic by:*
> (a) *Repeating the key terms of the topic throughout the **Aims Section** and throughout other sections of the narrative.*
> (b) *Avoiding synonyms and most pronouns when referring to the topic.*
> (c) *Using abbreviations for selected, lengthy key terms in your topic.*
> (d) *Defining key terms.*

A few lengthy key terms constituting the topic (3 or 4) can be abbreviated to save space (see Chapter 8.6.3). The abbreviation is introduced in parentheses immediately after the topic, following its first mention, and is then used throughout the narrative. The abbreviation is not re-introduced when first used in a subsequent section of the narrative.

In sentence 1 of Case 2-1, *HCN* is introduced for *hyperpolarization-activated, cyclic-nucleotide-gated*; in Case 2-2, *Bp* is introduced for *Burkholderia pseudomallei*; in Case 2-3, *TE/RM* is introduced in sentence 1 for *tissue engineering/regenerative medicine*; and in sentence 1 of Case 2-4, *CNS* is introduced for *central nervous system*, *RRMS* for *relapsing remitting multiple sclerosis*, and *AON* for *acute optic neuritis*.

In addition to being abbreviated, some topical key terms are also defined (see Chapter 8.2).

2.2.2 TRACS: The review

The review consists of detailed sentence(s) that explain what is already known in your discipline about your topic or research area, and/or proposed methods. The reviews primarily comprise: (1) information that is common knowledge and is shared by experts in your proposed area of research, in your discipline, or in the public domain; this common knowledge includes established definitions of key terms; (2) published research of other investigators, the PI, and other members of the PI's research team; and (3) non-published preliminary research from the PI and other members of the PI's research team.

Review sentences in the **Aims Section** do not typically have citations identifying the source of information, unlike review sentences in other sections of the narrative. However, some funding agencies may require citations in the **Aims Section**, especially if the **Aims Section** for that funding agency includes the **Background Section**, such as NSF. Other funding agencies that have a separate section for background information often do not require citations in the **Aims Section** but only in the **Background Section**. Note, however, that if you start using citations in Passage 1, you need to be consistent and use citations where appropriate throughout Passages 2 and 3 of the **Aims Section** (see Chapter 8.1).

The tone of a review should reflect that of a neutral, unbiased observer who is giving facts, data, and methods from previous research. Sentences that review information are composed of an optional identifier and a ***precise generalization***, as illustrated in Examples 1–3. Examples 4–5 are review sentences that need revision (see discussion below).

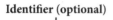

Identifier (optional)   Precise generalization

Example 1. Smith and Jones (53) found that

Example 2. In 2005, we found that

the canine parvovirus and its close relative feline panleukopenia virus bind to the host transferrin receptor type-1 to infect cells.

Example 3. Smith and Jones (53) have shown that

Example 4. [3]‡It has been found that

Example 5. [4]*In 2005, researchers have found that

● **Identifier.** The identifier names the source of the reviewed information, as shown in Examples 1–3. The present perfect, such as *have shown* in Example 3, as opposed to the simple past tense (e.g., *found* in Examples 1 and 2), is used when the reviewed information is particularly noteworthy or relevant to proposed research. The impersonal phrasing of the identifier illustrated in Example 4 (*It has been found that*) is currently out of favor in scientific English (indicated by ‡) unless the generalization is so ubiquitous that the source of information may be difficult to trace. Also note that, if the identifier includes a specific past time (e.g., *In 2005* in Example 2), its verb needs to be in the simple past (e.g., *found* in Example 2) due to grammatical constraints in English. Thus, *have found* in Example 5 is ungrammatical (indicated by *) and needs to be rephrased as *found* if a past date (e.g., *In 2005*) is used.

● **Precise generalization.** A precise generalization is not an oxymoron. A precise generalization presents facts and data (e.g., a characteristic, condition, pattern, or process) that are believed to be ***constant over time***. The precise generalization in

---

[3] The symbol ‡ indicates that the passage is grammatical but not appropriate or not commonly used.
[4] The symbol * indicates that the passage is ungrammatical.

Examples 1–5 describes a feature of the canine parvovirus and the feline panleuko-penia virus that is constant over time – a feature that was characteristic of the viruses yesterday and today and, assuming that laws of nature hold, will be characteristic of them tomorrow. Generalizations and their details are derived from: (a) common knowledge, (b) others' research, (c) your previous research, and (d) your experience. If the generalization has not yet been disproved, its verb is in the simple present tense (illustrated by *bind* in Examples 1–5), in active or passive grammar. If the generalization has been disproved, the simple past form is used.

- **Definition.** In deciding what information to review, you should not overlook definitions of key terms. In the **Aims Section**, reviews in Passages 1 and 2 often include definitions of key terms.

   In the narrative to a grant proposal, a definition is a type of precise generalization that reviews the meaning of an established term. A new term is rarely coined, introduced, and newly defined in the narrative; new terms are introduced in research papers. For example, in Case 2-2, *Bp bacterium* is defined in sentence 2; in Case 2-3, the definition for *tissue engineering/regenerative medicine* is explained in sentences 1–2; and in Case 2-4, the definition for *acute optic neuritis* is composed of information in sentences 1–3. Chapter 8.2 discusses definitions in detail.

2.2.3   TRACS: The analysis

As earlier explained (Chapter 1.9), you should not just describe (i.e., not just review) what is known about your research topic or area of study. You need to give your educated opinion about the information that you review. This educated opinion is termed *analysis* in scientific argumentation.

   An analysis is a statement that characterizes a weakness, limitation, signifi-cance, benefit, and/or strength of information you have just reviewed. By means of this characterization, you imply the novelty and significance of the issue that you propose to investigate. Also, by injecting your analyses into the proposal, you can indicate the problem-solving character of your proposed research, which enhances your credibility. The analysis typically follows a review, in an arrange-ment indicated by *review-analysis*. The analysis is much shorter than the review in terms of textual space.

> **Guideline 2.5**   *After reviewing previous research related to your proposed research area and topic, briefly analyze the reviewed research in order to identify what is not yet known.*

   As illustrated by the analyses identified in the **Aims Section** of Cases 2-1 to 2-4, an **Aims Section** can have one or multiple analyses. The analysis should follow clearly and logically from the information you review. In this way, the analysis creates a logical transition from the historical research context that you characterize in your review to your proposed research (as indicated in the connection component of a scientific argument). Thus, the analysis is critically important to your credibility because it demonstrates your critical thinking skills and reflects your understanding of scientific methodology.

There are different types of analysis statements in a scientific argument. The main types are: (1) problem statements, (2) technical questions, (3) speculations, (4) hypotheses, (5) predictions, (6) extension statements, and (7) conclusions.

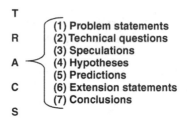

T
R
A
C
S

(1) Problem statements
(2) Technical questions
(3) Speculations
(4) Hypotheses
(5) Predictions
(6) Extension statements
(7) Conclusions

(1) Problem statements. A common way to evaluate your research area is through a problem statement, which explicitly identifies a shortcoming – a gap – in the research that you have reviewed. As illustrated in Examples 6 and 7, a problem statement consists of 2 or 3 components: an optional problem identifier, and a non-optional limitation and technical area of investigation.

Problem identifier (Optional)	Limitation	Technical area of investigation
↓	↓	↓

Example 6
On the other hand,   relatively little is known   about how kinematic information is extracted from the population of proprioceptive afferents.

Example 7
____   Little is known   about the mechanisms by which maternal diabetes affects cardiac development during early pregnancy, when maternal hyperglycemia is thought to induce cardiac malformations.

- **Problem identifier**. The problem statement often begins with a word or phrase that provides a transition from the neutral review to the critical, but civil, analysis of the information you have just reviewed. Problem identifiers include *on the other hand* (Example 6, above), *however* (Example 8), *while* (*while* meaning adversative; Examples 9 and 10), *although* (Example 11), *but* (Example 12), and *despite* (Example 13).[5] However, as shown in Example 7, a problem statement can lead with a limitation, such as *Little is known.*

Example 8
However, the exact role of inflammation in atherosclerosis and the risks it poses to patients – both systemically and within a specified plaque – are *not well understood*.

Example 9
While the binding of FXI has been well-described on the molecular level in purified systems, the role that these interactions play in mediating FXI binding to clots is *unclear*.

Example 10
While there are much data showing that S4 is the major voltage sensor, there still exists *controversy* over *how* S4 moves.

Example 11
Although there has been increased interest in postural preparation in the standing human (refs), APAs have *not yet been tested* in this context after an active manipulation of the SMA.

Example 12
Electrical stimulation of sensory afferents and somatosensory discrimination training have been shown to improve haptic perception, *but* relatively *little attention* has been paid to the benefits of sensory stimulation on motor coordination.

Example 13
Despite this cleavage being of *unknown function*, it is beneath the threefold region of the capsid, and its greatest effect would likely be to alter the stability of the capsid or the DNA packaging.

- **Limitation**. A necessary component of a problem statement is an identified shortcoming in the previous research. This shortcoming is usually phrased negatively and is related to undesirable research situations, such as: (a) insufficient or problematic research; (b) unclear or unknown phenomena; or (c) a controversy.

Examples 11 and 12 above use the expressions *not yet tested* and *little attention*, which signal insufficient research. However, even if a particular topic or research area

---

[5] These terms have other functions in addition to introducing problem statements. Therefore, just because a sentence uses a problem identifier does not mean that the sentence is a problem statement. The other 2 components (limitation and technical area of investigation) are also needed for the sentence to be unambiguously identified as a problem statement.

has been extensively studied, certain aspects or features may still be unknown or unclear. Phrases signaling this limitation are: *little is known* (Examples 6 and 7), *not well understood* (Example 8), *unclear* (Example 9), and *unknown* (Example 13). A controversy deals with a topic or research area about which researchers disagree in some respect. This type of problem statement uses such terms as *controversy, little consensus, no consensus, disagree,* and *disagreement,* illustrated in Example 10. When using a controversy problem statement to analyze previous research, you need to briefly review each side of the controversy.

As illustrated in Examples 6 to 13, the limitation is phrased in non-technical vocabulary.

● **Technical area of investigation.** Another necessary component of a problem statement is the specific technical area of investigation that needs to be researched. In the **Aims Section,** the technical area of investigation should be limited to only one or 2 sentences, but broad enough to cover the proposed aims. The technical area of investigation can be phrased with or without an interrogative, such as *whether, what, why, when, how, which, by/in/for which,* and *to what extent.* Interrogatives are used in Examples 6 (*how*), 7 (*by which*), and 10 (*how*).

Precise phrasing requires problem statements that have only one possible interpretation. However, problem statements sometimes lend themselves to multiple interpretations due to problems in phrasing the technical area of investigation. Without the interrogative, the technical area of investigation could have multiple interpretations, in which case the reviewers are left to infer the precise technical area of investigation. Consider the variations in Example 14:

Example 14

a. *Despite* the prevalence of this perceptual basis, there is little or no consensus ***about its occurrence or its significance*** for human vision.

b. *Despite* the prevalence of this perceptual basis, there is little or no consensus ***about whether it occurs or what significance it has*** for human vision.

c. *Despite* the prevalence of this perceptual basis, there is little or no consensus ***about how or why it occurs or what significance it has*** for human vision.

These 3 problem statements, each with a slightly different phrasing of the technical area of investigation, illustrate the important role that interrogatives play in phrasing precise problem statements. Example 14a uses no interrogative in the technical question and can be interpreted in multiple ways, shown in Examples 14b and 14c. Example 14c is the original version of this sentence – and also the most precise.

(2) **Technical questions.** A technical question seeks information about a particular technical area of investigation. By your posing a question that you do not immediately answer, reviewers assume that researchers have not yet answered the question. Unlike a problem statement, the technical question is not identified with

any negative evaluation; the technical question only identifies the technical area that needs investigation. Since no problem is included in the technical question, this type of analysis statement is particularly useful when you are reviewing your own previous research or that of another researcher whose relationship to you may warrant political expediency and no negativity, such as your mentor. In Case 2-2, sentences 13–15 and 17 are technical questions.

(3) **Speculations**. A speculation is a conjecture, an educated guess about a particular issue or topic in a research area that you have just reviewed. When a speculation is offered, the problem is implied: the problem is that there is no certainty about the issue and so researchers in the field are left to speculate. Similar to a technical question, a speculation avoids an explicitly negative characterization of the reviewed research. When a speculation addresses a predictor-outcome issue, it often rises to the level of a hypothesis (see Item #4 below and also Chapter 1.10).

A speculation is a common way to analyze reviewed information. Certain expressions distinguish a speculation from another type of analysis statement, and without these expressions, your statement is not a speculation. Indicators of speculations include *might* (Example 15), *possibility* (Examples 16 and 17), *seem* (Example 17), *unlikely* (Example 17), and *likely*. The adverb *likely* in sentence 2 of Case 2-1 also indicates a speculation, as does *suggesting* in the last half of sentence 3 in Case 2-1.

Example 15
The differences in phenotypes of T cells recruited simultaneously in 2 different organs *suggest* that the microenvironment *might* also play a role in determining lymphocyte homing.

Example 16
The functional consequence of this substitution is not clear, but one intriguing *possibility* is that syntaxin 3, but not the other syntaxins, interacts with the unique calcium channels of retinal ribbon synapses, to regulate transmitter release at these synapses.

Example 17
While other *possibilities* such as differences in the affinity of binding of the neutralizing and non-neutralizing Fabs *seem unlikely* due to the similar affinities of most of the antibodies examined, we will also be able to test that *possibility* directly in the studies proposed by selecting for increased affinity variants.

(4) **Hypotheses**. A powerful way to analyze previous research in the **Aims Section** is to present your hypothesis that emerges from the research that you have just reviewed. A hypothesis, as a type of speculation, implies a lack of certainty about a particular issue in the previous research and so warrants additional investigation. Some professionals in grant writing note that a hypothesis that does not

pursue a mechanistic issue is trivial.[6] Descriptive and applied studies arguably might not have hypotheses.[7]

A hypothesis, when presented in Passage 1 of the **Aims Section**, is the primary claim that you expect to confirm[8] by investigating your proposed aims. This hypothesis is termed the ***major hypothesis***. A ***minor hypothesis*** relates to a particular aim and so is identified in an aims paragraph. However, if you present a minor hypothesis in one of the aims paragraphs, reviewers sometimes expect each of the aims paragraphs to offer a minor hypothesis.

> **Guideline 2.6** *If you include a minor hypothesis in an aims paragraph, consider including a minor hypothesis in each aims paragraph.*

As illustrated in Examples 18, 19, and 20a, a hypothesis has 2 necessary components: an identifier and a claim.

Hypothesis identifier   **Claim**

↓   ↓

Example 18
We hypothesize that   Rac1 activation *is regulated* by the mTOR/S6K1 signaling pathway in platelets.

Example 19
Our hypothesis is that   chronic unilateral peripheral vestibular loss *results in* persistent central vestibular asymmetries.

Example 20
a. We hypothesize that   the song-induced expression of zenk in the NCM, an early component of a regulatory cascade, *leads to* long-lasting modifications of NCM neurons.

b. ‡It is hypothesized that   NsrR-dependent regulation and the dual . . .

Both the identifier and the claim are required in a hypothesis. The identifier labels the claim (a speculation) as a hypothesis in a verb (*hypothesize*) or noun (*hypothesis*) form. It is this identifier that distinguishes the hypothesis from other types of analysis statements. The verb in the identifier is in the present tense since the hypothesis is current to the proposed research. A hypothesis is not identified with an impersonal *It is hypothesized* (see Example 20b) since this passive phrasing does not identify who is performing the mental exercise of hypothesizing.

---

[6] As Ogden and Goldberg note, "Important hypotheses relate to basic mechanisms, the understanding of which advances science. With some thought, it is almost always possible to transform a study based on trivial or phenomenological hypotheses into one involving basic mechanisms and scientifically important hypotheses." Ogden, Thomas E. and Goldberg, Israel A. *Research Proposals, a Guide to Success*, 3rd Ed. New York, NY: Academic Press; 2002, p. 89.

[7] Hulley, Stephen B. *et al. Designing Clinical Research: An Epidemiologic Approach*, 2nd Ed. Philadelphia, PA: Lippincott Williams & Wilkins; 2001, p. 57.

[8] You expect to prove the hypothesis based on preliminary studies that already suggest its claim will be proven.

The claim in the hypothesis is what you intend to prove through your proposed methods. The verb in the claim of a hypothesis is phrased, in most instances, with a verb in the simple present tense. This verb captures a predictor–outcome or impactor–impactee relationship (see Chapter 1.10.2). The verb in the claim of a hypothesis is not phrased with any speculation terms, such as *may, might,* or *possibly.* Notice that none of the verbs in the hypotheses of Examples 18, 19, and 20 are phrased as speculations; rather, they express definite predictor–outcome or impactor–impactee relationships: *is regulated* (Example 18), *results in* (Example 19), and *leads to* (Example 20). The hypotheses in Cases 2-1, 2-3, and 2-4 also have verbs in their claims that express predictor–outcome or impactor–impactee relationships: *underlies* and *stabilize* (Case 2-1, sentences 4 and 14), *evaluate* (Case 2-3, sentence 11), and *to reduce, to preserve,* and *to improve* (Case 2-4, sentence 19).

In Passage 1 of an **Aims Section**, the hypothesis is an analysis logically derived from information that you have reviewed:

Reviewed information   —————————————▶   Analysis:
                                                        Hypothesis

In Case 2-1, the review in the first half of sentence 3 is followed by a speculation in the second half of the sentence, and then the hypothesis follows in sentence 4. In Case 2-4, the review of literature extends from sentences 9 to 18 and is immediately followed by the hypothesis in sentence 19.

Reviewed information ——▶ Analysis: ——▶ Analysis:
                                 Speculation   Hypothesis

The analysis needs to logically follow from the reviewed information. Because of this logical relationship between the review and the analysis, ***it is especially important that you review information that will lead your reviewers to consider your hypothesis as plausible enough to be researched***. As a result, you need to select information for the review that reviewers will readily, logically relate to the hypothesis.

Guideline 2.7   *Select information to review so that reviewers will consider the hypothesis as logically relating to the reviewed information.*

(5) Predictions. A prediction identifies what the PI expects will happen upon the execution of a proposed method(s). As shown in Examples 21a and 22a, a prediction has 2 components: (1) an identifier and (2) a future result or outcome.

Prediction identifier   **Future result or outcome**
            ↓   ↓

Example 21
a. Our prediction is that   only one 3D volumetric image will be sufficient when the mirror moves either forward or backward.
b. ‡Our hypothesis is that   only one 3D volumetric image will be sufficient when the mirror moves either forward or backward.

Example 22
a. We predict that   our oral lipoic acid treatment, compared with a placebo,
                              will reduce optic nerve inflammation and neural loss,
                              thereby preserving the RNFL in our subjects, as
                              measured by optical coherence tomography.
b. ‡We hypothesize that   our oral lipoic acid, compared with a placebo,
                              will reduce optic nerve inflammation and neural loss,
                              thereby preserving the RNFL in our subjects, as
                              measured by optical coherence tomography.

PIs sometimes misphrase predictions as hypotheses (and sometimes vice versa), such as in Examples 21b and 22b. Both a prediction and a hypothesis are types of speculations. However, a prediction is not a hypothesis, and a hypothesis is not a prediction. A prediction speculates about the result or outcome of an experiment or a procedure; a hypothesis gives the underlying theory on which the prediction is based. Thus, a prediction and a hypothesis need to be phrased distinctly since they are distinct ways to analyze reviewed information. In a prediction, the claim is about a possible future result or outcome, and it uses a future verb, such as *will be* in Example 21a and *will reduce* in Example 22a. In a hypothesis, the claim is not reserved to a particular point in time and so includes a simple present verb to indicate a precise generalization that has not been disproved. Also, the identifier for a prediction further distinguishes a prediction from a hypothesis.

The verb *predict* and its associated noun *prediction* are usually reserved for predictions associated with one or a set of results from the proposed experiments or procedures. However, often in clinical research, the primary outcome is limited to the single, key result of the experiment. In Example 23a, the result is not labeled as *outcome*. To help draw the reader's attention to your outcome, you can rephrase your predicted result with *outcome*, as shown in Example 23b.

Example 23
a. We will use OCT images to reconstruct 4D images of the beating chick heart – images that will enable analysis and visualization of cardiac motion during early development.
b. We will use OCT images to achieve our major outcome: the reconstruction of 4D images of the beating chick heart – images that will enable analysis and visualization of cardiac motion during early development.

(6) Extension statements. An extension statement is an analysis that identifies how your research will build on the strengths of previous research. An extension can be variously phrased, but it uses a verb or a noun that suggests growth, forward movement, or an increase in scope, such as *advance, advancement, broaden, build upon, emerge, emergence, extend, extension, expand*, and *expansion*. The use of such positive terms, plus the absence of negative words, creates a positive tone to an extension statement. As a result, this type of analysis is often also used when it is politically feasible not to identify limitations in the research you have reviewed, such as when you are analyzing your own research or the research of any person whom you want to present in a neutral or positive light, or when you are submitting a competitive renewal of a grant proposal. The following examples illustrate extension statements that use verbs, adverbs, and adjectives that denote and connote growth and forward movement and that provide a positive tone.

Example 24
The present study *extends* the previous work on human muscle spindle Ia receptors, to *further* characterize the position- and velocity-dependence of Ia afferents.

Example 25
In an *extension* of our previous work, our study will characterize the movement and location of S4 in relation to the pore, in both shaker and HCN channels.

Example 26
An additional, important mechanism for regulating the excitability of spinal circuits and the control of muscle tone is beginning *to emerge*.

(7) **Conclusions**. A type of analysis statement is a conclusion – a logical inference drawn from facts. Conclusions are very common in scientific arguments in discussions of preliminary studies, where you discuss your preliminary findings in ways that help reviewers understand your thinking processes and the implications and relevance of your preliminary research to your proposed research. However, conclusions are rarely used to evaluate reviewed information in the **Aims Section**.

● **Clusters of analyses**. A review of literature can be analyzed by a group of sequential analysis statements. Clusters of analyses can help reviewers understand your reasoning processes that led you from the reviewed research to your proposed research. In Case 2-1, a speculation is presented in the second half of sentence 3, and sentence 4 recasts the speculation into a major hypothesis.

Guideline 2.8   *Consider clustering analyses to help reviewers better understand your reasoning processes that led you to your proposed research.*

2.2.4   TRACS: The connection – research purposes, methods, and outcomes

Connection information in a scientific argument identifies the relevancy of the reviewed information and analysis to: (1) the proposed research purposes, (2) the proposed methods, and/or (3) the anticipated results or outcomes of your proposed research.

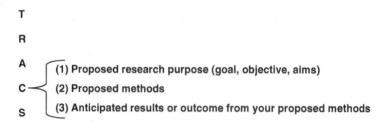

T

R

A ⎡ (1) Proposed research purpose (goal, objective, aims)

C ⎨ (2) Proposed methods

S ⎣ (3) Anticipated results or outcome from your proposed methods

● **Proposed research purpose**. The *proposed research purpose* is what you intend to achieve in your proposed research project. Two levels of research purpose, and an optional third level, need to be associated with your proposed research in the narrative, as shown in Figure 2-2: an optional long-term goal, one research objective, and the aims.

Different funding agencies use different terms when referring to these 3 levels of research purposes. For example, the National Multiple Sclerosis Society[9] uses the terms *specific aims* and *objectives* to refer to the purposes of your proposed methods; the terms that the Society for Family Planning uses for *specific aims* include the *long-term objectives*;[10] and NSF appears to use the term *objective* for all 3 levels of research purpose.[11] We differentiate the terms *long-term goal*, *research objective*, and *aims* in this book (Figure 2-2).

(1) Long-term goal (optional). The long-term goal is the research purpose that identifies what the PI intends to accomplish in a line of research over an extended period of time, such as 10–15 years. This time period might cover multiple research projects that are eventually funded through multiple grants. Figure 2-2 illustrates one long-term research goal that is being supported, for example, by 4 grants over a 10–15-year period.

The long-term goal is not usually required content in an **Aims Section**, but is more common in longer narratives, such as those that are 12 or 15 pages long. In an **Aims Section** that presents the long-term goal, it is located in Passage 1, illustrated in sentence 2 of Case 2-3.

(2) Research objective. The research objective is the research purpose that the PI intends to accomplish in one grant proposal. In other words, one grant proposal has one research objective (see Figure 2-2). The research objective is a logical subset of the long-term goal, but is more limited in scope since it relates to only one proposed research project. Your research objective should be broad enough to cover all of the proposed aims, but specific enough to cover only the proposed aims (and no others).

---

[9] http://www.nationalmssociety.org/ms-clinical-care-network/researchers/get-funding/research-grants/index.aspx. Instructions for online submission of research grant applications (page 10, accessed November 28, 2014).

[10] http://www.societyfp.org/research/applying.asp; Large and small research proposals; Download research grant RFP; 2013 REQUEST FOR PROPOSALS (pdf; page 4; accessed February 6, 2013)

[11] http://www.nsf.gov/pubs/policydocs/grantsgovguide0111.pdf (accessed November 28, 2014).

In the **Aims Section**, the research objective occurs in Passage 1, usually in a focus position of a paragraph (see Chapter 1.8.1). In Case 2-1, the research objective is in sentence 1. It is also in sentence 1 of Case 2-2, except that there it is called the *overall objective*. If the research objective is not in one of the focus positions, you can place it in emphatic font to make it more noticeable to reviewers, such as the research objective in sentence 20 of Case 2-4. Because of the importance of the research objective in pinpointing what you intend to achieve in your proposed research, it is commonly repeated not only in the **Aims Section** but also later in the narrative, such as in the introduction to the **Methods Section** where it is the organizing principle for the **Methods Section** (see Chapter 5.2) and sometimes again in the ending section to the narrative (see Chapter 6.12).

(3) **Aims**. An aim is the research purpose of a related group of proposed methods. The aims occur in Passage 2 of an **Aims Section** and are presented in a numbered list, in emphatic font to make them readily identifiable. The aims are the themes that tie together the proposed methods.

The list of aims usually has a lead-in, which announces the upcoming aims and functions as a transition between Passages 1 and 2. In Case 2-1, the lead-in is in sentence 5; in Case 2-2, sentence 20; in Case 2-3, sentence 14; and in Case 2-4, sentence 22.

It is difficult to say how many aims are too many in a grant proposal. However, it appears that 2 may be appropriate for a research grant spanning 2–3 years and up to 6 may be appropriate for a grant spanning 5 years. NIH notes that, "Most successful applications have 2–4 specific aims."[12] Figure 2-2 shows grant proposals with 2 to 3 aims each.

Just as the research objective should fit logically within the scope of the long-term goal, so too should all of the aims fit logically within the scope of the research objective. When there is not such a fit, you need to either: (a) broaden the research objective to cover all of your proposed aims and no other aims, (b) narrow the research objective to encompass only the proposed aims and no others, or (c) revise the aims so that their scope, collectively, matches the scope of the research objective. Because grant proposals have more funding success when the scope of their proposed objective and aims is relatively narrow and well-defined, you may want to narrow, not broaden, your research objective and aims.

In addition, grants seem to have more success when their aims are interrelated, but not fully dependent on each other. For example, if your second aim depends fully upon the data from the first aim, and if your first aim fails, then you cannot reach your second aim. When your aims are fully dependent on each other and you have no other way to design your proposed research, then in the **Potential Problems and Possible Solutions** subsection of the **Methods Section** (see Chapter 6.10), you should explain that your research will produce novel and significant results in any case, and then identify these results.

---

[12] NIH. Quick Guide for Grant Applications, Revised September 2010. http://deainfo.nci.nih.gov/extra/extdocs/gntapp.pdf (accessed November 28, 2014).

The aims need to be sequenced logically in order for readers to understand their relationship to each other and to the research objective. Where there are different types of aims in one proposed research project, the usual sequence is descriptive (if any) before analytical aims (if any); and analytical aims (if any) before applied aims (if any); and applied aims (if any) before clinical aims (if any), as indicated here:

Descriptive Aims → Analytical Aims → Applied Aims → Clinical Aims

When sequencing multiple aims of *one type*, you sequence them chronologically, that is, in the order that you intend to begin their respective methods. For example, if you have 2 analytical aims, you sequence them chronologically according to which you will start first:

Analytical Aim (Time 1) → Analytical Aim (Time 2)

If there are multiple applied aims, the logical sequence is to develop or refine a product or procedure, then verify it, and then validate it:

Developmental Aim → Verification Aim → Validation Aim

In a translational research project, the sequence is to conduct basic research, develop a potential intervention, and then test the intervention on non-humans, and eventually, on humans in clinical research:

Basic Aims → Developmental Aims → Verification Aims → Validation Aims

However, it is not common in one grant proposal, to propose research that spans bench to bedside. You or another researcher will have likely performed the basic research in a separate study. Thus, the sequence of aims would be similar to the sequence of multiple applied aims:

Developmental Aims →Verification Aims → Validation Aims

● **Phrasing of the research purpose statement**. Whether drafting the long-term goal, the research objective, or an aim, the research purpose statement has 3 components: (1) an identifier, which explicitly names the level of research purpose (*goal*, *objective*, or *aim*); (2) a research purpose verb (e.g., *elucidate* in Case 2-1, sentence 1); and (3) the technical area of investigation. The research objectives in Examples 27, 28, and 29 reflect this 3-part structure.

Level of research identifier	Research purpose verb	Technical area of investigation
↓	↓	↓
Example 27 Our research objective is	to determine	how the interaction between normal blood flow and cardiac tissue affect early development of the chick heart.

Example 28
The objective of   to elucidate   the molecular mechanism of
this project is   voltage hysteresis in
                                                     hyperpolarization-activated,
                                                     cyclic-nucleotide-gated (HCN)
                                                     ion channels.

Example 29
The overall   to better understand   how the Gram-negative bacterial
goal of this   pathogen *Burkholderia*
proposal is   *pseudomallei* (*Bp*) initiates and
                                                      sustains infections in the GI
                                                      tract.

The research purpose verbs in these examples – *determine, elucidate,* and *better understand* – are all common in research purpose statements of basic, applied, clinical, and translational research.

● **Research purpose verbs.** Research purpose statements – whether the long-term goal, research objective, or aims – have research purpose verbs. A research purpose verb focuses on what you intend to accomplish upon the successful completion of the proposed research. The following verbs are examples of research purpose verbs:

assess	*determine*	*refine*
characterize	*develop*	*understand*
create	*elucidate*	*validate*
define	*identify*	*verify*
describe	*improve*	

When choosing a research purpose verb, you can select a verb from the above list, but another source for a research purpose verb is in the RFP that you are targeting, for it will usually use key research purpose verbs in its explanations. There are certain patterns of usage with verbs in statements of your research purpose:

(1) The long-term goal and research objective are frequently phrased with such verbs as *to (better) understand, to determine,* and *to elucidate.* The verb *elucidate* is usually used only for the long-term goal, not for the research objective.

(2) If one aim is *to characterize* (or *to describe*), it is usually followed (***not*** preceded) by an aim beginning with *to determine.*

(3) If the purpose of an aim is *to develop* or *to create,* you may likely also be proposing *to validate* the efficacy of your creation, using appropriate methods from your discipline. The validation aim necessarily comes after development or modification of the product or intervention.

(4) Training aims, such as in educational grants, commonly begin with *to develop*, such as *to develop postdoctoral fellows' multidisciplinary research skills in the neurological sciences.*

(5) The research objective and aims of a study that proposes to continue research in a previously funded line of study (commonly termed a **competitive renewal**) sometimes use the verbs *to refine, to extend,* or *to continue,* and the adverb *further,* such as *to further refine.*

● **Technical area of investigation**. The technical area of investigation in a research purpose statement specifies the specific topic or scope of the proposed studies. Similar to the technical area of investigation in a problem statement (see Chapter 2.2.3), the technical area in a research purpose statement can be phrased as a technical question, beginning with an interrogative (e.g., *whether, what, where, how,* and *to what extent*). The technical area of investigation needs to be described in key terms.

Three common problems in phrasing a clear research purpose statement are: (1) the level of research purpose is not identified; (2) the technical area of investigation is too general or too specific; and (3) the verb is not a research purpose verb. Sometimes research purpose statements have all 3 problems, such as Example 30a, which is revised in Example 30b.

Example 30
a. ‡Our objective is to *investigate* subjects with chronic unilateral vestibular loss.
b. The objective of our proposed research is to ~~investigate~~ determine ~~subjects with chronic unilateral vestibular loss~~ whether subjects with chronic unilateral vestibular loss have underlying postural asymmetries.

Example 30a is not only too general with the verb *investigate*, but it focuses only on methods, with the actual technical area of investigation not specified. In its revision in Example 30b, the verb *investigate* is changed to the research purpose verb *determine*. In addition, the technical area of investigation, which begins with the question word *whether* in Example 30b, still presents methodological information but does so in the context of a technical area of investigation (i.e., a research issue, such as *underlying postural asymmetry*).

> **Guideline 2.9**   *Use a research purpose verb in a statement explaining the research purpose of your proposed research and methods.*

● **Connection: Methods.** In the **Aims Section**, another way to connect your review of literature and the analysis to your proposed research is to characterize your proposed methodological approach and methods briefly, and to identify expected methodological outcomes. In fact, NIH now requires its **Specific Aims Section** to include a summary of the expected outcome(s).[13]

---

[13] NIH. Quick Guide for Grant Applications, Revised September 2010. http://deainfo.nci.nih.gov/extra/extdocs/gntapp.pdf (accessed November 28, 2014).

The characterization of your proposed methodological approach involves identifying your object of study (e.g., humans, a bacterium, or an algorithm) and the primary means you will use to investigate it. The basic methodological approach is explained in Passage 1 if the approach is similar across your proposed aims or in Passage 2, in each aims paragraph, if the approach differs for each aim. For example, in Case 2-1, the methods are basically different for each aim so the methods for each aim are briefly described in each aims paragraph. In Case 2-2, the *mouse model of enteric Bp infection* (sentence 12) is mentioned in Passage 1 since it will be used in the methods for each of the 3 proposed aims.

The expected major results or outcomes of your methods, taken together, are identified in Passage 1 or Passage 3; and anticipated minor results or outcomes – those relating to a particular aim – are identified within the aims paragraph for the particular aim. For example, in Case 2-3, the outcome of the OCT approach is identified in Passage 3, sentences 29–31.

● **Connection: Anticipated results or main outcome.** Another way to connect the review of literature and the analysis of your proposed research is to identify the results or main outcome that you expect from executing your proposed research. This information is useful in the **Aims Section** to help reviewers begin to understand the significance and novelty of your proposed research.

2.2.5 TRACS: The significance

The significance is a type of analysis that identifies the importance of your proposed research. When explaining significance, you need to convince reviewers that your research will yield important discoveries, *as judged against current knowledge, norms, and values in your discipline and in society*. In other words, you need to argue that the outcomes from your research will ultimately have a substantial, positive impact on your scientific discipline, in particular (technical significance), and/or on the larger domain of science and society (social significance). Significance statements can identify the importance of your proposed research from several perspectives, such as:

● **Scientific, technical, and/or medical perspective**. Importance of your proposed research from the perspective of advancing the body of knowledge in your discipline or across disciplines. The NSF term *intellectual merit* to a certain extent aligns with this particular category of significance.

● **Health perspective**. Improvements in medicine and other health-related fields. This improvement can be in terms of prevention, diagnosis, treatment, containment, or cure of diseases, disorders, or injuries; or in terms of the maintenance or enhancement of natural conditions or processes.

● **Societal perspective**. Benefits that may eventually flow from your proposed research as they relate to values that the populace of a country holds dear, such as convenience and efficiency, which are valued in the United States.

• **Educational perspective**. How your proposed research might improve training, learning, and teaching at particular educational levels. This benefit speaks to NSF's *broader impacts*.[14]

• **Economic perspective**. Importance of your proposed research in financial, monetary, or budgetary terms.

• **National security perspective**. Importance of your proposed research in terms of military operations, and the safety and security of citizens and soldiers in a particular country or international alliance.

It is particularly important for you to express significance in Passage 1 of the **Aims Section**, either implicitly or explicitly, to help convince reviewers, from the outset of the narrative, the pressing need for your proposed research and the beneficial impact your research results can ultimately have. In applied, clinical, and translational research (as opposed to basic research; see below), the significance of the proposed research is easier to address since significance is usually obvious. Example 31 is an excerpt from Case 2-4 (sentences 19-20), which expresses significance through a discussion of a therapeutic agent (lipoic acid), which is to reduce inflammation in the optic nerve, preserve neural integrity, and improve visual outcomes in persons with multiple sclerosis who have acute optic neuritis (AON).

Example 31
Based on our preclinical studies and initial clinical trials, we hypothesize that lipoic acid (LA) is an effective therapy to reduce inflammation in the optic nerve, preserve neural integrity, and improve visual outcome in AON subjects. To test our hypothesis, we propose a randomized, controlled trial to study the effects of oral LA on retinal nerve fiber layer thickness in the affected eye of AON subjects.

In basic research, pinpointing significance usually takes more writing effort since the consequence of researching a basic topic needs to be explained. In describing significance of basic research, you can ask yourself such questions as:

> *Who cares if X is studied or not?*
> *Why is it important that I investigate X?*
> *What does X have to do with the prevention, diagnosis, treatment, containment, or cure of X disease, X disorder, or X injury?*
> *What does the X have to do with the maintenance or enhancement of the natural condition X or the natural process X?*

The **Aims Section** in Case 2-1 is from a narrative addressing basic research. Sentence 2 explains the significance of HCN channels in the *prevention of arrhythmic behavior in the heart and brain*.

---

[14] www.nsf.gov/pubs/policydocs/pappguide/nsf3001/nsf13_1.pdf (II-8, II-11; accessed November 28, 2014).

In addition to locating significance information in Passage 1, if space allows, you can also include significance in Passage 3, in an ending paragraph, to reiterate the importance of your proposed research. In Case 2-3, the last half of sentence 29 identifies the technical significance of the proposed research: *provide extremely high-resolution testing results without removing the vascular construct from the bioreactor or otherwise damaging the construct.* Sentence 30 also identifies the technical significance of the proposed research: *track the development of the engineered tissue … while providing on-line quality control measures,* and sentence 31 notes the medical significance: *to consistently generate the engineered tissue substitutes for surgical implantation.*

2.2.6   Why is scientific argumentation important?

Why should you care whether your narrative, in general, or your **Aims Section**, in particular, conveys at least one scientific argument? The components of a scientific argument are the ***minimally necessary components*** (***prima facie elements***) of the conceptual framework of your proposed research. Without including information comprising at least one scientific argument (**TRACS**), you might describe your proposed research, but you will not be able to sell it – that is, to convince your reviewers that your proposed research is so significant and novel that it merits funding.

Two types of information that PIs sometimes inadvertently omit in the **Aims Section** are the **A** and **S** of **TRACS**: analysis and significance. Although PIs usually include a brief review of research in the **Aims Section** (and a longer review in the **Background Section**), they sometimes do not adequately analyze the reviewed research. Maybe you are one of these PIs. Chapter 2.2.3 provides different ways that you can analyze previous research. You should choose the type of analysis that allows you to do justice to your proposed research.

Also, sometimes PIs do not adequately identify the significance of the proposed research in the **Aims Section**. Significance, a type of analysis, is so critical to convincing reviewers of the merit of your proposed research that one major United States funding agency (NIH) likely renamed the **Background Section**, the **Significance Section** (among other reasons).

What is noteworthy is that both analysis and significance are ultimately derived from the review of previous research, which is presented in the **Aims**, **Background**, and **Preliminary Studies/Progress Report Sections**, and at least in a rationale subsection in the **Methods Section**. Thus, in all of these (sub)sections you need to review the previous research and also to analyze it and to identify significance.

2.3   Novelty and scientific argumentation

Novelty can be implicit or explicit in the **Aims Section**. When submission requirements require its explicit discussion, you need to identify novelty through such terms as: *for the first time, novel, novelty, original,* and *originality*. In most instances, novelty is not explicit but rather an inference derived from the analysis. Novelty is usually expressly discussed in the **Background Section**. (See Chapter 3.9.)

2.4   Additional perspectives on the Aims Section

Thus far, the **Aims Section** has been discussed in terms of content that comprises a scientific argument, but there are at least 3 other ways to view an **Aims Section**.

- **Known and unknown information.** The first additional perspective of an **Aims Section** is in terms of your proposed research advancing scientific, technical, and/ or medical knowledge. The **Aims Section** presents information that moves from *what is known* about your specific topic or research area through your review of literature, to *what is not known*. The gap in knowledge between what is known and what is not known is identified through the PI's analysis of what is known.

The information that is known is located primarily in Passage 1, before the aims, but can also be included in the aims paragraphs of Passage 2.

By identifying what is and is not known about your proposed research topic or research area, you can help reviewers understand the novelty and significance of your proposed research.

- **Specific overview.** The **Aims Section** can be considered a specific overview[16] of your proposed research and of the scientific, technical, or medical context from which it emerges – an overview that is further detailed in subsequent sections of the narrative. Similar to a precise generalization (see Chapter 2.2.2), *specific overview* is not an oxymoron. The **Aims Section** is specific; thus, the narrative moves from *specific* information in the **Aims Section** to *equally specific or even more specific* information in subsequent sections of the narrative:

---

[15]   The layout showing the relationship between what is and what is not known is adapted from the argumentation scheme in Toulmin, Stephen. *The Uses of Argument*. Cambridge: Cambridge University Press; 1958.
[16]   In business writing, the specific overview is sometimes called the *executive overview*.

An **Aims Section** that is specific is important for many reasons. It is the reviewers' first encounter with your understanding of the scientific method, your ability to design research, and your ability to review and analyze previous research. You will hurt your credibility and the reviewers' perceptions of the merit of your proposed research if you are not specific and detailed; if you are not clear or accurate; if you do not present a specific, unambiguous picture of your research objective or of others' research; and if you have grammatical or even typographical problems. Depending on the extent of damage to your credibility, you may not be able to recover in the remaining sections of the narrative.

● **Problem-solution features**. Yet another perspective on the **Aims Section** is that it sets the stage for the problem-solving character of your proposed research. The **Aims Section** identifies the problem in the current research that you are trying to solve with your proposed research. The solution that you are trying to achieve might be, for example, through the collection of data that address a hypothesis or through the development or refinement of a piece of equipment or of an intervention.

Because of the importance of the **Aims Section**, you should plan on continually reviewing and possibly revising it while drafting the other sections of the narrative.

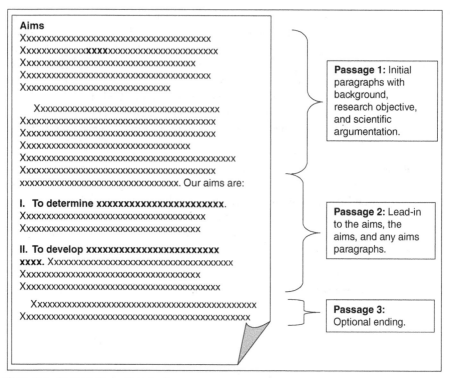

Figure 2-1. Passages in an Aims Section.

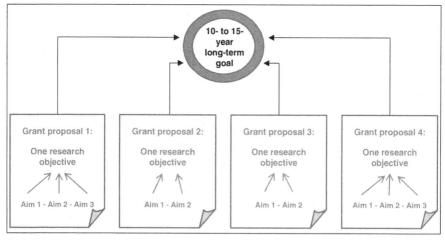

Figure 2-2. Relationship among a long-term goal, research objective of proposed research, and aims. This figure shows a line of research that the PI intends to investigate for 10–15 years. The purpose of the line of research is the long-term goal. Over the 10–15 years, the PI's research investigating the long-term goal might be supported by 4 grant proposals, each of which has a research objective. In the line of research illustrated in this figure, Grant proposal 1 has 3 aims, Grant proposals 2 and 3 have 2 aims each, and Grant proposal 4 has 3 aims.

Cases

Case 2-1. Aims Section from a narrative proposing basic research. Passage 1: Sentences 1–4; Passage 2: Sentences 5–14. **Bold** = Topic, **A** = Analysis, **C** = Connection.

Specific Aims

[1]The objective of this project is to elucidate the **molecular mechanism of voltage hysteresis in hyperpolarization-activated cyclic-nucleotide (HCN) ion channels.** [2]HCN channels undergo a shift in their voltage dependence (hysteresis) during pacemaker activity that is likely critical for the prevention of arrhythmic behavior in the heart and brain (refs). [3]Hysteresis occurs in the voltage dependence of HCN channels even in excised, cell-free patches, and it is not modulated by second messengers, such as cAMP, suggesting that the mechanism is intrinsic to the core of the channel. [4]We hypothesize that a conformational change intrinsic to the HCN channel underlies hysteresis in HCN channel voltage dependence. [5]We propose the following specific aims:

C: Purpose & Topic
A: Review

Significance

Review

A: Speculation

A: Hypothesis

Specific Aim 1: To identify the regions of HCN channels that change conformation during voltage hysteresis. [6]Using voltage-clamp fluorometry (VCF), we will introduce cysteine residues one-at-a-time to each region of the HCN channel that is believed to reside extracellularly based on structure-function studies and homology with voltage-gated potassium (Kv) channels. [7]These mutant channels will be expressed in Xenopus oocytes and exposed to a cysteine-reactive fluorophore. [8]The insertion of the fluorophore will induce an increased fluorescence signal that we will detect with a confocal microscope or photo-multiplier tube. [9]Following verification of the fluorophore attachment, voltage hysteresis will be induced using a standard, two-electrode voltage clamp (TEVC). [10]If the region to which the fluorophore is attached undergoes conformational changes during voltage hysteresis, these changes should alter the fluorescence signal (spectral shift or amplitude change) of the attached environment-sensitive fluorescent probe.

C: Purpose

C: Methods

A: Prediction

Specific Aim 2: To determine specific amino acid interactions that are critical for HCN channel voltage hysteresis. [11]Amino acids in the region(s) of the channel that undergoes conformational change during voltage hysteresis (as reported by VCF measurements in Aim 1) will be mutated to identify residues that are critical for HCN channel voltage hysteresis. [12]For example, if electrostatic interactions between charged residues are altered during hysteresis, we will identify these interactions by neutralizing or reversing the critical charges. [13]Further, we weill attempt to disrupt or enhance voltage hysteresis by cross-linking the regions that undergo conformational changes. [14]Our hypothesis is that specific amino acids interactions stabilize the voltage sensor of the HCN channel in the open or closed stated during voltage hysteresis.

C: Purpose

C: Methods

A: Hypothesis

Case 2-2. Aims Section proposing basic research with a scientific argument and without aims paragraphs. Passage 1: Sentences 1–19; Passage 2: Sentences 20–23. **Bold** = Topic, **A** = Analysis, **C** = Connection.

2. Specific Aims. ¹The overall goal of this proposal is to better understand **how the Gram-negative bacterial pathogen** ***Burkholderia pseudomallei* (*Bp*) initiates and sustains infections in the GI tract.** ²The *Bp* bacterium is normally found in soil and water, but is also a deadly pathogen in humans, where it can cause a variety of difficult-to-treat infections ranging from acute sepsis to chronic abscesses. ³While *Bp* is endemic in southeast Asia and northern Australia, infections are now being diagnosed with increasing frequency around the world, including in Central and South America. ⁴Therefore, it is likely that *Bp* infections will soon be identified in the U.S., as the result of either deliberate or accidental introduction. ⁵Thus, the proposed studies to gain a better understanding of the pathogenesis of *Bp* infection can be justified based on both national and international health concerns.

⁶Little is known regarding how infection with *Bp* develops, though inhalation or cutaneous inoculation are currently considered the most likely routes of infection. ⁷*However, our new studies indicate that Bp is actually a primary enteric pathogen that can readily establish acute or persistent GI tract infection following oral inoculation in mouse models.* ⁸*Furthermore, our findings also suggest that GI tract is the primary reservoir for maintenance and dissemination of Bp during chronic infection.* ⁹Thus, re-defining *Bp* as a primary enteric pathogen will have major implications for understanding how humans are infected with *Bp* and the risks posed by *Bp* contaminated food, soil, and water. ¹⁰However, at present essentially nothing is known regarding the pathogenesis of enteric infection with *Bp*.

¹¹Therefore, the studies proposed here are intended to fill a critical void in our understanding of pathogenesis of infection with this important and emerging bacterial pathogen. ¹²To address these knowledge gaps, we will use a mouse model of enteric *Bp* infection developed in our lab to answer three key questions. ¹³First, is enteropathogenicity a property of all isolates of *Bp*, or are only certain isolates virulent after oral inoculation? ¹⁴If highly virulent enteric strains of *Bp* are identified, will in vitro assays of invasion correlate with the virulence phenotype? ¹⁵Secondly, what are the target cells for *Bp* infection in the intestine during acute and chronic infection? ¹⁶This information could be very important for developing new vaccination or treatment strategies. ¹⁷Third, how does *Bp* disseminate from the intestine to other organs following enteric infection, since widely disseminated infections are a key feature of *Bp* infection? ¹⁸For example, if dissemination were found to be primarily cell-associated, then different classes of antimicrobials could be used to treat chronic, as opposed to acute, infection. ¹⁹The information generated from these studies may substantially alter our view of *Bp* as a pathogen and lead to a reassessment of the risks

C: Purpose
Topic

Review

A: Problem

A: Prediction

C: Aim &
Significance

A: Problem &
Speculation

Review

A: Speculation

Significance

A: Problem

A: Purpose

C: Methods

A: Technical
Questions

Significance

A: Technical
Question

Significance

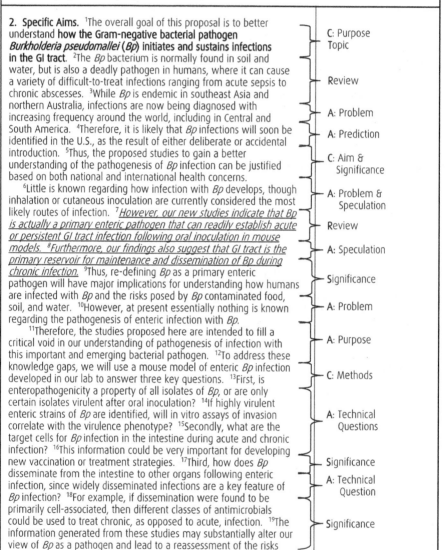

posed by oral *Bp* infection. [20]The questions raised above will be addressed by means of 3 Specific Aims.

Aim 1. [21]Determine whether enteropathogenicity is a general feature of all or only some *B. pseudomallei* isolates.

Aim 2. [22]Identify intestinal target cells for *B. pseudomallei* during acute and chronic enteric infection.

Aim 3. [23]Determine how *B. pseudomallei* disseminates from the GI tract following oral inoculation.

C: Purpose

64

Cases

Case 2-3. Aims Section from a narrative proposing applied research. Passage 1: Sentences 1–13; Passage 2: Sentences 13–14; Passage 3: Sentences 29–31. **Bold** = Topic, **A** = Analysis, **C** = Connection.

A. SPECIFIC AIMS

[1]The field of **tissue engineering/regenerative medicine** (TE/RM) is a rapidly expanding discipline with the potential to revolutionize multiple medical disciplines. [2]While the ultimate goal of TE/RM is the production of fully functional tissues and organs readily available for transplantation, another potential application of TE/RM is the use of tissue-engineered products to study the pathology of disease and its treatments in vitro. [3]In both cases, it is vital to reproducibly manufacture morphologically and functionally consistent tissues. [4]This manufacturing requires non-invasive, real time monitoring and on-line assessment of the engineered tissue construct.

Topic, Review, & Significance
Review
Significance
A: Conclusion

[5]An **application of tissue engineering** that has been the subject of intensive research is the **treatment of cardiovascular disease**. [6]A functional and durable small-caliber vascular graft would have enormous implications for the **treatment of cardiovascular disease**. [7]Unfortunately, despite many efforts, the clinical utility of conventional vascular graft materials, such as ePTFE, remains limited by adverse blood reactions (thrombosis) and healing responses (intimal hyperplasia). [8]The engineering of fully biologic vascular constructs that simulate natural vessel properties is a strategy that has many important merits, although several key issues remain to be addressed. [9]Foremost among these is the ability to **assess/screen these tissue-engineered constructs in real time, in a non-invasive and non-destructive manner**. [10]Optical coherence tomography (OCT) and optical elastography (OE), individually and combined, are viable technologies that have the potential to monitor the morphological and mechanical characteristics of these developing tissues and to provide sensory feedback for bioreactor control.

Topic & Review
Significance
A: Problem
Review
Topic & A: Problem
A: Speculation

[11]The hypothesis of this proposal is that OCT and OE non-invasively and non-destructively evaluate the changing functional properties of TE constructs. [12]Specifically, we propose that OCT, combined with OE, will be able to measure the bulk and local mechanical properties and morphological characteristics of developing tissue-engineered constructs, while also providing feedback for bioreactor control. [13]To investigate our hypothesis, we will conduct the project with a common, tissue-engineered scaffold for vascular graft applications using vascular cells embedded in a collagen gel. [14]Our specific aims are to:

A: Hypothesis
A: Prediction
C: Methods

1. [15]Determine time-dependent morphological changes in TE constructs. [16]As the seeded cells populate, attach, migrate, and deposit extracellular matrices within the model constructs, the depth-resolved morphology of the construct becomes an important, dynamic parameter that is closely related to cell-growth profile and tissue turnover. [17]We predict that our optical techniques will successfully monitor the dynamic morphology of a developing

C: Purpose
Review
A: Prediction

tissue-engineered vascular scaffold during a lengthy culture period. [18]We will use OCT to measure the active production of collagen and will document the degree of cell coverage for smooth muscle cells (SMCs) at different depths, in a longitudinal time scale. [19]We will test a number of seeding concentrations of SMCs and the influence of different chemical stimuli on cell proliferation.

⟩ **A: Prediction**

⟩ **C: Methods**

2. [20]**Identify time-dependent changing mechanical properties.** [21]SMCs embedded in collagen gels reorganize the collagen, compact the gel, and alter the mechanical properties of the developing tissue. [22]We predict that these changes can be effectively documented in a non-destructive and non-invasive manner, using our optical techniques. [23]Specifically, OCT combined with OE will measure the bulk and local mechanical properties of the tissue-engineered construct as the tissue develops over time in our established bioreactor system.

⟩ **C: Purpose**

⟩ **Review**

⟩ **A: Prediction**

⟩ **C: Methods**

3. [24]**Provide real time sensory feedback control on the biophysical state of construct.** [25]We predict that OCT, combined with OE, will perform real-time assessment of the developmental progress of the tissue-engineered vascular constructs. [26]This real-time assessment will allow us to control the environmental parameters of the constructs in order to produce the desired functionality. [27]We will use the morphological and mechanical properties of the construct measured in real time, using our optical techniques, to iteratively alter the chemical and mechanical environment of the bioreactor.

⟩ **C: Purpose**

⟩ **A: Prediction**

⟩ **C: Methods**

[28]For all of the measurements we perform, we will directly compare our results with standard end-point evaluations (e.g., histology and scanning electron microscopy).

[29]The primary advantage of our approach over traditional microscopy (Aims 1 & 3) and mechanical testing (Aims 2 & 3) is that we can provide extremely high-resolution results without removing the vascular construct from the bioreactor or otherwise damaging the construct. [30]Thus, we will be able to track the development of the engineered tissue in the bioreactor while providing on-line quality control measures and time-resolved measures of the functional properties of the tissue-engineered construct. [31]In turn, this real-time assessment will allow tissue engineers to adjust the environmental parameters (e.g., flow rate, nutrient supply) to consistently generate the engineered tissue substitutes for surgical implantation.

⟩ **Significance**

⟩ **C: Outcome**

Case 2-4. Aims Section from a narrative proposing translational research.
Passage 1: Sentences 1–21; Passage 2: Sentences 22–31; Passage 3: Sentence 32.
Bold = Topic, **A** = Analysis, **C** = Connection.

A. Specific Aims

[1]Demylinating diseases are inflammatory diseases of the central nervous system (CNS), which include relapsing remitting multiple sclerosis (RRMS) and monophasic conditions, such as **acute optic neuritis (AON)**. [2]AON, which is most commonly associated with multiple sclerosis (MS), is one of the most frequent inflammatory causes of visual disability among young and middle-aged adults [1,2]. [3]AON involves subacute, painful loss of vision due to inflammation of the optic nerve. [4]While most patients with AON experience significant recovery of vision, many have some residual, visual symptoms, and most have evidence of permanent demyelination and axonal loss, despite the single currently available treatment of corticosteroids. [5]High-dose corticosteroids, given intravenously for 3-5 days, have been shown to hasten the recovery of vision acutely, but do not affect the long-term visual outcome in AON [3,4]. [6]In addition, corticosteroids can cause significant side-effects, including gastric irritability, mood disorders, insomnia, hyperglycemia, electrolyte imbalance, and increased susceptibility to infections. [7]There is, therefore, a need to develop a **non-steroidal treatment of AON** that is well-tolerated and decreases permanent injury of the optic nerve.

[8]**One possible treatment to reduce visual impairment in AON is lipoic acid (LA)**. [9]LA is a natural antioxidant available as an oral dietary supplement. [10]We discovered that LA was an effective therapy in a murine mouse model of MS (experimental autoimmune encephalomyelitis [EAE]) – a finding that has since been replicated by others [3,4]. [11]More recently, we demonstrated that LA is also effective in treating a mouse model of AON: experimental autoimmune optic neuritis (EAON). [12]Importantly, we discovered that LA dramatically reduces axonal loss in the spinal cords of EAE mice and in the optic nerve of EAON mice. [13]We also discovered a novel mechanism of action in LA, related to its therapeutic effects in EAE and EAON. [14]LA stimulates cAMP in CD4+ T cells and natural killer cells via the prostaglandin EP2 and EP4 receptors, thereby activating protein kinase (driven by the immunoregulatory pathway) [5]. [15]Based on our preclinical studies, we conducted 2 pilot trials to determine the pharmacokinetics (PK) of LA and the safety of a 1200-mg dose of LA on MS subjects [6,7]. [16]The subjects tolerated well the LA dose, and LA levels in their serum were comparable to those obtained in mice receiving a therapeutic dose of LA [7]. [17]Further, oral LA administration stimulated cAMP in peripheral blood mononuclear cells (PBMC) of the subjects (unpublished results). [18]Using a 2-week placebo-controlled trial of 3 different doses of LA in MS subjects, we also found that LA was well-tolerated and was associated with a change in 2 serum markers of inflammation: matrix metalloproteinase-9 (MMP-9) and soluble intercellular adhesion molecule (sICAM-1 [6]).

Topic

Review

A: Problem

A: Speculation
Topic

Review

[19]Based on these studies, we hypothesize that LA is a safe, effective therapy to reduce inflammation in the optic nerve, to preserve axonal integrity, and to improve visual outcome in AON subjects. [20]To test these hypotheses, we propose a 24-week, randomized, placebo-controlled, 2-arm pilot trial to determine the effects of oral LA (treatment period of 6 weeks at 1200 mg LA dose per day) on retinal nerve fiber layer (RNFL) thickness in the affected eye of AON subjects since RNFL thinning has been found to occur after the onset of AON [8,9]. [21]We predict that oral LA, compared with a placebo, will reduce optic nerve inflammation and axonal loss, thereby preserving the RNFL in our subjects, as measured by optical coherence tomography (OCT). [22]Specific aims of this clinical trial are:

— A: Analysis

— C: Methods

— A: Prediction

— C: Purpose

[23]Specific Aim 1: To determine whether AON subjects who receive LA have less permanent optic nerve injury than those receiving a placebo. [24]The primary outcome measure will be the difference from baseline in RNFL thickness of the affected optic nerve, as determined by OCT, at 12 weeks post LA treatment. [25]Secondary outcome measures to assess optic nerve injury will be changes from baseline in the RNFL thickness at week 24, and changes from baseline in low- and high-contract visual acuity, contrast sensitivity, and visual field changes at weeks 12 and 24.

— C: Outcome

[26]Specific Aim 2: To determine the safety of the LA treatment regimen on AON subjects. [27]Safety will be determined by assessing LA side effects (e.g., gastrointestinal tolerability and liver function).

— C: Purpose

— C: Methods

[28]Specific Aim 3: To determine whether serum LA levels in AON subjects treated with LA affect PBMC cAMP levels and whether they affect serum inflammatory immune markers. [29]We will conduct PK studies to quantify serum concentrations (Cmax) of LA. [30]We will collect blood to determine cAMP, MMP-9, sICAM-1, and pro-inflammatory cytokine (e.g., interferon-γ; IFN-γ), tumor necrosis factor-α (TNF-α), and interleukin (IL-2) levels. [31]We will then determine whether LA Cmax is related to: a) changes from baseline in cAMP levels in PBMCs, b) changes from baseline in serum MMP-9, sICAM-1, IFN-γ, TNF-α, and IL-2 at week 6, c) expression of mRNA in PBMC for INF-γ, TNF-α, and IL-2 at baseline and at week 6, and d) clinical outcomes at weeks 12 and 24 (e.g., RNFL changes, visual acuity, contrast sensitivity, and visual field).

— C: Purpose

— C: Methods

— C: Outcome

[32]The results from this research will provide preliminary data and a rationale for a phase II trial to determine whether oral LA is an effective treatment to reduce disability in AON subjects.

— C: Outcome

The Background Section

The **Background Section** provides an historical research context for your proposed research. When explaining this context, you need to review the previous research on your topic or research area, and then evaluate its strengths and weaknesses.

A strong **Background Section** helps reviewers understand that your proposed research is not arbitrary, but the *logical* outgrowth of previous research. A strong **Background Section** also helps you persuade reviewers that your proposed research is *significant* and *novel*. It is not surprising that NIH has divided its former **Background Section** into 2 major sections: **Significance** and **Innovation**.[1] In contrast to NIH, NSF typically provides little guidance in structuring the **Background Section** (or the **Aims Section**), but does identify the need to address the state of knowledge in the field and significance:

> The Project Description should provide a clear statement of the work to be undertaken and must include: objectives for the period of the proposed work and expected significance; relation to longer-term goals of the PI's project; and *relation to the present state of knowledge in the field*, to work in progress by the PI under other support and to work in progress elsewhere. (emphasis added)[2]

---

[1]   http://grants.nih.gov/grants/funding/phs398/phs398.html (I-46; accessed November 28, 2014).

[2]   http://www.nsf.gov/pubs/policydocs/pappguide/nsf11001/gpgprint.pdf (accessed August 4, 2012).

Perhaps due to NSF's general guidance on the **Background Section** (and the **Aims Section**), it is not unusual for the initial section of an NSF narrative to be composed of information equivalent to information in both the **Aims** and **Background Sections** (see Chapters 2 and 3). However, most funding agencies still require separate **Aims** and **Background Sections**. Before writing the **Background Section**, you need to review submission requirements from your targeted funding agency and follow them.

A strong **Background Section** can also underscore your credibility by revealing how well you are familiar with and understand the previous research leading to your research topic – its strengths and its shortcomings. Further, a strong **Background Section** helps reviewers evaluate your understanding of scientific methodology, in general, and of the scientific method in the research context for your proposed research, in particular.

Other sections of a narrative that also can include background information are the **Aims Section** (Chapter 2.2.2), the **Preliminary Studies Section** (Chapters 4.5 and 4.7), and the rationale subsection of the **Methods Section** (Chapter 5.6). Heading outlines of **Background Sections** are in Cases 3-1 and 3-2, and examples of subsections from **Background Sections** are in Cases 3-3 to 3-12.

3.1 Scope and basic structure of the Background Section

3.1.1 Scope

The **Background Section** expands on the background information that is first presented in the **Aims Section** and addresses the major topics in the **Aims Section**.

Two major problems tend to plague the **Background Section**. One is that its scope is either too inclusive or too narrow. The **Background Section** sometimes covers background information that does not relate to major topics raised in the **Aims Section**, or it fails to cover major topics that are mentioned in the **Aims Section**. In the **Background Section**, you need to present background information on the key topics in your proposed research objective, aims, and hypothesis, as articulated in the **Aims Section**. The other major problem is that the **Background Section** uses different terminology from that used in the **Aims Section**. To help reviewers follow topics across sections, you need to use consistent terms. You can introduce additional terms for additional topics in the **Background Section** if you need them, but you need to make sure that you use key terms from the **Aims Section**, in the **Background Section**.

> **Guideline 3.1** *Discuss key topics from the **Aims Section** in the **Background Section**, and use key terms from the **Aims Section** in this discussion.*

Because of the dependency of the **Background Section** on the **Aims Section**, the **Background Section** is easier to plan and to write after the content and key terms in the **Aims Section** are stable.

> **Guideline 3.2** *Draft the **Background Section** after the **Aims Section** is stable.*

3.1.2   Basic structure and heading outline

The **Background Section** includes an introduction (i.e., a paragraph), body subsections, and an ending paragraph (Figure 3-1). The headings to the body subsections provide the topics of the subsections and help reviewers focus their attention on the topics. To conserve space, a run-in format (Chapter 8.6.2) is usually used for the headings to the body subsections.

An alternative organization to the one shown in Figure 3-1 has the introduction as the first subsection. It is followed by paragraphs without headings, but the elements are the same as in Figure 3-1. The paragraph style is used in short **Background Sections**, such as one page or less. Case 3-3 shows a short **Background Section** from an NIH narrative (which NIH terms **Significance**) without subsections.

Before drafting a **Background Section**[3] with headings, it is useful first to list the headings that you intend for the body subsections. Such a list (termed *heading outline*; see Chapter 1.11.3) allows you to test phrasing, organizational, and layout alternatives before drafting the section. Cases 3-1 and 3-2 give examples of heading outlines to 2 **Background Sections**, showing their subsections, headings, and itemization schemes. The heading outline allows you to evaluate the hierarchical relationship of content before you start writing, and once you begin writing, it helps you stay organized. In most cases, funding agencies specify the required heading to the **Background Section** but not for its subsections. For **Background** subsections, you create unique headings, such as **A.1 Impediments to identifying the vestibular contribution to balance control** in Case 3-1. Guideline 3.3 gives suggestions for phrasing and itemizing unique headings for **Background** subsections:

> ### Guideline 3.3
> (a) *Phrase unique headings to **Background** subsections with key terms from the proposed research objective, aims, methods, and hypotheses in the **Aims Section**.*
> (b) *Phrase unique headings to **Background** subsections as nouns with modifiers.*
> (c) *Consider using an alphanumeric itemization scheme with the subsection headings.*

The heading outlines in Cases 3-1 and 3-2 present headings with nouns that are variously modified with adjectives and prepositional phrases, and both use itemization schemes. Itemization schemes can help reviewers follow the organizational hierarchy of the section and can help you achieve concise cross-references.

3.2   Informal and formal introductions to the Background Section

A **Background Section** needs an introduction, whether the section is divided into subsections or is a series of paragraphs.

---

[3] It is also useful to draft an outline before drafting a **Background Section** without subsections, in which case the heading outline becomes a topic outline.

In a **Background Section** with subsections, sometimes the introduction is a paragraph, located before the first body subsection in the **Background**. This type of introduction, without a heading, is termed an *informal* introduction. However, sometimes the introduction is the first subsection in a **Background Section**. An introduction with a heading is termed a *formal* introduction. In a **Background Section** without subsections, the introduction usually corresponds to the first paragraph, as shown in the first paragraph of Case 3-3. Case 3-5 presents a formal introduction. You can give a formal introduction a standard heading, such as **Introduction**, or a unique heading, such as *Staphylococcus aureus* **as a human pathogen** in Case 3-5.

The **Background** introduction needs to convey significance and, optionally, definitions and fundamental concepts.

● **Significance of the proposed research.** The **Background** introduction needs to identify the significance (importance) of the proposed research in terms of science or technology, medical issues, society, education, economics, or national security (Chapter 2.2.5).

Significance can be explicitly stated or can be implied through statistics that identify the extent and impact of a problem. You can present statistics on: (a) the extent or prevalence of a problem, (b) an issue associated with the research topic, or (c) resources, such as monetary costs, associated with the problem. These statistics help underscore the significance of the proposed research by implying that the proposed research will ultimately, favorably impact large numbers of persons (or other entities) and may help conserve resources. Case 3-3 presents detrimental consequences through a conclusion in sentence 8 and through statistics in sentence 3, both of which serve to underscore the significance of the issue that the PI is proposing to research. In Case 3-4, sentence 8 provides the prediction that the application of GVS (galvanic vestibular stimulation) will enhance a vestibular-impaired subject's ability to control balance. In Case 3-5, in sentences 10 and 11, significance is explicitly addressed in the discussion that identifies the relevance of the proposed research to the development of antimicrobial agents.

Even though significance is mentioned in the introduction, it can again be mentioned in the **Background** subsections. However, the **Background** subsections often do not again address significance, in which case the scope of significance in the introduction covers the entire **Background Section**. Therefore, to achieve *significance by association* in the **Background** body subsections, it is critically important that the **Background** introduction addresses significance explicitly or includes statistics from which significance can be readily implied. However, even if significance is addressed in the **Background** introduction, significance is so critical that it does not hurt to mention it in the **Background** subsections.

> **Guideline 3.4** *Address significance in the introduction to the **Background Section**, and consider addressing it again in subsequent **Background** body subsections.*

- **Definitions of key terms.** Definitions are useful, not only to help clarify key terms for lay reviewers, but also to help technical reviewers assess the PI's understanding of key concepts. The **Aims Section** can include definitions, but oftentimes the **Background** introduction expands on these earlier definitions, especially if the terms (a) are used throughout all or most body subsections of the **Background** or (b) have not been fully defined in the **Aims Section**. The **Background** introduction in Case 3-3 leads with a definition of *Burkholderia pseudomallei*; in Case 3-5, the **Background** introduction leads with a definition of *S. aureus*; and in Case 3-6, the subsection leads with a definition of lipoic acid. Definitions are further discussed in Chapter 8.2.

- **Fundamental features of the topic.** In addition to significance and definitions, a **Background** introduction can offer fundamental features of the proposed research topic and common knowledge about the topic, thus helping reviewers understand the context for the proposed research. Also, a **Background** introduction with fundamental features of the topic and common knowledge is particularly useful to lay or semi-technical reviewers who do not have a basic understanding of critical issues surrounding the topic. The **Background** introductions in Cases 3-3 to 3-5 present fundamental features of, and common knowledge about, the proposed research topic.

- **Scientific argumentation.** Besides significance, a **Background** introduction can present other information comprising a scientific argument, as shown in the right-hand columns of Cases 3-3 to 3-5.

3.3   Body subsections

A **Background Section** comprises *at least* 2 categories of content that are typically formatted into subsections. Guideline 3.5 gives a strategy for dividing the **Background** body into content categories.

> **Guideline 3.5** *Determine the categories (**Background** subsections) of **Background** content by reviewing your research objective, aims, methods, and hypotheses, as phrased in the **Aims Section**.*

3.3.1   Sequencing body subsections

Deciding on the sequencing of the **Background** subsections can be difficult. One strategy is to sequence them in the order that the key topics are introduced in the **Aims Section**.

> **Guideline 3.6** *In the **Background Section**:*
> (a) *Present body subsections that discuss previous research relating to the proposed research objective, then body subsections that relate to more than one aim, and then body subsections that relate to individual aims.*
> ***Exceptions:*** *Proposed research that seeks to develop a new or to improve an existing method or therapeutic intervention. For these exceptions, you explain: (i) the standard method or standard of care, (ii) existing non-standard methods or therapies, and then (iii) briefly, any proposed methods or therapies.*
> (b) *Sequence the body subsections in the order that key concepts in the aims are presented in the **Aims Section**.*

As noted in Guideline 3.6(a), there are notable exceptions to Guideline 3.6. For these exceptions, you first explain any standard method or standard of care, then any existing non-standard methods or therapies, and then, briefly, any proposed methods or therapies. This organization can be seen in the heading outline of Case 3-2(I), where B.2, B.3, and B.4 each begin with a subsection describing the *standard techniques*, which is then followed by a subsection describing OCT (optical coherence tomography). An alternative to Case 3-2(I) is shown in Case 3-2(II), which has all of the standard techniques grouped together in a subsection (B.2) and then OCT in another subsection (B.3).

3.3.2   Background body subsections

The **Background Section** needs to persuade reviewers that you know the relevant background research to your proposed research, that your proposed research is related to previous research, and that your proposed research is important and novel. To achieve these purposes, each **Background** subsection needs *at least* one scientific argument and needs to be associated with the proposed research objective or at least one of the proposed aims.

> **Guideline 3.7** *Include **at least** one scientific argument in **each Background** body subsection: topic, review, analysis, connection, and significance (**TRACS**).*

3.4   Topic in a Background subsection

A heading to a **Background** body subsection should indicate the main topic of the subsection – that is, what the **Background** body subsection is about. Strategies for establishing a clear topic in a subsection are given in Guideline 3.8:

> **Guideline 3.8**
> (a) *Using key terms from the **Aims Section**, create headings to **Background** subsections to indicate the main topics of each subsection.*

(b) *For each **Background** body subsection, integrate the topics into the first or second sentence of the subsection.*

(c) *Repeat the key terms from the heading throughout the **Background** subsection to help reviewers track the topics in the subsection.*

As noted in Guideline 3.8(a), the heading to a **Background** subsection should repeat key terms from the **Aims Section**. This repetition helps reviewers track key concepts from the **Aims Section** into the **Background Section**. In the heat of drafting, it is not uncommon for writers to use synonyms in the text for the key terms in the headings, or not to address the topics until well into the subsection. Both of these writing problems need to be avoided. After reading through the heading, the reviewers should encounter the topic within the first or second sentence of the subsection.

The heading to the **Background** body subsection in Case 3-6 consists of key terms from the **Aims Section**, and the very first sentence uses a key term from the heading: *LA* (lipoic acid), which is repeated a total of 17 times in the subsection. In the **Background** body subsection presented in Case 3-7, the heading identifies the topics *mutations, α_{1F} calcium channel,* and *X-linked congenital stationary night blindness.* Examining α_{1F} *calcium channel,* for example, the words that comprise this term occur multiple times in the subsection: the word α_{1F} occurs 8 times, beginning with the first sentence; and the term *calcium channel* also occurs in the first sentence and is repeated a total of 4 times in the subsection. In Case 3-8, the heading presents the topics *regulation, hypothalamic-pituitary gonadal axis* (*HPG axis*), and *autoimmunity. Regulation* is mentioned in sentence 6, in its verb form *regulating;* and in sentence 16, regulation is mentioned through its antonym *dysregulation.* The first sentence mentions the *HPG axis* twice, and sentences 6, 9, 10, 12, 14, and 17 also mention the *HPG axis.* Sentences 1, 10, 15, and 16 mention *autoimmunity* in its adjective form *autoimmune.*

Headings to subsections are also very important in helping reviewers anticipate what will be discussed. By repeating key terms from the heading throughout the subsection, you can establish a topic, which can help reviewers more readily understand your discussion.

3.5   Review of previous research

You did not conceive of your proposed research in a vacuum. Your mind was prepared to pursue your selected research topic, to formulate your research objective and aims, and to design your methods, to some extent as the result of your evaluating other researchers' studies and your own. You need to pay homage to these previous, relevant studies, and you need to identify how your proposed research relates to the strengths and short-comings of other researchers' studies (and to a lesser extent, to your previous studies).

The significance and novelty of your proposed research can only be identified against a backdrop of known information. You need to select previous research that will help you articulate why your proposed research is important and how it is original or distinctive.

Although the term *review of literature* is sometimes used synonymously with the **Background Section**, you might misunderstand the scope of the **Background Section** if you construe it to be equivalent to a review article. It is not. You need to be highly selective when choosing which studies and facts to review in the **Background Section** (see Chapter 3.1.1 on scope). To this end, Guideline 3.9 offers criteria for selecting studies and facts to review.

Guideline 3.9 *In a **Background** subsection, review studies and facts that:*
(a) *Are directly relevant to the topic of the subsection and to your proposed objective, aims, methods, and/or hypotheses.*
(b) *Come primarily from published research of other researchers.*
(c) *Are up-to-date.*
(d) *Help pinpoint both the significance and the novelty of your proposed research.*
(e) *Are politically expedient.*
(f) *Logically relate to and reinforce your analysis of the research.*

Background subsections are not surveys of research; they should be limited to facts from previous research that are relevant to your particular proposed research, as you characterize it in the **Aims Section**. It is because of this direct relevance that you should be able to cross-reference to *the research objective, hypothesis, or at least one of the proposed aims* in each **Background** subsection (see Chapter 3.7 for a discussion of connection).

Four types of information typically comprise the review in a **Background** body subsection:

(1) Common knowledge.

(2) Other investigators' previous research that is relevant to your proposed research and proposed methods.

(3) Your preliminary research that is relevant to your proposed research topic and/ or your research skills that will be needed to execute the proposed methods.

(4) Fundamental concepts underlying the research topic.

● **Common knowledge.** Common knowledge is information that experienced professionals in a specific field of study consider acceptable and that no longer needs substantiation. Common knowledge does not usually need citations (see Chapter 8.1). However, it may be difficult for some PIs, especially those early in their professional careers, to identify what information is common knowledge. If you are such a PI, you can seek out the information that you suspect is common knowledge in review articles and recent textbooks, and then cite to them. Keep in mind that common knowledge is dynamic; what was considered common knowledge in the 1980s might have been refuted in 1999 and might have become folklore by 2014. Therefore, it does not hurt to provide citations for common knowledge. Common knowledge is illustrated in sentence 2 of Case 3-6, for example, which describes how lipoic acid is available over the counter in the United States and by prescription in Germany.

- **Other researchers' and your previous research.** The review primarily consists of facts from and about other researchers' published studies, and it needs to fairly, accurately, and objectively represent the previous research. There are many different facts from which to select, such as others' previous research purposes, methods, collected data, findings/results, conclusions, speculations, and hypotheses. You should select those facts that are most relevant to your proposed research. All information and facts that you review from published sources, except for common knowledge, need citations (see Chapter 8.1).

The most credible source for reviewed information is from published papers of original research since the vast majority of them have been subjected to *peer review*. You can use relevant, sound, up-to-date research from peer-reviewed papers, in hard-copy and electronic formats, as long as the journals in which the papers are published, are considered reputable by professionals within your field.

Information from reputable, peer-reviewed research papers is usually preferable to information from review articles. Two problems with review articles are that: (1) you are viewing the previous research through the eyes of the author writing the review, and (2) the reviews usually do not include enough details about the research for you to assess the quality of scientific methodology and the degree to which the reviewed research is relevant to your proposed research. However, review articles can be invaluable in synthesizing research in a field and in directing you to the relevant original articles via citations.

Another source of facts that you can review is from first-person, personal conversations with other investigators. However, facts offered from *personal communication* should be kept to a minimum because they are hearsay and carry little credibility, having not been scrutinized in the peer-review publication process. Sentences 10 and 17 in Case 3-3 show personal communication as a source of information.

Reviews are not limited to other researchers' work. Reviews can include information from *your* (the PI or a research team member's) research that is relevant to the proposed research, even if the narrative has a **Preliminary Studies/Progress Report Section**. If the narrative does not have a **Preliminary Studies/Progress Report Section**, the discussion of your previous research in the **Background Section** can be lengthier, but if it does have a **Preliminary Studies/Progress Report Section**, the discussion of your previous research should be brief.

Case 3-7 provides an example of a **Background** body subsection with a review that includes a brief discussion of the PI's previous research. Notice that sentence 8 opens with *We have localized*, and sentence 9 leads with *Preliminary data indicate*. Likewise, Case 3-8 briefly provides a discussion of the PI's relevant research in sentences 12–15.

The information that you review also needs to be up-to-date. What does *up-to-date* mean? It means whatever your panel of reviewers decides. Thus, to anchor your proposed research into an appropriate historical research context, you might need to review not only recent research but also research that you might consider dated. If there is an older study that is directly relevant to your proposed research or includes a result that is similar to what you expect to achieve in your proposed research, you can leave yourself open to charges of lack of novelty or lack of familiarity with the research in your field, if you fail to cite it and to distinguish it from your proposed work. However, if you extensively cite to older, original articles, you could give the appearance that you are not reading enough current literature in your research area.

A strong review is politically expedient – that is, it includes information from research performed by investigators whom you expect might be reviewing your grant proposal. Since you do not know for sure who these reviewers will be, this criterion may be particularly hard to apply. If you suspect a particular researcher will likely be a reviewer and you are discussing this person's research in a **Background** subsection, you need to take particular care in reviewing the research with as little bias or subjectivity as possible and in analyzing the research without negative criticism, such as with problem statements. Instead, you could analyze this reviewer's research through speculations and technical questions. (See Chapters 2.2.3 and 3.6.)

A few words need to be said about the organization of a review within a **Background** subsection. Although the background establishes the historical research context for your proposed research, the review is seldom organized chronologically, from early studies to the most recent studies. The review is usually organized by topics, as shown in Case 3-2A.

3.6 Analysis in a Background subsection

A strong **Background** subsection (whether an introductory or a body subsection) does not just review the previous research – it also analyzes it. This analysis is central to your demonstrating your acumen in science and your ability to evaluate previous research. Without the analysis, the **Background** subsection cannot express significance or novelty, for it is only through limitations in previous research, captured in analysis statements, that your proposed research promises to contribute to scientific knowledge in novel ways.

As discussed in Chapter 2.2.3, there are different ways to analyze reviewed research in a scientific argument, including the 7 ways listed here:

T
R
A
C
S

(1) **Problem statements**
(2) **Technical questions**
(3) **Speculations**
(4) **Hypotheses**
(5) **Predictions**
(6) **Extension statements**
(7) **Conclusions**

- **Proportion of review to analysis.** The analysis usually takes much less space than does the review. For example, Case 3-7 has one *review-analysis* pair (see below): sentences 1–9 serve to review information that is analyzed in less space, in sentences 10 and 11. Case 3-3 also has a more extensive review than an analysis: sentences 1–7 review background literature, and sentence 8 analyzes it; then sentences 9–11 review information that is analyzed in sentence 12; and sentences 13–18 review information that bolsters the analysis in sentence 12.

> **Guideline 3.10** *Within a **Background** body subsection, include more review than analysis, and locate the analysis immediately after the review of literature, creating a **review-analysis** pair.*
> *Exception: A review that bolsters an analysis (see below) follows the analysis.*

- **Review-analysis relationships.** The analysis follows the review of literature and evaluates it. Combined, the reviewed literature and its analysis form the following content relationship:

review-analysis

The *review-analysis* pair comprises the minimum, logical unit of a **Background** subsection, for the analysis needs to logically follow from the reviewed information (however, see bolstering below).

The examples in Cases 3-4, 3-5, 3-6, and 3-7 show the most common organization, where a **Background** subsection (including a **Background** introduction) has either one or 2 *review-analysis* pairs. A less common organization is shown in Cases 3-8 and 3-9, where each **Background** subsection has 3 or 4 *review-analysis* pairs. Clarifying the logical relationships across many *review-analysis* pairs in the same **Background** subsection is difficult.

You need to draft the review so that reviewers can readily perceive the logical relationship between the review and analysis. In other words, when writing a **Background** subsection, you need to strive to articulate a clear, logical *review-analysis* pair and to relate *review-analysis* pairs to each other.

> **Guideline 3.11** *Within every **Background** body subsection, draft a **review-analysis** pair, making sure that the analysis clearly and logically follows from the reviewed information.*

For example, in Case 3-6, sentences 6 and 7 review information about *LA* (lipoic acid), and sentence 8, an analysis statement, uses a problem statement to identify a gap in the previous research, through the term *unknown*. Had the preceding sentences reviewed research that determined whether or not the *R* isoform of *LA* is beneficial in comparison to racemic, the analysis of *unknown* would not logically fit the context.

A review needs to clearly and fairly describe the previous research, and the analysis needs to be a logical inference that flows directly from the reviewed information in order for reviewers to perceive an analysis as logical.

- **Bolstering the analysis.** If you cannot *draft the review so that reviewers can readily perceive the logical relationship between the review and analysis*, you should consider: (1) re-selecting the literature to be reviewed so that the subsequent analysis logically follows, (2) rephrasing the analysis so that the analysis logically follows from the reviewed literature, and/or (3) adding another review after the analysis in order to bolster the analysis and to make the logical relationship between the reviewed literature and the analysis more transparent. This last strategy – adding another review to bolster an analysis – creates a 3-pronged *review-analysis* relationship:

review-analysis-review

There may be times when you will need to add a review after a *review-analysis* pair, creating a *review-analysis-review* relationship, to bolster the logical relationship between the first review and the analysis. As shown in the second paragraph of Case 3-3, sentences 9 to 11 review information, and this information is analyzed in sentence 12 with a speculation. This analysis is supported by a second review in sentences 13 to 18. The content pattern in the scientific argument in this second paragraph of Case 3-3 is:

Pattern: *review-analysis-review*
Sentences: 9–11   12   13–19

The content patterns of the review and analysis of **Background** body subsections are captured in the following guideline:

> **Guideline 3.12** *In **Background** subsections, to establish the novelty of your proposed research, organize the content according to one of 2 **review-analysis** relationships:*
> (a) ***Review-analysis***
> (b) ***Review-analysis-review***

- **Summary analysis statements.** In addition to analysis statements that follow reviews, they can also begin and end **Background** body subsections. In these positions, the scope of the analysis statements is the entire subsection; they serve to analyze *all* of the reviews and analyses in the subsection.

> **Guideline 3.13** *Consider using a summary analysis to synthesize all of the information in a **Background** subsection.*

In Case 3-8, sentence 1 in the **Background** body subsection presents a *summary analysis* (and a review of information) that opens the subsection and presents the topic (*disruption of the HPG axis*) that is continued through to the last sentence of the subsection. The summary analysis is followed by a series of *review-analysis* combinations; the subsection ends with a statement that connects the topic to Aim 3 of the proposed research:

Pattern 1:	***summary analysis***
Sentence:	1
Pattern 2:	***review-analysis***
Sentences:	2–5 5
Pattern 3:	***review-analysis-review***
Sentences:	6–7 7 8
Pattern 4:	***review-analysis***
Sentences:	9 10
Pattern 5:	***review-analysis***
Sentences:	11–13 14–16
Pattern 6:	***connection: research purpose***
Sentence:	17

The example in Case 3-8 is noteworthy in many respects, not only because of its summary analysis in sentence 1 and the clear pattern of ***review-analysis*** pairs, but in addition: (1) the terms in the heading constitute the topics in the subsection and are repeated throughout the subsection (e.g., *regulation of the HPG axis*); (2) an individual sentence can comprise both a review and an analysis (such as sentences 5 and 7); (3) an individual sentence can comprise more than one analysis, such as sentence 10, which is a conclusion in the first half and a speculation in the second half; and (4) since a hypothesis is a type of speculation (Chapters 1.10.1 and 1.10.2), you can analyze information as a speculation and then identify it as your hypothesis (sentences 10 and 11) to help reviewers understand your logical thought processes.

● **Tactfulness.** If you are reviewing and analyzing research of an investigator whom you suspect might be a reviewer, you need to select an analysis statement that allows you to evaluate the previous research, but to do so without being negative. The analysis statements can present your assessment of the previous research in a neutral or a positive light – except for the problem statement. Example 1a presents 2 sentences from Case 3-6, the second one of which is an analysis in the form of a problem statement. Example 1b gives a more neutral revision of the analysis, with the problem statement changed into a technical question. Example 1c softens the analysis even further through a speculation.

Example 1
a. The form of LA most readily available on the market is a racemic mixture (50:50) of R and S forms [22–24]; the R form is known to be biologically active [21]. However, whether its administration is beneficial in comparison to the racemic form is unknown. (Case 3-6, sentences 7–8.)
b. The form of LA most readily available on the market is a racemic mixture (50:50) of R and S forms [22–24]; the R form is known to be biologically active [21]. We now ask: is the administration of the R form beneficial, in comparison to the S form?

c. The form of LA most readily available on the market is a racemic mixture (50:50) of R and S forms [22–24]; the R form is known to be biologically active [21]. The administration of the R form may be beneficial, in comparison to the S form.

3.7   Connection to your proposed work

A strong **Background** subsection connects the research that you have reviewed and analyzed with: (1) a proposed research purpose, such as your research objective or *at least* one of your proposed aims from the **Aims Section**, and/or (2) an aspect of your proposed methods. To identify the connection, you can ask the following question for every **Background** subsection of the narrative:

> *What does this information that I have just reviewed and analyzed*
> *have to do with my proposed research?*

The connection between your proposed research and the previous research can be expressed in a sentence or part of a sentence, or it can be tucked into a sentence via parentheses. Regardless of how the connection is expressed, it is located after the relevant *review-analysis* pair, usually towards the end of a **Background** subsection. As noted in Guideline 3.14, to connect previous research with your proposed research, you can identify the proposed aim that is relevant to the previous research.

Guideline 3.14   *At the end of each **Background** subsection, identify the proposed aim(s) that is relevant to your analysis of the previous research by:*
(a) Inserting a parenthetical cross-reference that identifies the proposed aim(s) by number, or
(b) Adding a sentence at the end of the subsection that identifies the related proposed aim(s) by number.
***Exception**: Consider not following (a) or (b) if the subsection includes only fundamental background information.*

In the **Background** subsection of Case 3-6, the connection statement is given in sentence 20 and serves to identify the relationship between the previous research and Aims 1 and 2 of the proposed research. In Case 3-7, the connection statement is in the last sentence of the subsection, in sentence 13. Case 3-8 has a connection statement in sentence 17, which associates Aim 3 with the information in the section.

If you cite to one of the aims by number in any **Background** body subsection, for consistency, you should cite to a relevant aim in all **Background** body subsections. Also, if you cite to one of your aims by number in any **Background** body subsection, you should cross-reference to each of your proposed aims at least one time, in an appropriate **Background** body subsection.

> **Guideline 3.15** *Refer to each of your proposed aims at least once in the entire Background Section.*

A connection statement can underscore the cohesiveness of the narrative and can contribute to your credibility by indicating that your proposed research is so well-conceived and structured that you can link each of your proposed aims to your discussion of previous research.

3.8 Significance in a Background subsection

Significance in scientific argumentation is the importance of your proposed research. However, just claiming that your proposed work is significant is not enough; you need to substantiate significance. Your proposed research acquires significance from research topics that address diseases, disorders, and other conditions that are not considered favorable on their face, from statistics in the **Background** introduction, and from an explicit identification of the significance.

Also as earlier mentioned, in a **Background Section**, significance does not need to be explicitly stated in each **Background** body subsection, as long as it is explicitly identified or implied through statistics in the **Background** introduction. Thus, for instance, the examples in Cases 3-6, 3-8, and 3-9 do not include significance, but all are associated with significance that is addressed in the **Background** introduction (not shown).

3.9 Novelty

Your research needs to be novel – that is, original or distinctive in some critical way – and you need to describe both the previous research and your proposed research in a way that reviewers can readily understand your novelty. As earlier mentioned (Chapter 1.6), NIH requires a major section in the narrative that addresses novelty, which it terms **Innovation**.[4] Whether or not your funding agency follows NIH submission requirements that specify a separate section for novelty, it is useful to draw the reviewers' attention to the novelty of your proposed research by including a **Background** subsection, paragraph, or sentences that address it.

Novelty can be explicitly or implicitly expressed. Explicit statements of novelty include such terms as:

authentic	*original*	*new*
authenticity	*originality*	*the first*
novel	*unique*	*the first time*
novelty	*distinct*	*the only*
unique	*distinctive*	*innovative*

---

[4] http://grants.nih.gov/grants/funding/phs398/phs398.html (I-46; accessed November 28, 2014).

Cases 3-9, 3-10, and 3-11 present 3 very different examples showing how novelty can be presented in a **Background** subsection. Case 3-9 presents a **Background** subsection in which one of the connection statements (sentence 11) contrasts previous studies to the proposed research, indicated by *Unlike previous studies*. Case 3-10 presents an **Innovation Section** that explains the novelty of the proposed research in a numbered list. Notice that in the first numbered item, each sentence uses the term *new*; the second item uses *unique* in its heading and in its first sentence; and the third item uses *novel* in the heading. Case 3-11 presents an **Innovation Section** in paragraph style. Notice that the first sentence does not just state that the proposed research is *highly innovative* (sentence 1), but it goes on to explain why. Sentence 3 addresses novelty by stating that *no one* has demonstrated a function for the proteins.

Novelty can be implicit in **Background** subsections, through a *review-analysis* pair of a scientific argument. The analysis identifies a shortcoming in the reviewed research, and through this identification, the implication is that the proposed research will do something that no other researcher and no other research has yet accomplished – which is to eliminate or to reduce the shortcoming. For example, in Case 3-7, novelty is implied in the speculation of sentence 11: α_{1F} *may also be involved in synaptic transmission in the cone pathway*. Implicit in the discussion is that results from the proposed research will provide certainty, either that α_{1F} is involved in synaptic transmission in the cone pathway or that it is not involved.

3.10 The ending to the Background Section

The **Background Section** does not usually have an ending in order to save space for other information in the narrative. However, if space permits, an ending is useful as a transition, to refocus the reviewers' attention from a review consisting primarily of other investigators' research, back to the purpose, significance, and innovation of your proposed research.

> **Guideline 3.16** *Consider including an ending subsection, to identify the purpose, importance, and innovation of your proposed research.*

The ending to the **Background Section** can be presented as a paragraph or a subsection. It presents at least 3 types of information from a scientific argument:

- Topic of the proposed research.

- Review (optional).

- Analysis (optional).

- Connection information: the purpose of the proposed research, in terms of the long-term goal and/or research objective.

- Significance of the proposed research.

Case 3-12 gives endings to 3 **Background Sections**. In Case 3-12A, the ending is brief, with the first sentence presenting the topic of the proposed research from the **Aims Section** (*the movement of S4*) and connection information – the proposed research objective (*characterization* of the S4 movement) and the long-term goal (*further our understanding*) from the **Aims Section**. The second sentence identifies the significance by associating the proposed research with the development of treatments for particular neurological disorders.

Similar to the ending in Case 3-12A, the ending in Case 3-12B identifies the topic (sentence 3, *control mechanisms*) and purpose (sentence 3, *understanding* the control mechanisms) of the proposed research. However, unlike Case 3-12A, Case 3-12B offers a ***review-analysis*** pair, to establish the underlying need for the proposed research. Case 3-12B gives a more extensive discussion of significance than does Case 3-12A, with significance addressed in sentences 4–6. Case 3-12C presents an ending section[5] that focuses on significance in applied research. Similar to the other 2 examples of ending sections, this ending consists solely of all content that comprises a scientific argument.

The relationship between reviewed information and its analysis in the **Background Section** has been emphasized throughout this chapter. For a review to work for you – that is, for a review that not only describes the previous research but also shows how the previous research is related to the novelty and significance of your proposed research – you need at least one ***review-analysis*** pair in each **Background** subsection. Further, you need to explain to reviewers which proposed aim or method that the previous information relates to, by identifying the connection between the ***review-analysis*** and a proposed aim or method.

---

[5] The grant proposal from which this example was taken, presents 2 **Significance** subsections (the other one entitled **Public Purpose**) as ending subsections to its **Background Section**.

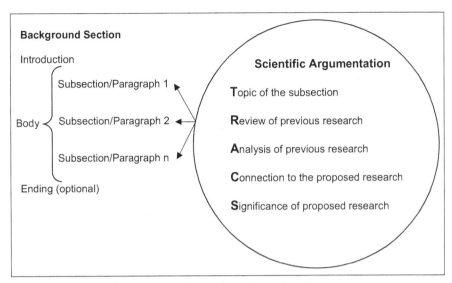

Figure 3-1. Content and organization of a generic Background Section.

Case 3-1. Heading outline of a Background Section.

A. Significance

A.1 Impediments to identifying the vestibular contribution to balance control.

A.2 Methods to reveal the vestibular contribution.
Comparisons between BVL and normal subjects.
Model-based interpretation of experimental data.
Use of GVS.

A.3 Clinical significance.

B. Innovation

B.1 Model-based interpretation of vestibular contribution to balance.

B.2 Unique application of GVS to manipulate vestibular variability in order to test the hypothesis of optimal sensory integration.

B.3 Novel use of GVS to manipulate the vestibular contribution to balance control.

Case 3-2. Heading outline of a Background Section. (I) Subsections B.2, B.3, and B.4 each have 2 sub-subsections, the first of which gives background on standard techniques and the second of which gives background on OCT. **(II)** Standard techniques are now grouped together in the subsection B.2, followed by background on the proposed OCT grouped together in B.3.

I.

B. BACKGROUND AND SIGNIFICANCE

B.1 THE PROCESS OF ENGINEERING
   TISSUES

B.2 MEASUREMENT OF
   MORPHOLOGICAL PROPERTIES
   OF ENGINEERED TISSUES
   B.2.1 Standard techniques to
       measure morphological
       properties in tissue
       engineering
   B.2.2 OCT imaging of
       morphological parameters
       in tissue engineering

B.3 MECHANICAL PROPERTIES OF
   ENGINEERED TISSUES
   B.3.1 Standard techniques to
       measure mechanical
       properties in tissue
       engineering
   B.3.2 Optical elastography
       techniques

B.4 MEASUREMENT OF FLUID
   DYNAMICS IN ENGINEERED
   TISSUES
   B.4.1 Standard techniques to
       measure fluid dynamics in
       tissue engineering
   B.4.2 OCT measurements of flow

B.5 MODELS OF TISSUE-ENGINEERED
   VASCULAR GRAFTS

II.

B. BACKGROUND AND SIGNIFICANCE

B.1 THE PROCESS OF ENGINEERING
   TISSUES

B.2 STANDARD TECHNIQUES
   B.2.1 Measurement of
       morphological properties
       of engineered tissues
   B.2.2 Mechanical properties of
       engineered tissues
   B.2.3 Measurement of fluid
       dynamics in engineered
       tissues

B.3 OCT IMAGING TECHNIQUES
   B.3.1 Measurement of
       morphological properties
       of engineered tissues
   B.3.2 Mechanical properties of
       engineered tissues
   B.3.3 Measurement of fluid
       dynamics in engineered
       tissues

B.4 MODELS OF TISSUE-ENGINEERED
   VASCULAR GRAFTS

Case 3-3. Example of a Background Section without subsections.
Bold = Topic, **A** = Analysis.

(a) Significance. [1]*Burkholderia pseudomallei (Bp)* infection is a Gram-negative bacterial pathogen that normally survives as a saprophyte in **soil and water**, but is also capable of infecting most mammals and causing serious infections (1-5). [2]*Bp* infection is a major cause of bacterial sepsis and chronic disseminated infections (meliodosis) in humans in Thailand and northern Australia (4-8). [3]The fatality rate for patients with *Bp* infection, even with prompt and aggressive treatment, still ranges from 20% to over 50%. [4]Moreover, *Bp* is an emerging pathogen and infections have been increasingly reported in many regions of the world, including Central and South America (9-13). [5]In fact, *Bp* infection is now considered endemic in regions of China and India, and in Brazil (11,12,14). [6]Infections with *Bp* are particularly dangerous because the organism is intrinsically resistant to many antimicrobials, can persist for years in the soil and in water, and can cause a wide array of clinical symptoms, ranging from acute sepsis, to chronic recurrent infection, to clinically silent infection (5-8, 15-17). [7]Meliodosis is also an increasing problem in travelers who have visited regions of the world where *Bp* is endemic (18). [8]Thus, *Bp* is a dangerous bacterial pathogen with high potential for spread into new regions of the world, including the US, via deliberate or accidental introduction in soil, food, or water.

    [9]Currently, *Bp* is not considered a primary enteric pathogen for infection in humans. [10]At present, infection with *Bp* is presumed to occur following inhalation or cutaneous inoculation, though the actual link between cutaneous exposure and infection is weak (Dr. Jane Smith, see Letter of Support). [11]Thus, current treatment and prevention efforts for human meliodosis do not consider the impact of oral infection or persistent fecal carriage and shedding of the organism (5, 7, 8). [12]There is, however, epidemiological evidence to suggest that oral infection with B*p* does occur in humans. [13]For example, outbreaks of meliodosis in villages in Indonesia have been linked directly to drinking water supplies contaminated with *Bp*, which can survive for years in water (19). [14]Infections with *Bp* increase significantly during times of greater exposure to very wet conditions (e.g., rice farming during the monsoon season), which would be consistent with oral exposure to a later-borne agent (20-22). [15]Outbreaks of meliodosis have also been associated with tsunami events (3, 24). [16]In addition, patients with meliodosis have been misdiagnosed as having typhoid (enteric fever) (25). [17]In fact, clinical observations (Dr. Smith, personal communication) suggest that oral infection may be a much more important route of infection with B*p* in humans than previously assumed. [18]Since *Bp* can persist in water or soil for years, enteric infection of humans with *Bp* would have major public health consequences (26, 27).

    [19]Virtually nothing is known regarding enteric infection with *Bp*. [20]Development of an animal model of enteric *Bp* infection proposed here would therefore be valuable for several reasons. [21]For one, a mouse enteric *Bp* model would be essential to help understand the pathogenesis of enteric meliodosis in humans. [22]A new mouse model of an enteric *Bp* infection would also add an important animal model to study enteric pathogens in general. [23]From a clinical perspective, an enteric *Bp* infection model in mice would also be critical for the development of new vaccines for *Bp* and for the development of new antimicrobial treatments and preventions strategies.

Annotations (right margin):
- Topic
- Review
- A: Conclusion
- Review
- A: Speculation
- Review
- A: Problem
- Significance

Case 3-4. Example of an informal introduction to a Background Section. Right-hand column: Analysis of the scientific argumentation. **Bold** = Topic, **A** = Analysis, **C** = Connection.

Background [1]Although the vestibular contribution to eye movement control is well understood [ref], much less is known about the **vestibular contribution to balance control**. [2]The fact that the vestibular system does contribute to balance is irrefutable. [3]Subjects without vestibular function cannot maintain stance in environmental conditions where they are deprived of accurate visual and proprioceptive orientation cues in contrast to subjects with intact vestibular function, who can [refs]. [4]While complete absence of vestibular function is a rare disorder, a recent large-scale study reported that 35% of Americans over the age of 40 years show balance deficits attributable to vestibular dysfunction [ref]. [5]This study also documented a potential link between vestibular dysfunction and falls, reporting that vestibular-dysfunction subjects with self-reported dizziness had a 12-fold increase in the odds of falling (6-fold increase in vestibular-dysfunction subjects with no dizziness; both odds ratios adjusted for other risk factors). [6]Falls have been identified as a leading cause of death in older individuals [refs], and falls are a substantial contributor to healthcare costs, with direct medical costs in 2000 of about $19 billion for fatal and non-fatal falls in adults 65 years of age and older [refs]. [7]These public health studies motivate our desire to obtain a clearer understanding of how the vestibular system contributes to balance control. [8]We further propose that this improved understanding will inform our application of GVS to enhance a vestibular-impaired subject's ability to utilize vestibular function for balance control.	Topic & Review A: Conclusion Review A: Problem & C: Purpose Significance

Case 3-5. Formal introduction to a Background Section. Right-hand column: Analysis of the section in terms of scientific argumentation. **Bold** = Topic, in sentence 1; other bolding in the original; **A** = Analysis.

Staphylococcus aureus as a human pathogen. [1]*S. aureus* is an opportunistic **pathogen** capable of causing diverse infections, ranging from superficial and relatively benign infections of the skin to serious and even life-threatening disease (41). [2]The most serious are the deep-seated infections that arise either after invasion of the bloodstream from primary sites of infection or after the direct introduction of *S. aureus* as a result of trauma. [3]Specific examples include osteomyelitis and endocarditis, both of which involve the colonization of a solid-surface substratum (41). [4]These infections are extremely difficult to resolve for two reasons. [5]**The first** is the continued emergence of *S. aureus* strains that are resistant to multiple antibiotics (34). [6]Indeed, an increasing number of cases, the only treatment option is the glycopeptide antibiotic vancomycin. [7]Moreover, reports describing the isolation of *S. aureus* strains that are relatively resistant to vancomycin emphasize the tenuous nature of our reliance on this antibiotic (refs). [8]**The second** complicating factor is the formation of a bacterial biofilm on the solid-surface substratum (Fig. 1). [9]Because the biofilm is an effective impediment to antibiotic delivery, resolution of deep-seated *S. aureus* infections typically requires surgical intervention **to debride the infected tissue and/or remove the offending implant.**	Topic Review
[10]We believe our proposal has relevance with respect to the development of new therapeutic agents and with respect to the delivery of those agents to the site of infection. [11]Specifically, we believe that *sar* may be an appropriate target for the development of antimicrobial agents capable of attenuating the virulence of *S. aureus* and that these agents may, by virtue of their ability to interfere with the coordinated regulation of *S. aureus* virulence factors (see below), inhibit biofilm formation and thereby increase the efficacy of conventional antimicrobial agents. [12]Moreover, recent evidence suggests that therapeutic strategies directed at *sar* may have a direct impact on the resistance of *S. aureus* to at least some antimicrobial agents. [13]For instance, Smith et al. (2) suggested that transcription from the *sar* P_3 promoter may be dependent on the *S. aureus* stress-response sigma factor σ [B]. [14]That is a significant observation since Jones et al. (1943) demonstrated that inactivation of the sigB operon in the homogenously resistant *S. aureus* strain COL results in a 64-fold increase in the susceptibility to methicillin (i.e., a 64-fold decrease in the methicillin MIC). [15]The observation that *sar* mutants exhibit a small but reproducible increase in the susceptibility to methicillin (44) supports the hypothesis that the inability to express *sar* may contribute to the decline in methicillin resistance.	Significance A: Speculation & Review A: Speculation Review Significance & Review Review A: Hypothesis

Case 3-6. Background body subsection. Significance was previously given in this **Background Section**. **Bold** = Topic; **A** = Analysis, **C** = Connection. Bolding, except in sentence 1, from the original; underscoring from the original.

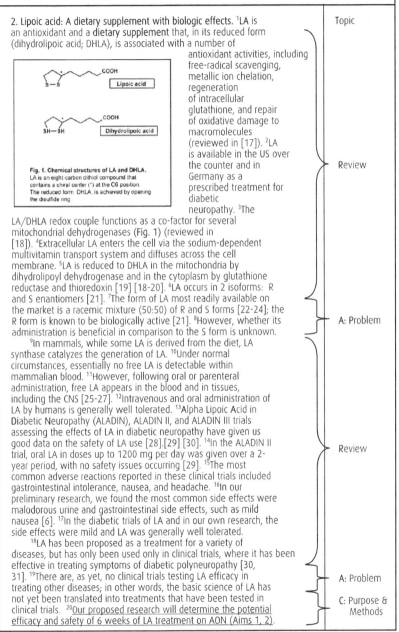

2. Lipoic acid: A dietary supplement with biologic effects. [1]LA is an antioxidant and a dietary supplement that, in its reduced form (dihydrolipoic acid; DHLA), is associated with a number of antioxidant activities, including free-radical scavenging, metallic ion chelation, regeneration of intracellular glutathione, and repair of oxidative damage to macromolecules (reviewed in [17]). [2]LA is available in the US over the counter and in Germany as a prescribed treatment for diabetic neuropathy. [3]The LA/DHLA redox couple functions as a co-factor for several mitochondrial dehydrogenases (**Fig. 1**) (reviewed in [18]). [4]Extracellular LA enters the cell via the sodium-dependent multivitamin transport system and diffuses across the cell membrane. [5]LA is reduced to DHLA in the mitochondria by dihydrolipoyl dehydrogenase and in the cytoplasm by glutathione reductase and thioredoxin [19] [18-20]. [6]LA occurs in 2 isoforms: R and S enantiomers [21]. [7]The form of LA most readily available on the market is a racemic mixture (50:50) of R and S forms [22-24]; the R form is known to be biologically active [21]. [8]However, whether its administration is beneficial in comparison to the S form is unknown.

[9]In mammals, while some LA is derived from the diet, LA synthase catalyzes the generation of LA. [10]Under normal circumstances, essentially no free LA is detectable within mammalian blood. [11]However, following oral or parenteral administration, free LA appears in the blood and in tissues, including the CNS [25-27]. [12]Intravenous and oral administration of LA by humans is generally well tolerated. [13]Alpha Lipoic Acid in Diabetic Neuropathy (ALADIN), ALADIN II, and ALADIN III trials assessing the effects of LA in diabetic neuropathy have given us good data on the safety of LA use [28].[29] [30]. [14]In the ALADIN II trial, oral LA in doses up to 1200 mg per day was given over a 2-year period, with no safety issues occurring [29]. [15]The most common adverse reactions reported in these clinical trials included gastrointestinal intolerance, nausea, and headache. [16]In our preliminary research, we found the most common side effects were malodorous urine and gastrointestinal side effects, such as mild nausea [6]. [17]In the diabetic trials of LA and in our own research, the side effects were mild and LA was generally well tolerated.

[18]LA has been proposed as a treatment for a variety of diseases, but has only been used only in clinical trials, where it has been effective in treating symptoms of diabetic polyneuropathy [30, 31]. [19]There are, as yet, no clinical trials testing LA efficacy in treating other diseases; in other words, the basic science of LA has not yet been translated into treatments that have been tested in clinical trials. [20]Our proposed research will determine the potential efficacy and safety of 6 weeks of LA treatment on AON (Aims 1, 2).

Topic

Review

A: Problem

Review

A: Problem

C: Purpose & Methods

Fig. 1. Chemical structures of LA and DHLA. LA is an eight-carbon dithiol compound that contains a chiral center (*) at the C6 position. The reduced form, DHLA, is achieved by opening the disulfide ring.

Case 3-7. Background subsection. Bold = Topic; **A** = Analysis; **C** = Connection.

Mutations in the α_{1F} calcium channel and X-linked congenital stationary night blindness

[1]In 1998, the CACNA1F gene was discovered and found to encode the α_1 subunit of a novel voltage-activated **calcium channel**, α_{1F}, expressed exclusively in the retina (refs). [2]Sequence analysis confirms that α_{1F} is a member of the L-type family of **calcium channels** displaying greatest amino acid identity, 62%, to the α_{1D} subunit of brain L-type **calcium channels**. [3]CACNA1F is located on the X-chromosome and was found to be the genetic locus for incomplete X-linked congenital stationary night blindness (CSNB2). [4]CSNB2 is a recessive, non-progressive visual disease characterized by poor night vision and decreased visual acuity. [5]The electroretinograms (ERGs) of patients with CSNB2 display a nearly normal a-wave and a markedly reduced b-wave under scotopic conditions (refs). [6]CSNB2 is also marked by abnormalities in the photopic ERG (refs). [7]The electrophysiological and psychophysical phenotype of the disease can be explained by a defect in synaptic transmission within the retina (refs). [8]We have localized α_{1F} immunoreactivity to both the inner and outer plexiform layers of the retina in the mouse and rat (refs). [9]Preliminary data indicate that α_{1F} is present in rod photoreceptor and rod bipolar cell terminals (refs), consistent with the symptom of night blindness associated with CSNB2. [10]Thus, α_{1F} is a candidate **calcium channel** for mediating transmitter release at rod photoreceptor and rod bipolar cell ribbon synapses in the retina. [11]The abnormalities in photopic vision that are characteristic of CSNB2 suggest that α_{1F} may also be involved in synaptic transmission in the cone pathway. [12]Therefore, it is important to determine whether α_{1F} is also present at cone photoreceptor and cone bipolar cell terminals. [13]The localization of α_{1F} at cone photoreceptor and cone bipolar cell terminals will be investigated in Aim 1.1.

Topic

Review

A: Conclusion
A: Speculation
Significance
C: Purpose

Case 3-8. Background body subsection that opens with a summary analysis.
Bold = Topic, **A** = Analysis, **C** = Connection.

Regulation of the hypothalamic-pituitary gonadal axis in autoimmunity. [1]Although **disruption of the HPG axis** has been described for a variety of **autoimmune** disorders, including systemic lupus erythematosus (46), ankylosing spondylitis (47), RA (48), and gout (49), studies of the HPG axis in arthritis have by far received the most attention. [2]Androgen levels during the course of rheumatoid arthritis are reported to be at their lowest during exacerbations of disease (50). [3]An inverse relationship has been described between androgen levels and indices of disease activity, including erythrocyte sedimentation rate, C-reactive protein concentration and rheumatoid factor titers (50). [4]In addition, the induction of experimental adjuvant-induced arthritis results in a dramatic decrease in serum testosterone levels in rodents (51). [5]Also, since levels of LH were found to be elevated (51), the inhibition of testosterone production in this model appears to be attributed to a testicular deficiency. [6]Increases in serum cytokine levels, including IL-1 and IL-6, have been shown to accompany exacerbations of arthritic symptoms (52); and an inverse relationship between inflammatory cytokines and testosterone levels in arthritis is further evidence that cytokines are involved in regulating the HPG axis (53). [7]Since androgens are known to be immunosuppressive, the reduction in androgen levels during an arthritic attack may serve to augment the progression of disability. [8]Studies assessing the efficacy of androgen replacement therapy in male RA patients have demonstrated a significant reduction in joint inflammation and various other indices of disease activity (54).

  [9]There is ample evidence demonstrating a disruption of the HPG axis and a deficiency in androgen production during exacerbations of rheumatic diseases in males. [10]Because of the similarities between this family of autoimmune disorders and MS, there is a strong likelihood that the HPG axis is disrupted during exacerbations of MS in males. [11]The high prevalence of sexual dysfunction in men with MS is an additional clue that supports this hypothesis. [12]Preliminary experiments presented in the following paragraphs have revealed that a disruption of the HPG axis does need to occur in EAE. [13]Moreover, an inverse relationship was found to exist between disease severity and testosterone levels. [14]There is also evidence that the HPG axis may be disrupted in men with MS, but the extent to which this occurs is unknown. [15]These preliminary data raise the possibility that a disruption of the HPG axis is a consequence of autoimmune demyelination. [16]Dysregulation of sex hormones in patients with autoimmune demyelination may ultimately lead to the increased severity or duration of disease symptoms experienced in males. [17]The primary objective of Aim 3 is to determine whether **a disruption of the HPG axis** occurs during attacks of MS.

Annotations (right-side brackets):
- Topic
- A: Problem
- Review
- A: Speculation
- Review
- A: Speculation
- Review
- Review
- A: Conclusion & Speculation/ Hypothesis
- Hypothesis
- Review
- A: Speculation & Problem
- C: Purpose

Case 3-9. Background subsection expressing novelty. Bold = Topic; **A** = Analysis, **C** = Connection.

B.5 Biomechanical Modeling of the Developing Heart.
[1]Biomechanical models of the developing heart[23, 26] have generally focused on stresses on the walls of the ventricle and the effect of wall stress on heart growth[23, 26, 67]. [2]Those studies did not account for effects of blood flow (other than through blood pressure) on the walls. [3]Other biomechanical models have been used to characterize blood-flow and wall shear stress patterns in the developing chick heart. [4]Most of these studies assumed blood to be Newtonian and solved Navier-Stokes equations assuming a static heart geometry (reconstructed from frozen heart sections[19, 68] or based on idealized geometrical representations of the heart[69]). [5]In these models, however, the effects of wall motion and pressure pulsatility were not considered.

    [6]A recent study used a geometrically simple model of a tubular heart[70]. [7]That study found that when cardiac cushions are included in the OFT and in the atrio-ventricular channel, a traveling compression wave along the heart – a wave that reproduces the peristaltic-like compression of the myocardium at early stages of development – qualitatively reproduces changes in ventricular blood pressure during the cardiac cycle. [8]The study concluded that cardiac cushions are essential for blood pulsatility in the early developing heart and that wall shear stress in the regions of the cardiac cushions are distinct from those in the rest of the heart, suggesting that wall shear stress may play an important role in valve development. [9]In preliminary studies, we found a similar result in our models of the OFT that included cardiac cushions (Section **C.6**). [10]In our proposed study, we will determine how wall shear stress changes under normal and altered blood-flow conditions (OTB and VVL) through FEM models of the OFT that will include cardiac cushions (**Aim 2**). [11]Unlike previous studies, our models will include the motion of the OFT wall and the pulsatility of blood pressure (**Aims 1** and **2**).

Topics & Review
A: Problem

Review

A: Problem

Review

A: Speculation & Review

C: Purpose & Methods

Case 3-10. Innovation Section with a list.

B. Innovation

1. Model-based interpretation of vestibular contribution to balance. [1]New experimental data will be used to extend previous models of balance control in order to explain how the nervous system integrates multiple sensory orientation cues to control multi-segmental body motion. [2]These new models will be used: a) to identify the vestibular contribution to balance and how it changes as a function of environment and test conditions, b) to begin revealing how loss of vestibular function affects balance control, and c) to determine how BVL subjects compensate for vestibular loss. [3]Clarifying balance control as proposed in this application can be used to inform the development of new clinical balance function tests to assess underlying causes of balance deficits.

2. Unique application of GVS to manipulate vestibular variability in order to test the hypothesis of optimal sensory integration. [4]The unnatural pattern of vestibular activation resulting from GVS affords a unique opportunity to manipulate vestibular variability in vestibular sensors encoding AP motion without evoking AP body sway. [5]This manipulation will allow us to test the hypothesis of optimal sensory integration as it applies to balance control. [6]Because increased variability in sensory signals may occur with aging and disease (refs), better understanding how the balance system changes when vestibular variability increases will provide insight into how the nervous system compensates for age-related changes in vestibular function.

3. Novel use of GVS to manipulate the vestibular contribution to balance control. [7]The proposed research will exploit our model-based analysis methods to identify how the nervous system interprets the unnatural pattern of vestibular activation produced by GVS. [8]We propose that we can manipulate the GVS signal to compensate for this unnatural pattern in order to deliver GVS feedback that can effectively eliminate the vestibular contribution to balance control (in normal subjects) or, more importantly, to restore a vestibular contribution in BVL subjects. [9]Demonstration of this capability will establish the feasibility of our methodology for use in a future vestibular prosthesis to improve balance control in patients with vestibular deficits.

Case 3-11. Innovation subsection in paragraph style.

Innovation. [1]The proposed studies are **highly innovative because** they propose to test the hypothesis that the *Ifitm* proteins function as intracellular shuttles, moving cargo proteins from one intracellular site to the next, and in doing so, altering the state of activation of cells. [2]We propose that this shuttle mechanism is not only important in the control of normal immune cell activation (BCR for B cells, TCR for T cells and LPS/Tlr for macrophages) but is critical for the interferon-dependent anti-viral state that requires *Ifitm* proteins to block virus entry and replication. [3]Although the *Ifitm* proteins in have been studied for decades, **no one** has demonstrated a function for them. [4]A subset of these proteins is constitutively expressed while others clearly showed elevated expression following interferon treatment. [5]Upon cellular activation, these proteins move from a cytoplasmic location to the membrane of the cell. [6]We propose this movement of the *Ifitm* proteins shuttles a set of cargo proteins from a cytoplasmic location into the lipid rafts of the cell. [7]These cargo proteins then modify/degrade the proteins at that site, thus down-regulating cell activation and blocking virus entry and replication. [8]Our analyses will thus be from two perspectives: the response due to IFN-signaling to enhance the innate cellular immune response, and the second as part of normal signal transduction pathway.

Case 3-12. Endings to Background Sections. Bold = Topic; **A** = Analysis; **C** = Connection.

A.

[1]Our proposed characterization of the **precise movement of S4 in** Shaker and the HCN channels will further our understanding of how voltage causes these channels to open and close. [2]Since mutations have been found in voltage-activated ion channels in patients with diseases such as epilepsy, episodic ataxia, and periodic paralyses, an understanding of the structure and function of these channels can take science closer to developing treatments for a number of neurological disorders.

Topic
C: Purpose

Review

Significance

B.

Clinical Significance. [1]The proposed research should significantly advance our understanding of **how the CNS integrates and uses sensory information from multiple sources** to generate appropriate corrective postural reactions that are needed to control balance. [2]Clinical tests of quiet stance control, as well as most of the research on quiet stance, have focused on characterizing body sway in a given test condition or in a set of conditions [92], without understanding the nature of the underlying control mechanisms. [3]In contrast, our proposed research focuses exclusively on understanding these **control mechanisms**. [4]Fundamental knowledge of these mechanisms will contribute to understanding clinically significant postural deficits, where inadequate regulation of sensorimotor control is likely to leave a subject vulnerable to falls and to the increased morbidity and mortality associated with falling [64]. [5]The proposed research will also identify factors that promote or limit the ability of subjects to compensate for deficits affecting postural control, such as losing orientation information from sensory system associated with aging and various diseases, or delaying access to sensory information due to peripheral neuropathies. [6]The results should: (1) lead to a better understanding of balance disorders in humans, (2) suggest new and more effective rehabilitative methods for correcting those disorders, and (3) motivate the development of new tests of human balance function.

Topic
C: Purpose

Review

A: Problem

C: Purpose

Significance

A: Speculation

C: Purpose,
Significance

Significance

C.

Significance. [1]There are 2 to 2.5 million Americans who have such severe communication disorders that they cannot communicate effectively, using only their natural speech and handwriting (refs). [2]These individuals include young children with developmental disabilities, teenagers, and young adults with traumatic brain injury, middle-aged adults with multiple sclerosis and brain stem strokes, and older adults with amyotrophic lateral sclerosis, Parkinson's disease, and cortical strokes. [3]For these individuals, **AAC devices** – in theory – allow them to speak in a variety of situations: on the street, in stores, in classroom settings, and in restaurants. [4]In theory, this allows them to function as closely as possible to individuals with no speech impairments. [5]In practice, however, many individuals are discouraged from using AAC devices that will output a personalized, synthesized voice – that is, will output a synthetic voice that will actually sound like the voice of the user. [6]By **modeling the voice of the AAC user**, as we propose, the AAC device will generate speech that is gender and age appropriate, and will sound more personal and human, and less generic than the standard synthetic voices that come with current state-of-the-art AAC devices.

Review

Topic &
A: Conclusion

C: Purpose

Significance

The Preliminary Studies/Progress Report Section

As mentioned in Chapter 3.5, background research presents 4 types of information: (1) common knowledge; (2) other investigators' previous research that is relevant to your proposed research and proposed methods; (3) your previous research that is relevant to your proposed research and to the research skills needed in the execution of your proposed methods. Chapter 3 described the **Background Section** as it relates to the first 2 types of information. This chapter focuses on the third type, your previous research and your research skills, that is presented in a **Preliminary Studies/Progress Report Section** or as part of a **Background Section**.

In the **Preliminary Studies/Progress Report Section**, your credibility comes sharply into focus. Reviewers look to this section to evaluate your preparedness (e.g., your training, competence, and experience) to perform your proposed methods; to assess the likelihood of your successfully achieving your proposed research objective and within the funding period; and to assess the relationship between your preliminary research and your proposed aims.

The **Preliminary Studies Section** is sometimes termed the **Progress Report Section** when describing your previous research that was funded by a funding agency, the *same* funding agency from which you are seeking additional funds in order to continue the line of research. Unless otherwise noted in this chapter, all of the guidelines for the **Preliminary Studies Section** also apply to the **Progress Report Section**, so in this and subsequent chapters of this book, the term **Preliminary Studies Section** will be used to refer to either of these sections. However, Chapter 4.12 discusses a **Progress Report Section** and distinguishes it in a few respects from the **Preliminary Studies Section**.

4.1 Scope of the Preliminary Studies Section

Unless specified otherwise in submission requirements, in the **Preliminary Studies Section** you include only preliminary research that is relevant to your proposed research and to your preparedness to undertake your proposed research – whether or not your preliminary research has been published and whether or not it was funded by another funding agency. At least one funding agency, NSF, specifies that the **Preliminary Studies Section** needs to include previous research performed within the last 5 years, even if it is not directly relevant to your proposed research.[1] This requirement is likely to assist reviewers in evaluating your previous productivity and your research skills. Once again, you need to follow the submission requirements of your targeted funding agency.

> **Guideline 4.1** *Unless otherwise specified by your targeted funding agency, include in the **Preliminary Studies Section** only information about your preliminary research that is relevant to your proposed research objective, aims, methods, and hypotheses, and to your preparedness to perform your proposed research successfully.*

If you have preliminary research that is partially, but not completely, relevant to your proposed studies and preparedness, then in the **Preliminary Studies Section**, you discuss only those aspects that are relevant to your proposed study. Also, in the **Preliminary Studies** introduction, you mention this point to your reviewers (see Chapter 4.5). For instance, Example 1 (modified from sentence 3 in Case 4-2A) limits the scope of the **Preliminary Studies Section** to preliminary research of only the *lower extremity*.

Example 1
Since the proposed studies focus on the treatment of the lower extremity, we mainly report on preliminary studies that focus on our subjects enrolled for lower extremity treatment and who completed the protocol (11 of the 15).

---

[1] NSF requires that you include results from prior NSF support within the past 5 years, even when your previous studies are not directly related to your proposed research (http://www.nsf.gov/pubs/policydocs/pappguide/nsf13001/nsf13_1.pdf; II-9; accessed November 28, 2014).

4.2   Number of Preliminary Studies to include

It is difficult to suggest *a priori* how many preliminary data and how many preliminary studies you will need to discuss in order for reviewers to believe that: (a) you (or members of your research team) have the requisite skills to execute your proposed methods successfully and on time, (b) you will be able to achieve your proposed research objective, (c) your proposed research is feasible, and (d) your proposed studies are reasonable. However, here are a few factors that can influence how many preliminary data and studies to discuss:

● **Experience.** In general, the less established you are as an independent PI, the more you will need to convince reviewers that you are sufficiently experienced with the proposed methods to execute your proposed research successfully. With this said, some funding agencies make allowances for less established, independent PIs. For example, NIH notes that "for R01 applications, reviewers will be instructed to place less emphasis on the preliminary data in applications from Early Stage Investigators than on the preliminary data in applications from more established investigators."[2] However, regardless of whether funding agencies require preliminary studies, your application will be strengthened if you include them.

● **Controversy and contradiction.** The more controversial your proposed hypothesis, research design, or methods, or the more your preliminary data contradict or call into question common knowledge or other researchers' published findings, the more preliminary studies you will likely need to include in the **Preliminary Studies Section**. You need to convince reviewers that: (a) you have sound reasons to pursue your research as proposed, in light of any controversy or contradiction, (b) your pursuit will resolve or account for any real or imagined controversy or contradiction between your and other investigators' research, and (c) your research will bring your discipline substantially closer to resolving the controversial issue.

● **Selection of preliminary research.** The strongest applications have preliminary data to support each of the proposed aims. When selecting which preliminary data to include, you should try to support each proposed aim with preliminary data that underscore the feasibility of your proposed aims.

● **Completed studies.** You should not provide so many preliminary studies that reviewers get the impression that you have already performed a substantial part of your proposed studies. With this said, you should include preliminary studies that show you have the skills to perform your proposed methods and that you have preliminary evidence that warrants your strong suspicion that you will achieve significant outcomes.

---

[2] NIH notes that "preliminary studies, data, and/or experience pertinent" to the grant application is not needed for Exploratory/Development Grants (R21/R33), Small Research Grants (R03), Academic Research Enhancement Award (AREA) Grants (R15), and Phase 1 Small Business Research Grants (R41/43)" (www.nih.gov/grants/funding/phs398/phs.398.html; I-47; accessed November 28, 2014).

● **Relationship to methods.** Sometimes your preliminary studies involve methods that are similar or identical to those you are proposing, or sometimes the results from your preliminary studies justify your proposed methods in particular ways. In such cases, you have an option: to describe these methods in the **Preliminary Studies Section** or in the **Methods Section**.

4.3   Relationship between the preliminary research and Methods Sections

Before drafting the **Preliminary Studies Section**, you need a clear understanding of precisely how your preliminary research relates to your proposed studies. Questions that will help you clarify this relationship, first to yourself and then to the reviewers, include:

> *Have my previous research design, methods, and/or results influenced the design or methods of my proposed research?*
>
> *If so, which particular aspects of my proposed research have been so influenced?*
>
> *How have my training, competence, and previous experience prepared me to undertake my proposed research?*

You need to answer these questions clearly in order to increase your awareness of the significance and novelty of your proposed research and the role that your previous research has played in bringing you to this point in your line of research and in your research career.

If you draft the **Preliminary Studies Section** after you draft your proposed **Methods Section**, you will likely need to revise both of these sections less than if you first draft the **Preliminary Studies Section**.

Guideline 4.2 *Consider drafting the Methods Section first and then the Preliminary Studies Section. Then revise the Methods Section to make sure it complements the Preliminary Studies Section.*

The need to track the relationship between preliminary research and the **Methods Sections** will become clearer in Chapters 4.5 and 4.7, where we discuss the connection between your previous and your proposed research.

4.4   The heading outline: Preliminary Studies subsections

Similar to the **Background Section**, the **Preliminary Studies Section** is usually divided into subsections unless it is short, such as one page or less, in which case it is presented as paragraphs.

A first step before writing the text of a **Preliminary Studies Section** is to determine its subsections and their hierarchical relationship, and to draft a heading outline of the section. In this heading outline and in the text, you use the itemization scheme used in the **Background Section** (Chapter 3.1.2) in order to promote readability and to underscore the unity across all sections of the narrative.

> **Guideline 4.3** *Draft a heading outline that reflects a generic organization of the*
> ***Preliminary Studies Section** and that includes:*
> *(a) Headings to its subsections and*
> *(b) An itemization scheme that matches the one in the **Background Section**.*

The left-hand side of Figure 4-1 gives a generic organization of the **Preliminary Studies Section**: an optional but recommended introduction, at least 2 body subsections, and an optional ending.[3]

4.4.1   Location of the Preliminary Studies subsections in the narrative

Many submission requirements from funding agencies specify that the **Preliminary Studies Section** should follow the **Background Section** and precede the **Methods Section**, a basic organizing principle also used for sequencing the chapters in this book.

> **Aims Section**
> **Background Section**
> **Preliminary Studies Section** ◄——————
> **Methods Section**
> **Ending Section (optional)**

Some funding agencies do not provide much advice on where to locate **Preliminary Studies** subsections in the narrative. A case in point is NIH, which notes, "The Research Strategy . . . is composed of 3 distinct sections – Significance, Innovation, and Approach. Note the Approach section also includes Preliminary Studies for new applications and a Progress Report for renewal and revision applications."[4]

Perhaps the reason why NIH and some other funding agencies are not more helpful regarding where to locate preliminary research is that, in the narrative, the location of **Preliminary Studies** subsections is a function of their scope. This means that where you locate **Preliminary Studies** subsections depends on which proposed methods (and research design) they support, which varies per project. With this said, however, a basic strategy can be offered to organize **Preliminary Studies** subsections.

---

[3] However, a **Preliminary Studies Section** that is relatively short, such as a page or less, can be presented in a series of paragraphs.

[4] NIH PHS 398, page I-42 (http://grants.nih.gov/grants/funding/phs398/phs398.html; accessed November 28, 2014).

The generic heading outlines in Tables 4-1A, 4-1B, and 4-1C show alternative locations for **Preliminary Studies** in an NIH narrative, all 3 of which follow Guideline 4.4. In Table 4-1A, the **Preliminary Studies Section** is the first subsection (C.1) in the **Approach**, immediately before all of the proposed procedures for the proposed aims. This location indicates that the scope of the preliminary studies is *broad*, relating in some similar way to all of the proposed methods (and research design) for Aims 1, 2, and 3. In Table 4-1B, the **Preliminary Studies** subsections are included in the **Approach**, with a **Preliminary Studies** subsection immediately preceding the relevant proposed procedures for each of Aims 1, 2, and 3. The scope of each **Preliminary Studies** subsection in this arrangement is *narrow* since each **Preliminary Studies** subsection relates only to the procedures it immediately precedes. Table 4-1B presents a generic heading outline for an NIH narrative. In this heading outline, a **Preliminary Studies** subsection is the first subsection in the **Approach** (similar to Table 4-1A), yet **Preliminary Studies** subsections also immediately precede relevant procedures for each aim (similar to Table 4-1B). In Table 4-1C, the scope of the **Preliminary Studies** subsection in C.1 is broad, covering C.2, C.3, and C.4, and the scope of each **Preliminary Studies** subsection in C.2, C.3, and C.4 is narrow, relating only to the relevant proposed procedures that it precedes.

If methods for 2 aims, such as Aims 1 and 2 in Table 4-1B, share the same preliminary research, but a third aim is related to different preliminary research, then a possible organization would be the following:

> **C. Approach**
>   **C.1** Aim 1 Methods
>   **Preliminary Studies**
>   Proposed Procedures
>
>   **C.2** Aim 2 Methods
>   Proposed Procedures
>
>   **C.3** Aim 3 Methods
>   **Preliminary Studies**
>   Proposed Procedures

In other words, you do not repeat the same preliminary research in Aim 2 (C.2) that you include in C.1. Instead, in your introductory discussion of **Preliminary Studies** in C.1, you mention that this preliminary research also applies to the methods in C.2. And in C.2, you mention that the preliminary research in C.1 also applies to the methods in C.2.

Table 4-2 shows 3 heading outlines of a narrative that follows a generic format, with **Preliminary Studies** subsections in locations similar to the locations in Table 4-1. If your funding agency does not specify the location of preliminary research, you can safely include it just before the relevant methods (following Guideline 4.4), using one of the organizational alternatives in Table 4-2. However, a caveat is in order for an NSF narrative. Since NSF requires that you include your previously funded research for the last 5 years – even if the previous research is not related to the proposed research – you can locate the **Preliminary Studies Section** *after* the **Methods Section**.

4.4.2   Headings as topics to Preliminary Studies body subsections

Each **Preliminary Studies** body subsection has a heading that identifies the topic of the subsection.

● **Key terms in headings**. As much as possible, you need to phrase headings (and the text) of the **Preliminary Studies** subsections with key terms that match both the preliminary research and the proposed research so that reviewers will associate the **Preliminary Studies** topics more closely with your proposed research.

> **Guideline 4.5**   *Phrase headings to **Preliminary Studies** subsections with key terms that reflect the key terms in the proposed research, especially those in the **Aims Section**.*

Case 4-1 shows headings to 3 different **Preliminary Studies Sections**. In Case 4-1A, the heading of C.1 is the term *displacement of the OFT wall*. If the **Aims Section** used the term *movement* instead of a *displacement*, you should use *movement* in the **Preliminary Studies** heading in order to more closely associate this topic with the **Aims Section**. All of the subsections are itemized to help the reviewers understand the hierarchical relationships and to facilitate any cross-referencing.

● **Grammar of headings to Preliminary Studies body subsections**. The headings to **Preliminary Studies** subsections are phrased as short sentences or nouns with modification. Case 4-1A presents headings phrased as short sentences. Short sentences allow you to be more precise in identifying your major preliminary results as the topics of the **Preliminary Studies** subsections. The verbs in sentence headings are in the simple present form, such as ***provides*** in C.1 and ***provides*** in C.2 of Case 4-1A, even though the preliminary research is of the past.

> **Guideline 4.6**   *Use the simple present form of verbs in **Preliminary Studies** headings that comprise sentences.*

Cases 4-1B and 4-1C show **Preliminary Studies** headings phrased as nouns with modifiers. Common types of nouns that are used in headings are: (1)

nominalizations – that is, nouns that are derived from verbs and that have such endings as + **tion**, + **ment**, + **ence**, and + **ness**, and (2) *gerunds*, which are nouns ending in + **ing**, which are derived from verbs (see Chapter 8.6.3). In Case 4-1B, the headings to the **Preliminary Studies** subsections use the following gerunds and nominalizations, shown here with the verbs from which the nouns are derived:

cloning (from the verb *to clone*)   *confirmation* (from *to confirm*)
expression (from *to express*)   *generation* (from *to generate*)
purification (from *to purify*)   *construction* (from *to construct*)

In Case 4-1C, the headings are phrased with the following nominalizations and gerund (also shown here with the verbs from which they are derived):

characterization (from *to characterize*)   *development* (from *to develop*)
preconditioning (from *to precondition*)   *endothelialization* (from *to endothelialize*)

Frequent use of nominalizations in prose can lower the readability of the text, but as key nouns in **Preliminary Studies** headings, nominalizations do not usually introduce readability problems.

In contrast to headings to **Preliminary Studies** subsections, headings to **Preliminary Studies** (sub)subsections are usually phrased as nouns with modification, not as short sentences. As shown in Case 4-1C, the headings to the sub-subsections under C.5 are all phrased as nouns with modification.

● **Grammatically parallel headings.** Headings to **Preliminary Studies** subsections at the same level in the organizational hierarchy need to be grammatically parallel to each other as much as possible: either all short sentences or all nouns with modification.

> **Guideline 4.7** *At each level in the organizational hierarchy, draft **Preliminary Studies** headings that are grammatically parallel.*

In the set of headings shown in Case 4-1A, all of the headings are short sentences, so all are considered to be grammatically parallel to each other. As shown in Case 4-1B, some headings are phrased as nominalizations and one, as a gerund, yet the headings are still considered grammatically parallel since both nominalizations and gerunds are types of nouns. A gerund, such as *cloning*, is used when the underlying verb (in this case, *to clone*) has no nominalized form (e.g., **clonement* and **clonization* are not grammatical nouns for the verb *to clone*).

Another related point to Guideline 4.7 is that a set of headings at a different level in the organizational hierarchy does not have to match the grammar of the heading at the next higher level. For example, in Case 4-1C, the headings to the subsections under C.5 are all phrased as nouns. However, as shown below, the subheadings under C.5 could all be phrased as short sentences:

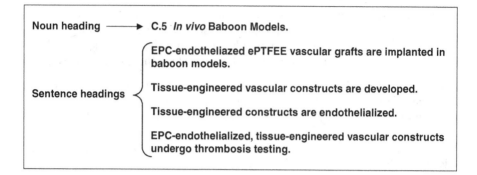

Noun heading ⟶ C.5 *In vivo* Baboon Models.

Sentence headings

EPC-endotheliazed ePTFEE vascular grafts are implanted in baboon models.

Tissue-engineered vascular constructs are developed.

Tissue-engineered constructs are endothelialized.

EPC-endothelialized, tissue-engineered vascular constructs undergo thrombosis testing.

4.5   Introduction to the Preliminary Studies Section

An introduction to the **Preliminary Studies Section** is recommended. A **Preliminary Studies** introduction provides you with an initial location where you can present your preliminary research in a favorable light and can highlight whatever information you want your reviewers to notice in the discussion that follows. A **Preliminary Studies** introduction also presents content comprising a scientific argument. Case 4-2 presents 2 examples of **Preliminary Studies** introductions.

● **Format and length.** The introduction to the **Preliminary Studies Section** is usually one or 2 brief, untitled paragraphs before the first **Preliminary Studies** body subsection.

● **Topic.** The main research area – that is, the topic of your preliminary research – is usually identified in the first or second sentence of the **Preliminary Studies** introduction and then is repeated in subsequent sentences of the introduction and in the upcoming **Preliminary Studies** body subsections. Notice in both of the introductions in Case 4-2, the topic is identified in the first sentence through key terms, and the key terms are repeated in the introduction.

● **Review of preliminary research.** The **Preliminary Studies** introduction reviews information that characterizes important aspects from (a) your preliminary research and (b) your training, competence, and experience. Guideline 4.8 identifies typical information that is commonly reviewed in the introduction.

> **Guideline 4.8**   *In the **Preliminary Studies** introduction, consider reviewing:*
> *(a) The primary research purpose of your preliminary research.*
> *(b) The research design of your preliminary research.*
> *(c) Significant features of the methods, major findings, major results, and/or major conclusions from your preliminary research.*
> *(d) Your research training, competence, and experience.*

In Case 4-2A, the **Preliminary Studies** introduction identifies 3 significant features of the methods: the subjects, the equipment (*AMES devices*), and where the subjects used the equipment. In Case 4-2B, the **Preliminary Studies** introduction characterizes methods (*cysteine accessibility* and *voltage clamp fluorometry*) and the major conclusion: *S4 is a voltage sensor in HCN channels*. Also illustrated in Cases 4-2A and 4-2B, the review of methods in the **Preliminary Studies** introduction serves to connect the preliminary studies to the proposed research through cross-references to the proposed research objective or aims. Also, if the preliminary methods are closely similar to your proposed methods, the inclusion of major findings and results in the **Preliminary Studies** introduction by implication indicates that the PI has the requisite skills to execute the proposed methods successfully.

If you have multiple major findings, results, and/or conclusions, rather than listing them in the **Preliminary Studies** introduction, you can identify the total number and then summarize them, indicating the **Preliminary Studies** subsections in which they are explained. However, if you list your major preliminary findings, results, and/or conclusions in the **Preliminary Studies** introduction, you can present them in a numbered list, which serves to emphasize the total number. If you do not have many major findings, results, and/or conclusions from your preliminary research, you can list them with alpha characters, to downplay the total number. The drawback of listing multiple major findings, results, and/or conclusions in the **Preliminary Studies** introduction is that the list considerably lengthens the introduction.

The examples that follow illustrate how you can: (a) summarize the major preliminary findings, results, and/or conclusions (Example 2, modified from sentence 1 in Case 4-2B), (b) forecast the upcoming discussion about them (Example 3), or (c) indicate the broad topic to which the preliminary findings, results, and/or conclusions relate (Example 4).

Example 2
To summarize, our major finding is that S4 is a voltage sensor in the HCN channels.

Example 3
In Sections C.2–C.5, we present our 8 major results from our preliminary studies.

Example 4
We had successful preliminary studies, and our findings elucidate many critical functions of S4.

The phrase *to summarize, our major finding* in Example 2 indicates that you achieved other findings, but you are describing only the major one. In Example 3, mentioning the relatively large, total number of major preliminary findings implies that you are productive. In Example 4, such terms as *successful, elucidate*, and *critical* have positive connotations and denotations. If you use such terms to characterize your preliminary research, you need to make sure that the information you include in the **Preliminary Studies** subsections does, indeed, substantiate each characterization. If reviewers end up assessing such statements as mere puffery and as not accurately reflecting your preliminary results, you may lose credibility.

● **Review of professional experience**. In addition to reviewing preliminary research that is relevant to your proposed research, you can review facts about your professional experience and about members of your research team. This type of review serves to emphasize that you and your research team are well-prepared to execute your proposed methods and to achieve your proposed research objective and aims in a timely manner.

> **Guideline 4.9** *In the **Preliminary Studies** introduction, consider reviewing information from your professional background to substantiate your qualifications to perform and to complete your proposed research successfully.*

You can address your experience – and enhance your credibility – by explicitly discussing it. However, just claiming that you are experienced in the proposed research is insufficient; you need to support the statement in order to turn the claim into a reasonable conclusion. Example 5 is a general statement about the research team's experience, but it is immediately supported by a list of references. Although the sentence could be more specific, at least the PI identifies references to support the statement. Example 6 presents a statement from a **Preliminary Studies** introduction that does not just cite to references but provides details that serve to substantiate the conclusion *We have extensive experience*.

Example 5
We have extensive experience with the experimental methods of the proposed research (see refs. 35–41, 48–52).

Example 6
We have extensive experience and multiple publications in each of the three core activity areas of the proposed project: concatenative synthesis (refs), voice transformation (refs), prosody modeling (refs), and perceptual evaluation (refs).

● **Connection between the preliminary studies and the proposed studies**. The **Preliminary Studies** introduction usually includes information to identify how the preliminary research is related to the proposed studies. This association helps keep your proposed research design in the forefront while your reviewers read details of your preliminary studies. For example, in Case 4-2A, sentence 3 connects the methodological relationship between the proposed studies that focus on the lower extremity and the preliminary studies that also address the lower extremity. In Case 4-2B, sentences 2 and 3 connect the preliminary methods to the proposed aims and methods.

4.6   Organization of Preliminary Studies body subsections

Preliminary Studies body subsections are primarily organized in one of 2 arrangements, as explained in Guideline 4.10. In either arrangement, the first preliminary information discussed is the preliminary research design, if relevant to the proposed research.

Guideline 4.10 *Organize **Preliminary Studies** body subsections in one of 2 ways:*
*(a) **Descending importance**. (i) If the preliminary research design is relevant to the*
*proposed research design, first present a **Preliminary Studies** body subsection that*
addresses results relating to the preliminary research design. (ii) For the remaining
***Preliminary Studies** body subsections, lead with a subsection that addresses the*
major preliminary findings, results, and/or conclusions, moving to subsections that
address the relatively minor preliminary findings, results, and/or conclusions.
*(b) **Chronological order**. (i) If the preliminary research design is relevant to the*
*proposed research design, first present a **Preliminary Studies** body subsection that*
addresses results relating to the preliminary research design. (ii) For the remaining
***Preliminary Studies** body subsections, sequence them in the order in which the*
preliminary methods were executed.

If you organize the **Preliminary Studies** body subsections according to descending importance, a critical question is:

How do you determine importance?

Importance is determined by the impact of your preliminary research on your proposed research: the greater the impact, the more important.

The chronological arrangement of **Preliminary Studies** body subsections is often used when preliminary findings and results led to your conducting other studies.

4.7 Scientific argumentation in Preliminary Studies body subsections

Similar to a **Background** body subsection, a **Preliminary Studies** body subsection needs to present content that comprises at least one scientific argument. However, in contrast to a **Background** body subsection, in which the *review-analysis* pair plays a central role in establishing the significance and novelty of the proposed research, in a **Preliminary Studies** body subsection, the 3-pronged relationship among the *review-analysis-connection* plays a central role. Identification of the connection in a **Preliminary Studies** body subsection identifies the relevancy between the PI's preliminary and the proposed research. Identification of the connection also promotes the credibility of the PI by indicating that the PI (and other members of the research team) is adequately prepared to accomplish the proposed research, as characterized in the **Aims Section**, as shown by the **Preliminary Studies Section**, and as explained in the **Methods Section**.

Cases 4-3 to 4-6 give examples of **Preliminary Studies** body subsections, with their scientific arguments analyzed in the right-hand column.

● **Topic: Results from the Preliminary Research.** Similar to headings to **Background** body subsections, headings to **Preliminary Studies** body subsections identify the topics of the subsections and are phrased in key terms that are repeated throughout the subsections. These topics are typically the results from the preliminary research. However, in contrast to headings to **Background** body subsections, which are primarily phrased as nouns with modifiers, headings to **Preliminary Studies** body subsections are phrased either as nouns with modifiers or as short sentences with present-tense verbs. The first heading below is a heading to a **Preliminary Studies** body subsection, that is a noun with modification. The second heading is a short sentence.

| _Confirmation_ | _that the purified protein is SarA_ | (Heading from Case 4-3) |
| Noun | Modification (Adjective Clause) | |

| _DBS_ | _improves_ | _postural stability in quiet stance_ | (Heading from Case 4-4) |
| Subject | Verb | Direct Object | |

● **Review.** In **Preliminary Studies** body subsections, the review mainly consists of: (a) the research purpose of the preliminary study that is under review, (b) relevant features of the preliminary methods, and (c) relevant major preliminary findings and results. To a much lesser extent, the review can comprise common knowledge and research from other researchers' previous studies, which, among other purposes, serves to substantiate your analysis of your own preliminary research (see below for a discussion of the analysis).

> **Guideline 4.11**   _In **Preliminary Studies** body subsections, consider reviewing:_
> _(a) The research purpose of each preliminary study._
> _(b) Relevant features of the preliminary methods, and/or_
> _(c) Relevant major preliminary findings and results._

All of the **Preliminary Studies** body subsections in Cases 4-3 to 4-6 present reviews primarily consisting of: (a) the research purpose of the preliminary study (sentence 1 of Case 4-3 and sentence 1 of Case 4-5), (b) relevant features of the preliminary methods (Cases 4-3, 4-4, 4-5, and 4-6), and (c) relevant major preliminary findings and results (Cases 4-3, 4-4, 4-5, and 4-6). The reviews of major preliminary findings and results (data analyses) can be presented in full sentences and in visuals.

A problem in writing **Preliminary Studies** body subsections is that you might review too few preliminary data to support your analysis of them (see below). You need to draft a review that has enough information for reviewers to view the analysis in the scientific argument as a logical inference, given the preliminary research that you have reviewed.

If submission requirements allow, you can review additional information to substantiate the analysis. This additional review can follow the analysis, or can be included on websites or in an appendix to the narrative. If you relegate substantiating data to a supplementary location (website or appendix), you need to include representative data in your **Preliminary Studies** subsection, and you need to tell your reviewers that the information you are reviewing is representative. However, reviewers are under no obligation to go outside of the narrative to retrieve information.

● **Analysis**. The analysis in **Preliminary Studies** body subsections evaluates the preliminary research that you have reviewed, thus creating *review-analysis* pairs and *review-analysis-review* combinations, similar to those used in **Background** body subsections. Also similar to the analysis in **Background** body subsections, the analysis in **Preliminary Studies** body subsections needs to be a reasonable inference that is based on the reviewed preliminary research.

However, there are a few differences between *review-analysis* pairs in **Background** and **Preliminary Studies** subsections. First, as explained earlier (e.g., Chapter 2.2.3), there are different types of analyses: (1) problem statements, (2) technical questions, (3) speculations, (4) hypotheses, (5) predictions, (6) extension statements, and (7) conclusions.

Although problem statements are frequently used in analyses of previous research in the **Background** subsections (and in the **Aims Section**), they are not often used in analyses of **Preliminary Studies** body subsections because they cast your preliminary research in a negative light. *Review-analysis* pairs and *review-analysis-review* combinations in a **Preliminary Studies** body subsection need to convey a less negative tone and to imply the motivation or rationale for your proposed research.

Further, although conclusions are not often used in analyses of previous research in **Background** subsections (and in the **Aims Section**), they are frequently used in analyses of preliminary research since they represent your ability to derive reasonable, supportable conclusions from the preliminary results, and to suggest that the experimental or theoretical results of the preliminary results have guided you in formulating useful, tractable questions, appropriate techniques, and robust analyses.

> **Guideline 4.12** *Consider analyzing your preliminary research not with problem statements, but with conclusions, technical questions, speculations, hypotheses, and predictions.*

As shown in Case 4-3, the analyses in sentences 5 and 6 are conclusions: the results that are reviewed in the first half of sentence 5 are analyzed as confirming *that the purified protein is SarA*, and the results reviewed in sentence 6 are analyzed as *demonstrating that the E. coli-derived SarA preparation is appropriate for the proposed experiments*. Case 4-4 presents analysis statements that are speculations in sentence 8 and in the last half of sentence 9, both of which logically follow the review of preliminary results. Case 4-5 shows a **Preliminary Studies** body subsection with an analysis passage in sentences 5–11: the analysis in sentence 5 is a speculation, signaled by the verbs *suggests* and *may play*. Sentences 6–11 present a set of technical questions that logically emerge from the review of preliminary research in sentences 1–4. Case 4-6 shows a **Preliminary Studies** body subsection that has analyses in sentences 5 and 8, both of which are conclusions.

It is important for intellectual and professional honesty, that your analyses not overstate your preliminary research. This requires your **unbiased** evaluation of your preliminary research and your selection of an analysis that accurately evaluates your findings and results.

Guideline 4.13 *Use a type of analysis that matches and does not misrepresent the positive features of your preliminary research.*

For instance, if your preliminary research does not support a particular conclusion, you should not so conclude. Instead, you can select another type of analysis to evaluate your preliminary results. In general, especially with conclusions, to increase their accuracy you will need to focus on the verb in order to make the conclusion less conclusory or consequential, less certain, or more speculative.

Example 7a presents a conclusion that is rephrased in Example 7b to be less conclusory. Examples 7c and 7d illustrate how you can further revise the conclusion into a speculation and, ultimately, into a technical question.

Example 7

a. The results from our intracellular modification of S4 residues *lead to* the conclusion that S4 is a voltage sensor in HCN channels. (Type of analysis: conclusion.)

b. The results from our intracellular modification of S4 residues *are consistent with* our proposed hypothesis that S4 is a voltage sensor in HCN channels. (Type of analysis: a less definite conclusion.)

c. The results from our intracellular modification of S4 residues *suggest* that S4 *may* be a voltage sensor in HCN channels. (Type of analysis: speculation.)

d. The results from our intracellular modification of S4 residues *lead to* the question: Is S4 a voltage sensor in HCN channels? (Type of analysis: technical question.)

● **Connection.** The connection component of a scientific argument identifies how the preliminary research and your analysis of it is related to the proposed research.

Review of Preliminary Research – Analysis – Connection to Relevant, Proposed Studies

To indicate the close relationship between the preliminary and proposed research, each **Preliminary Studies** body subsection needs to relate the review and analysis to the proposed research in terms of: (a) a feature of the proposed method, and/or (b) a proposed aim. This relationship points to your preparedness to execute the proposed methods with skill and to achieve your proposed research purposes (research objective and aims) in a timely manner.

To relate the review and analysis in a **Preliminary Studies** subsection to one of your proposed methods or aims, you need to identify the proposed method and/or aim.

Guideline 4.14 *Include a connection statement in each **Preliminary Studies** subsection either in a sentence or in parentheses, to relate the preliminary research to the relevant, proposed method(s) and/or aim(s).*

In Case 4-3, the preliminary research is associated with the relevant, proposed aim at the end of sentence 6, with the relevant aim identified in parentheses. The connection statements in Cases 4-4 (sentence 10) and 4-5 (sentences 12–15) relate the preliminary studies to both the relevant, proposed methods and the proposed aims. In Case 4-4, the relevant, proposed aim is presented in parentheses, and in Case 4-5, the relevant, proposed aims are integrated within sentences that discuss the proposed methods. Sentence 9 in Case 4-6 connects the preliminary research to both the proposed method and the proposed Aim 1.

The connection statement needs to specify the relevant, proposed method and/or aim by number; such explicitness helps reviewers understand the precise connection between your preliminary and your proposed research. Also, cross-referencing by number to the related method or aim implies that you are organized. Further, if you specify by number the relevant, proposed method or aim, you should specify by number *all* of your proposed aims at least once in the entire **Preliminary Studies Section**.[5] Otherwise, you may leave a reviewer thinking that any aims that you have not explicitly identified by number are not supported by preliminary research. You need to show reviewers that all of your proposed aims are supported in some way by your preliminary studies.

Guideline 4.15 *Try to relate all of your proposed aims to your preliminary studies, and identify by number which aims relate to which preliminary studies.*

Connection statements can also relate your preliminary research to your proposed research through the use of nouns and verbs that denote advancement and growth, such as ***advance, broaden, build upon, emerge, emergence, extend, extension, expand, expansion, lead, pursue***, and ***pursuit***. In Examples 8 and 9, the verbs (in bold italics) connect the preliminary and the proposed research. The verbs in connection statements can be in present or future tense forms.

Example 8
The proposed research ***builds upon*** the physiological classes identified in preliminary studies, to provide a context for interpreting patterns in the cells from which we recorded (see proposed Exp. 4).

Example 9
Our proposed Specific Aim 1 ***will extend*** the characterization of S4 movement in the preliminary studies by introducing additional cysteines at more internally located positions and testing their intracellular accessibility.

- **Significance**. When we address significance in terms of the **Preliminary Studies** body subsections, we are really referring to 2 types of significance:

---

[5] However, sometimes it is not realistic to specify by number all of your proposed aims at least once in the entire **Preliminary Studies Section**. And sometimes you might need to cross-reference the same aim more than once.

(1) Preliminary research that contributes to or furthers scientific knowledge.
(2) Preliminary research that facilitates the proposed research.

Preliminary research acquires significance not only through substantive results (significance type 1) but also by enriching the PI's experience so that the PI can successfully execute the proposed methods with skill and can achieve the proposed research objective and aims (significance type 2).

By the time the reviewers have read the **Preliminary Studies** body subsections, the significance of your proposed research in terms of warranting the contribution of your proposed research to scientific knowledge should be clear since you will have addressed this type of significance in the **Aims Section**, in a **Background Section**, and possibly in the introduction to the **Preliminary Studies Section**. As a result, it is not critical that significance of the preliminary research to scientific knowledge be stated again (note significance is not stated in the examples, in Cases 4-3 to 4-6).

It is critical that reviewers understand that your preliminary research is significant in terms of its facilitating your preparedness to undertake the proposed research and to complete it successfully and on time. In most cases, this type of significance is implicit through your relating your preliminary research to your proposed methods and/or aims through the connection statement. However, you can make this point explicit, as shown in Example 10:

Example 10
These preliminary results show our ability to execute successfully the immunohistological studies that we will also use in our proposed research (see Methods 2.3).

Guideline 4.16 *Consider explicitly identifying the significance of your preliminary research in terms of your preparedness to execute the preliminary research successfully.*

4.8   Ending the Preliminary Studies Section

Due to length constraints, the **Preliminary Studies Section** often does not include an ending. But if you do have the space, an ending subsection can be useful in order to reiterate the connection between your preliminary and proposed research.

If you include an ending subsection, it typically consists of content indicating a scientific argument. It identifies the topic that you investigated in the preliminary research. It reviews, in summary fashion, key preliminary information that you discussed in the body sections – information that is particularly relevant to your proposed research. The ending analyzes the preliminary research that you have summarized. It connects the preliminary research to your proposed methods, research objective, and/or aims; and it optionally includes the significance of the proposed research.

Case 4-7 presents an ending to a **Preliminary Studies Section**, with its scientific argument analyzed in the right-hand column.

4.9   Referring to your publications in the Preliminary Studies Section

Since a **Preliminary Studies Section** gives background information, you need to include citations to your publications. You need to let reviewers know that you are publishing – to convince them that you are productive, that you get results, and that your peers have already scrutinized your preliminary research and have found it sufficiently scientifically meritorious, both in terms of significance and novelty, to be published.

   If you use the citation by number scheme, reviewers need to interrupt their reading and check the list of references to learn more about the publications. But if you use the author scheme for citations, you still run the risk of reviewers needing to check the list of references, if your name is not the first author cited. Regardless of the citation scheme that you choose, in the **Preliminary Studies Section** you need to maintain a focus on you and your research team so that reviewers can readily understand that you have been publishing and are productive. Citations are further discussed in Chapter 8.1.

   The most straightforward way to help maintain the reviewers' focus on you and your research team is to educate them about those publications stemming from your preliminary research that are relevant to your proposed research.

> **Guideline 4.17**
> *(a) In the **Preliminary Studies** introduction, explicitly mention your publications.*
> *(b) In the **Preliminary Studies** body subsections, explicitly refer to yourself or to members of your research team.*

   Examples 11 and 12 present sentences from introductions to **Preliminary Studies Sections**, both of which refer to publications by number and explicitly associate the publications with the PI and the research team via *We, the PI*, and the name of a collaborator.

Example 11
The collaboration takes advantage of the expertise of the PI in the molecular genetics of *S. aureus* and the biochemical expertise of Dr. John Smith in transcription factor structure and function (14,15).

Example 12
We have extensive experience with both the experimental methods and the underlying theoretical background that we will use to approach these issues (see refs 69–76 and Appendix 1.1–1.7).

4.10   A Preliminary Studies Section without subsections

The discussion of **Preliminary Studies Sections** has thus far focused on those that are formatted into subsections. However, if you have a relatively short **Preliminary**

Studies Section, such as one page or less, you can present it in a series of paragraphs. An example of a **Preliminary Studies Section** without subsections is shown in Case 4-8. A benefit of formatting this section into a series of paragraphs is that space that would have been used for headings is used for text. A drawback from your (the writer's) perspective is that the section may be harder to organize since there are no visual cues (i.e., headings) to help you delineate topics of discussion and scientific arguments. A similar drawback, this time from the reader's perspective, is that a section formatted as a series of paragraphs may result in lower readability since there are no headings to identify the topics of discussion. A work-around to this problem is to present a series of paragraphs as a bulleted list of paragraphs, with each paragraph receiving a run-in heading.

> **Guideline 4.18**   *In a relatively short **Preliminary Studies Section** that is formatted as a series of paragraphs:*
> *(a) Include at least one scientific argument.*
> *(b) Divide the section into an optional introduction, at least one body paragraph, and an optional ending.*
> *(c) Consider changing the series of paragraphs into a bulleted list of paragraphs, with each paragraph receiving a run-in heading.*

The introduction to a **Preliminary Studies Section** is formatted as a paragraph. This first sentence either: (a) stands alone as a one-sentence introductory paragraph followed by paragraphs that comprise the body of the section or (b) is the first sentence of a body paragraph.

In Case 4-8, the first sentence is the introduction, which identifies the purpose of the preliminary research, and sentences 2–4 present body information. The introductory sentence identifies the topic (*imaging the chick heart using FDOCT*) and significance (*feasibility*). The first and subsequent paragraphs also give content comprising a scientific argument, with 3 review passages and 3 connection passages.

4.11   Assorted issues relating to Preliminary Studies and Progress Report Sections

4.11.1   Distinguishing preliminary from proposed studies

For clarity, your reviewers need to readily understand when you are discussing your preliminary research and when you are discussing your proposed research. Through consistent and precise phrasing, you can help distinguish preliminary from proposed research, as explained in Guideline 4.19:

> **Guideline 4.19**   *Distinguish preliminary from proposed research through:*
> *(a) Strategic use of past and future verb tenses.*
> *(b) Adverbs that indicate sequence or chronology.*
> *(c) Adjectives and other noun modifiers that denote time.*

- **Verb tenses**. Verb tenses need to be controlled to signal previous and preliminary research from proposed research, and to distinguish reviews from analyses, connection statements, and significance statements.

The following chart summarizes the use of verbs in **Preliminary Studies** and **Progress Report Sections.**

Verb Tense and Form	Use	Examples
Simple past	Reviewing past research	found, were found
Present perfect	Reviewing past research, with subtle emphasis	have found, have been found, has found, has been found
Simple present	Generalizations that have not been disproved, analyses that have not been disproved	find, are found finds, is found
Simple future	Future research, connection statements	will find, will be found

(1) Simple past or present perfect. The review of completed previous research is primarily presented in the simple past (e.g., *found* or *were found*)[6] or the present perfect (e.g., *have found* or *have been found*)[7]. The present perfect is particularly useful to emphasize subtly the relevance of the previous research to the proposed research. However, the present perfect takes more characters than does the simple past, so if space is needed, you can use the simple past, but the subtlety in emphasis will be lost. In Case 4-3, most of the verbs in the sentences that comprise the review (sentences 1–4) are in the simple past (e.g., *performed* [sentence 1] and *was generated* [sentence 2]). However, in Case 4-5, the verb in sentence 14 is in the present perfect (*have now obtained*), which subtly emphasizes the currency and relevancy of the information to the proposed research.

(2) Simple present. Common knowledge, generalizations about previous research, and analyses of previous research – whether the previous research is yours or other researchers' – are in the simple present, if they have not been disproved. For example, in Case 4-4, sentence 1, a review sentence, uses the simple present verbs *stands* and *is* because the sentence generalizes about persons who stand quietly. Sentence 8 in Case 4-4 uses the simple present *suggest* because you are currently analyzing the reviewed research. Many sentences in Case 4-6 use the simple present to describe models that were developed in the preliminary studies since the review describes how the models function

---

[6] The simple past verb *found* is in active grammar and *were found* is in passive grammar.

[7] The present perfect verb *have found* is in active grammar and *have been found* is in passive grammar

in general. Verbs in the simple present in sentences 3 and 4 in Case 4-6 include: *expresses, is multiplicated, are, reflect,* and *has.*

(3) **Future.** Sentences that discuss proposed research, such as connection statements, use verbs in the simple future since the proposed research will be conducted in the future. Verbs in the present tense include *will need* in sentence 10 of Case 4-4 and *will be used* in sentence 15 of Case 4-5.

● **Adverbs.** Adverbs that communicate sequence or time can also be used to distinguish previous from proposed research. The use of the adverb *now* or *presently,* with the present perfect in particular, serves to pinpoint the relevancy of previous research to proposed research. Example 13 uses the phrase *have now obtained,* which emphasizes the relevancy of the previous research to the proposed work.

Example 13
We **have now obtained** cDNA clones of wild-type sodium channels from susceptible S-lab and HAmCq-S strains .... (from Case 4-5, sentence 14)

● **Chronological modifiers of verbs and nouns.** Another way to distinguish previous from proposed studies is to use adjectives and adverbs that denote chronology. Chronological adjectives include *preliminary, previous, proposed,* and *new,* as illustrated in Example 14.

Example 14
Since the **proposed** studies focus on treating the lower extremity, we mainly report on **preliminary** studies that focus on our subjects who used our **proposed** device for lower extremity treatment and who completed the protocol (from Case 4-2A, sentence 3).

Some chronological adjectives are problematic because they are context-dependent, such as *current.* In a description of your previously funded research in a **Progress Report Section,** *current* can mean previously funded research that you are still working on. However, in a **Preliminary Studies Section,** *current* can mean your proposed research. To avoid such ambiguity, it is best to use other chronological modifiers.

4.11.2   Focusing on investigators: personal pronouns, and active and passive grammar

One way to bring your credibility (and the credibility of your research team) into focus and to increase the readability of your text is to mention yourself in your descriptions of your preliminary studies. This inclusion is achieved by **humanizing** your preliminary research by: (1) using personal pronouns to refer to yourself and using names to refer to people on your research team, and (2) using active grammar.

> **Guideline 4.20** *Humanize the description of your preliminary research through personal pronouns and active grammar.*

● **Personal pronouns and names.** You, as PI, can subtly associate yourself with your preliminary research, especially when it is particularly noteworthy, by using first-person personal pronouns (e.g., *we, our*) to refer to yourself and/or the research team.

If you are the only investigator involved in your proposed research, you can use either *we, the PI,* or *the applicant.* The pronoun *we* can be used even when the research team consists of only the PI because a PI cannot submit a grant proposal without the PI's institutional pre-approval – an approval that signals the institution's commitment to the proposed research project. Thus, in this context, the pronoun *we* includes the PI's institution. The third-person noun *the PI* (or *the applicant*) can be used, but some people consider such usage too formal and are more comfortable with first-person personal pronoun references, such as *we.*

Note that the singular pronoun (*I, me, my, mine*) to refer to the PI is acceptable. However, it suggests the PI is researching in isolation.

Another way to humanize the description of your preliminary research is to identify, by name, which members of your research team were involved in which preliminary research projects or experiments. For example, in Case 4-8, sentence 1 not only mentions the PI but also identifies the Co-I (*Dr. Smith*) by name. The following chart identifies how to present a team member's name, depending on whether the person has been previously mentioned and identified in the text:

Use of a Researcher's Name within the Text of a Preliminary Studies or a Progress Report Section	Example: First Mention of the Researcher's Name	Example: Subsequent Mention of the Researcher's Name
First mention of an investigator	James Smith, MD (Co-Investigator, University of Southern California)	Dr. Smith or Dr. Smith (Co-I) or Co-I Dr. Smith
First mention of an investigator who does not have a role in the proposed research	James Smith, MD (University of Southern California)	Dr. Smith

In sentence 1 of Case 4-8, because the Co-I is identified without his first name (*Dr. Smith*), we can assume that this is not the first time he has been mentioned in the narrative.

● **Active and passive grammar.** To refer to yourself (the PI) with a pronoun, you need to use active phrasing. Active phrasing requires the identification of the agent. Example 15a shows a sentence with its verb in active grammar (*will use*) and the agent in first-person pronoun (*We*). Example 15b is a passive sentence, with the agent of the verb *will* *be used* not identified; this type of passive sentence is termed *an agentless passive*. Example 15c, a passive sentence, is problematic (note the symbol ‡) because the agent is identified by the first-person pronoun *us*. In scientific English, when the agent is the PI, passive sentences do not identify the first-person agent, and readers assume that the agent is the writer. Chapter 8.5 further discusses active and passive grammar, agentless passive, and the use of personal pronouns.

Example 15

a. **We will then use** segmented images to visualize and to quantify motion and morphology of the chick heart at HH1 to HH16, and to develop subject-specific FEM models of the developing Heart (Aim 3) (from Case 4-8, sentence 15).
b. Segmented images **will then be used** to visualize and to quantify motion and morphology of the chick heart at HH1 to HH16, and to develop subject-specific FEM models of the developing heart (Aim 3).
c. ‡Segmented images **will then be used by us** to visualize and to quantify motion and morphology of the chick heart at HH1 to HH16, and to develop subject-specific FEM models of the developing Heart (Aim 3).

4.11.3   Expected versus problematic preliminary research problems and results

You may have had complications or surprises in your preliminary methods or results. For example, your preliminary research may have yielded results that you did not expect (or did not want), your preliminary results might not have yielded clear patterns, or your preliminary methods may have been problematic in design or execution.

You must stay true to your preliminary research, which means mentioning problems if they are relevant to your proposed research. Thus, the issue is not whether to mention the problems but, rather, how to describe them. Guideline 4.21 gives strategies for describing problematic preliminary research:

> **Guideline 4.21** *When reviewing problematic preliminary research:*
> *(a) Describe the problem in neutral or positive terms, such as **surprising, unexpected**, or **exciting**, as appropriate.*
> *(b) Analyze unexpected preliminary result(s), such as by speculating on its possible source or cause, or on what the results may suggest.*
> *(c) Describe how the unexpected preliminary result(s) affected your preliminary studies, your proposed research methods, and/or design.*

You need to identify and describe your significant, problematic preliminary result(s) in a way to imply that you are optimistic – or at least are not worried – about it. You need to convey a confident tone: although unexpected or problematic results arose, your proposed research can still successfully build upon them. You can achieve this positive (or neutral) tone by careful selection of words. In Example 16, sentences 1–5 call a problem *a surprising finding* and analyze it in a way to suggest that it is actually a research opportunity that the PI will pursue in the proposed research. Sentence 3 in this example reviews the surprising preliminary result, sentence 4 analyzes it with a problem statement, and sentence 5 connects the surprising finding to the proposed research.

Example 16
[1]Simulation results indicate that an unchanging postural control system is not able to manage more than very modest load changes without the risk of instability (Fig. 6). [2]The obvious compensation for increased load would be to increase the stiffness of the system (increase K_p). [3]However, when stiffness is increased, damping must also be increased. [4]*A surprising finding of our preliminary results* was that when stiffness was increased as a result of sensory manipulations (Fig. 9), damping appeared to be controlled mainly by decreasing the control loop time delay (Fig. 10), rather than by increasing the neural controller damping factor, K_D. [5]It is unknown which of these mechanisms the postural control system uses to adjust for transient load changes. [6]Specific Aim I, Exp. 3 will investigate this issue.

If your research problem stems from issues involving your personal life, the quality of the research performed, or administrative complications, you can present the problem but focus more on the impact of these problems on your scientific methodology. Consider this scenario: You are writing a **Progress Report Section** of a narrative. Your previously funded research was significantly behind your originally proposed schedule because of problems with a research technician's work, which made you question the accuracy of the collected data. As a result, you reran a number of tests, which slowed your progress. One way to explain this situation is not to deny that you are behind the original schedule because that would be a misrepresentation of facts. Rather, you can focus on the consequences of this problem – a redesign of the schedule and improved scientific rigor.

Example 17a illustrates an unacceptable way to describe a problem with the execution of the research since the term *personnel issues* suggests you are not adequately managing your laboratory staff. Example 17b illustrates a sentence explaining a scheduling issue.

Example 17
a. ‡*Due to personnel issues in our laboratory*, we are behind the schedule identified in the previous grant application because selected tests were rerun to verify the reliability of data collected. We found …
b. We are behind the schedule identified in the previous grant application because we reran selected tests in order to verify the reliability of data collected. We found …

You might characterize a situation as a problem when it may not be actually a problem. Perhaps you are a new, independent PI, and most of your previous work was in the laboratory of your mentor. Example 18a is a slightly modified excerpt from a new PI's actual first draft, apologetically explaining his use of preliminary data from his post-doctoral studies. Example 18b is a rewrite, which focuses on the resulting publications.

Example 18

a. Although I was just a post-doctoral fellow in Dr. Smith's laboratory, he let me run the experiments myself without his supervision and all data from the resulting publication were from my own efforts (even though Dr. Smith is listed as a co-author).

b. We ran 4 tests, the results of which were published (see appendix).

4.12   The Progress Report Section

When you seek to renew a grant proposal or to supplement a previously funded research project, the funding agency may require that you explain the progress you have made in the previously funded research cycle. The section that explains these preliminary studies is the **Progress Report Section**. Reviewers read the **Progress Report Section** to evaluate your progress in the previously funded research, the chances that you will produce valid findings in your proposed research in a timely fashion, and the logical relationship between the previously funded research and the research that you are proposing.[8] Case 4-11 gives examples of introductions in **Progress Report Sections**, and Cases 4-12 and 4-13 give examples of **Progress Report** body subsections.

Both the **Preliminary Studies** and **Progress Report Sections** present preliminary results and are similar in many respects. In fact, almost all of the guidelines for the **Preliminary Studies Section** can be applied equally well to the **Progress Report** subsections. For example, the **Progress Report Section** is similar to the **Preliminary Studies Section** in terms of the location of the **Progress Report Section** in the narrative. The **Progress Report Section** precedes the relevant **Methods Section**.[9] The organizational alternatives in Tables 4-1 and 4-2 can also be used for the **Progress Report Section**. However, the **Progress Report** and the **Preliminary Studies Sections** are different in a few respects, as explained here.

- **Location within the narrative.**

- **Organization.** Unlike the **Preliminary Studies Section**, which is organized by topics relating to the preliminary research design and major results, the **Progress Report** body subsections are usually organized according to the aims of your previously funded research. The previous aims become the headings to the **Preliminary Studies** body subsections.

---

[8] Note, however, that NSF requires a progress report on any NSF previous research that it funded within 5 years before the current grant proposal, whether related to the proposed studies or not. (http://www.nsf.gov/pubs/policydocs/pappguide/nsf13001/gpg_2.jsp#//C2diii; II-9; accessed November 28, 2014).

[9] However, the progress report for NSF narratives can be located after the Methods Section.

Guideline 4.22   *Organize the **Progress Report** body subsections according to the aims from the previously funded research.*

The headings to the **Progress Report** subsections are usually the actual previous aims, or they cite to the previous aims by number. If any **Progress Report** body subsection is further divided into subsections, the heading for a (sub)section usually indicates major results.

Guideline 4.23
*(a) Include the numbers of the previous aims in the **Progress Report** subsection headings.*
*(b) For **Progress Report** sub-subsections, use topic- or results-based headings.*

Cases 4-9 and 4-10 present heading outlines that show how **Progress Report** headings can be phrased. Case 4-9 presents noun headings in alternative arrangements. In Case 4-9A, the noun headings to the body subsections are topics from the aims taken from the previously funded grant proposal. The noun headings in Case 4-9B are all at the same level in the organizational hierarchy. Their corresponding aims are indicated at the end of each heading. The heading outline in Case 4-9B implies greater productivity since the results are distributed over 8 subsections. In Case 4-10, the headings are sentences and their corresponding previous aims are listed in parentheses.

- **Introduction to a Progress Report Section**. The introduction to a **Progress Report Section** orients the reviewers to the research objective of the previous grant, the dates of the previous grant cycle, the significance of the findings, and the progress made towards achieving the previous aims. Optional information can include: (a) the previous aims, especially if they were not indicated in the subsection headings (e.g., see Case 4-11A), (b) the previous methodological approach, (c) scope of the **Progress Report Section**, and (d) a forecast statement (e.g., sentence 6 in Case 4-11A).

Case 4-11 presents introductions from 2 **Progress Report Sections**, from NIH grants. NIH requires certain information in its **Progress Report Section**: the dates of the grant cycle, a summary of the aims, the significance of the findings, and the progress made toward achievement of the aims.[10] Even though NIH does not specify where to locate this information in the section, PIs usually locate it in the introduction, as shown in Case 4-11A. In the introduction in Case 4-11B, the aims are not summarized, which is the case when the previous aims are used as headings to the body subsections of the **Progress Report** (e.g., see Case 4-9A).

---

[10] www.nih.gov/grants/funding/phs398/phs398.html (Section 5.5; accessed February 6, 2013).

Guideline 4.24   *When aims are used as headings to **Progress Report** body sections, do not give or summarize the aims in the **Progress Report** introduction.*

Depending on the submission requirements, the scope of the **Progress Report Section** may include other research that was not funded or was not carried out in the previous funding period. If this is the case, you need to indicate this inclusive scope in the **Progress Report** introduction. Otherwise, reviewers might assume that your research in the **Progress Report Section** is exclusively from your previously funded grant submission.

Guideline 4.25   *If the **Progress Report Section** covers research beyond that funded by your targeted funding agency, include a statement in the **Progress Report** introduction that identifies the scope of the section.*

Assuming that submission requirements allow, if you have previous research that is not directly relevant to your proposed research but is relevant to your preparedness to undertake your proposed research, you can include this previous research. However, you need to mention both of these points in the **Progress Report** introduction, as shown in Example 19:

Example 19
Although our past research on healthy human participants is not directly relevant to our proposed research objective, we offer our preliminary research findings from healthy participants in Section C.4, in support of our skills to successfully record microneurographically from single muscle spindle Ia afferents.

● **Scientific argumentation in a Progress Report Section.** Similar to the body subsections in the **Background** and **Preliminary Studies Sections**, each body subsection in the **Progress Report Section** needs to consist of at least one scientific argument.

Guideline 4.26   *In each **Progress Report** body subsection, include, at the very least, these components of a scientific argument:*
(a) The research topic.
(b) A review of your previous research.
(c) An analysis of your previous research.
(d) The connection between your previous research and proposed research.

Cases 4-12 and 4-13 give examples of a **Progress Report** body subsection, with information comprising scientific arguments identified in the right-hand column. Just as connection statements are important in **Preliminary Studies** subsections for your credibility and to indicate your preparedness to take on the proposed research, connection statements are also important in **Progress Report** subsections for these same reasons.

Thus, there are commonalities across the **Background, Preliminary Studies, and Progress Report Sections**. They all are types of sections that give information about *previous* research and that help define the novelty and significance of the proposed research. The **Background Section** gives information primarily consisting of others' research, the **Preliminary Studies Section** gives the proposed research team's previous research that is relevant to the proposed research, and the **Progress Report Section** also gives relevant information derived from the proposed research team's previously funded research. In addition, all 3 sections are usually divided into subsections, and each subsection gives information that comprises at least one scientific argument.

Through the logical relationship between the review and the analysis components of a scientific argument, you can help reviewers understand the significance and novelty of your proposed research. And by explicitly identifying the relationship of the analyzed information to the proposed methods (via connection information), you can help reviewers understand how your proposed methods will generate significant and novel information that is reliable and valid.

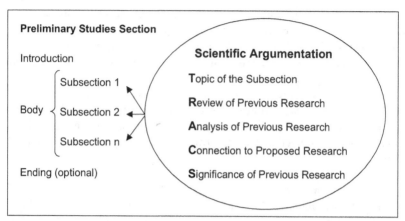

Figure 4-1. Components of a scientific argument in a generic Preliminary Studies/Progress Report Section comprising subsections. A scientific argument is included in each body subsection.

Table 4-1. Locations for preliminary studies in an NIH narrative.

A. One Preliminary Studies Section	B. Multiple Preliminary Studies Subsections with the Same Scope	C. Multiple Preliminary Studies Sections with Different Scopes
I. Specific Aims	I. Specific Aims	I. Specific Aims
II. Research Strategy	II. Research Strategy	II. Research Strategy
A. Significance	A. Significance	A. Significance
B. Innovation	B. Innovation	B. Innovation
C. Approach	C. Approach	C. Approach
C.1. Preliminary Studies	C.1. Aim 1 Methods	**C.1 Preliminary Studies**
C.2. Aim 1 Methods	• **Preliminary Studies**	C.2 Aim 1 Methods
• Proposed Procedures	• Proposed Procedures	• **Preliminary Studies**
C.3. Aim 2 Methods	C.2 Aim 2 Methods	• Proposed Procedures
• Proposed Procedures	• **Preliminary Studies**	C.3 Aim 2 Methods
C.4. Aim 3 Methods	• Proposed Procedures	• **Preliminary Studies**
• Proposed Procedures	C.3 Aim 3 Methods	• Proposed Procedures
	• **Preliminary Studies**	C.4 Aim 3 Methods
	• Proposed Procedures	• **Preliminary Studies**
		• Proposed Procedures

Table 4-2. Locations for preliminary studies in a generic narrative.

A. One Preliminary Studies Section	B. Multiple Preliminary Studies Subsections with the Same Scope	C. Multiple Preliminary Studies Sections with Different Scopes
I. Aims Section	I. Aims Section	I. Aims Section
II. Background Section	II. Background Section	II. Background Section
III. **Preliminary Studies Section**	III. Methods Section	**A. Preliminary Studies**
IV. Methods Section	A. Aim 1 Methods	B. Innovation
A. Aim 1 Methods	• **Preliminary Studies**	C. Approach
• Proposed Procedures	• Proposed Procedures	B. Aim 1 Methods
B. Aim 2 Methods	B. Aim 2 Methods	• **Preliminary Studies**
• Proposed Procedures	• **Preliminary Studies**	• Proposed Procedures
C. Aim 3 Methods	• Proposed Procedures	C. Aim 2 Methods
• Proposed Procedures	C. Aim 3 Methods	• **Preliminary Studies**
	• **Preliminary Studies**	• Proposed Procedures
	• Proposed Procedures	D. Aim 3 Methods
		• **Preliminary Studies**
		• Proposed Procedures

Case 4-1. Heading outlines for Preliminary Studies Sections.

(A)

C. PRELIMINARY STUDIES

C.1 Ultrasound images provide data on the displacement of the OFT wall during the cardiac cycle.

C.2 Variation of ventricular pressures during the cardiac cycle provides inlet boundary conditions.

C.3 FEM model of chick embryonic OFT allows quantification of hemodynamic forces.

C.4 Banding causes the heart of chick embryos to enlarge within a few hours.

(B)

C. PRELIMINARY RESULTS

C.1 Cloning and expression of sarA.

C.2 Purification of SarA.

C.3 Confirmation that the purified protein is SarA.

C.4 Generation of SarA-specific antiserum.

C.5 Construction of xylE reporter fusion vector.

C.6 Studies with *S. aureus* collagen adhesin gene (cna).

(C)

C. PRELIMINARY DATA

C.1 Vascular Grafts.

C.2 Characterization of EPCs.

C.3 The Role of Hemodynamic Preconditioning on Endothelial Function.

C.4 The Role of Extracellular Matrix Substrates on Endothelial Function.

C.5 *In vivo* Baboon Models.

    a. EPC-endothelialized ePTFEE vascular grafts in baboon models.

    b. Development of tissue-engineered vascular constructs.

    c. Endothelialization of tissue-engineered constructs.

    d. Thrombosis testing of EPC-endothelialized, tissue-engineered vascular constructs.

Case 4-2. Examples of introductions to Preliminary Studies Sections. Bold =
Topic, **A** = Analysis, **C** = Connection.

(A)

C. Preliminary Studies
[1]Preliminary studies have been carried out in which **AMES devices** have been placed in the homes of 27 chronic stroke subjects for treatment of the ankle (n =15), wrist/fingers (n = 9), and elbow (n = 3). [2]The **devices** are all similar in design, each assisting the movement of a paretic/spastic **joint(s)** and vibrating the tendons of the prime mover muscles crossing the treated **joint(s)**. [3]Since the proposed studies focus on treating the lower extremity, we mainly report on preliminary studies that focus on our subjects who used our proposed **device** for lower extremity treatment and who completed the protocol (11 of the 15).

Topic
Review

C: Methods

(B)

C. Preliminary Studies
[1]In support of this application, we have preliminary data measured with 2 **different methods**, indicating that **S4 is a voltage sensor in HCN channels**. [2]These **methods**, which will be used in the proposed study, are cysteine accessibility (Specific Aim 1) and voltage clamp fluorometry (Specific Aim 2). [3]In addition, we have preliminary data on the kinetics and voltage dependence of the gating currents in HCN channels (Specific Aim 3).

Topic
Review &
C: Purpose

Case 4-3. **Body subsection from a Preliminary Studies Section. Bold** = Topic, **A** = Analysis, **C** = Connection.

Confirmation that the purified protein is SarA. [1]To **confirm** that the **protein purified** from our *E. coli* lysates is **SarA**, we performed electrophoretic mobility shift assays (EMSA) with a 45 bp DNA fragment containing the heptad repeat cis to the *agr* P_3 promoter (see Fig. 5, Background and Significance). [2]The 45 bp fragment was generated by synthesizing and annealing complementary oligonucleotides. [3]After labeling the target DNA with ^{32}P, the fragment was allowed to equilibrate in solution with varying amounts of the purified protein. [4]The mixture was then resolved by native gel electrophoresis. [5]The fact that a mobility shift was observed with the agr-derived target DNA (Fig. 9) **confirmed** that the **protein** we purified from the *E. coli* lysates **is SarA**. [6]Moreover, these results, together with the results of our EMSA experiments employing DNA fragments derived from the region upstream of can (see below), demonstrated that our *E. coli*-derived **SarA** preparation is appropriate for the experiments aimed at the identification of additional SarA targets within the *S. aureus* genome (Specific Aim #3).

Topic

Review

A: Conclusion

C: Purpose & Methods

Case 4-4. Example of a Preliminary Studies body subsection. Bold = Topic, **A** = Analysis, **C** = Connection.

3. DBS improves postural stability in quiet stance
[1]When a person **stands** quietly, a continuous flow of **sensory information** from peripheral receptors is important for maintaining **balance** (22, 27). [2]In studies from our current NIH grant, subjects stood with eyes open on force plates for 3, 60-second trials. [3]We computed the excursion of the whole body CoP. [4]Figure 6A shows CoP displacements in the horizontal plane for a representative control subject and for a DBS subject post-surgery in each of the conditions. [5]In the OFF condition, postural sway area, quantified as the mean 95% confidence ellipse*s* (Fig. 6B) and the root mean square area (RMS; Fig. 6C), was larger in the PD subjects compared to the control subjects, and the sway area dramatically increased even more in the DOPA condition (p<.01). [6]In the DBS condition, postural sway decreased significantly (p<.05) to within the range of postural sway of the control subjects (see DBS ellipses, Fig. 6B). [7]For PD subjects, the BOTH condition resulted in postural sway that was not specific enough (Fig. 6B).
  [8]Preliminary analysis of the DBS sites suggests that DBS in both the STN and GPi sites, but not DOPA, reduces postural sway in quiet stance to within the range exhibited by control subjects (Fig. 6C). [9]However, after surgery, in the DOPA and BOTH conditions, the PD subjects with DBS in the STN, but not in the GPi, showed increased postural sway, as if the site of DBS may alter the response of the nervous system to levodopa. [10]We will need to compare the effects of DOPA (and the level of levodopa medication) before and after surgery, and the effects of DBS in the STN and GPi, to determine whether there are differential effects related to the site of stimulation (**D.2, Aim 1**).

Topic

Review

A: Speculation

Review

A: Speculation

C: Methods & Purpose

Case 4-5. Preliminary Studies body subsection. Bold = Topic, **A** = Analysis, **C** = Connection.

C.4 Mutations in the sodium channel of *Culex quinquefasciatus*. [1]To identify **mutations** in the entire **sodium channel** cDNAs of resistant *Culex* mosquitoes, we recently investigated SNPs among (a) susceptible S-Lab, (b) resistant parental HAmCq[G0] and MoAmCq[G0], (c) highly resistant offspring HAmCq[G8] and MoAmCq[G8] *Culex* mosquitoes, and (d) S-Lab[G8], the 8th generation of permethrin-selected offspring of S-Lab strain. [2]We isolated the full-length sodium channel cDNAs from these 6 *Culex quinquefasciatus* strains and compared the entire sodium channel cDNA sequences across these 6 strains. [3]We detected 30 cDNA clones from each mosquito strain and revealed 25 mutations: 7 nonsynonymous and 18 synonymous mutations (refs, in preparation). [4]The frequency of these mutations in each strain correlated with the strain's level of resistance. [5]This novel finding suggests not only that multiple mutations may co-exist in a single mosquito sodium channel that is involved in pyrethroid resistance, but also that synonymous mutations may play a significant role in insecticide resistance. [6]Five questions emerged from this study: 1) [7]Which of the 25 mutations are involved in insecticide resistance? 2) [8]How frequently do certain mutations co-occur in insecticide resistant mosquitoes? 3) [9]Is there a specific threshold of insecticide concentration at which a particular mutation or mutation combination occurs in a mosquito population? 4) [10]What effects do mutations or mutation combinations in the sodium channel have on sodium channel function? and 5) [11]What role do synonymous mutations play in sodium channel function? [12]Our proposed Aims 1-2 will address these areas by characterizing mutations in the mosquito sodium channel that are involved in insecticide resistance. [13]These 25 mutations will serve as valuable tools that will allow us to characterize the mutations, as proposed in Aims 1-2. [14]We have now obtained full-length cDNA clones of wild-type sodium channels from susceptible S-lab and HAmCq-S strains and full-length sodium channel clones from resistant HAmCq[G8] and HAmCq-R. [15]The wild-type full-length sodium channel clones from susceptible S-lab and HAmCq-S strains will be used in our Aim 2, to characterize both nonsynonymous and synonymous mutations, and to identify mutation combinations involved in sodium channel sensitivity, by introducing the 25 mutations and mutation combinations into these wild-type clones.

Topic

Review

A: Speculation

A: Technical Questions

C: Methods & Aims

Review

C: Methods & Aims

Case 4-6. Example of a Preliminary Studies body subsection. Bold = Topic, **A** = Analysis, **C** = Connection.

Duration modeling. [1]We have developed a **duration module** for the OGI Festival system that is based on van Santen's sums-of-products approach (refs). [2]At CSLU we implemented a multiplicative model, which is a simple case of a SoP model in which contextual factors have a multiplicative effect on the average duration of a phoneme (see Equation 2). [3]It expresses the predicted duration DUR (ref) of a phoneme p, the average duration of which $D_{mean}(p)$ is multiplicated with different parameters. [4]$S_i(c_i)$ are parameters whose values reflect the contributions of factor i when it has level c_i (i.e., the factor stress can have levels stressed and unstressed, and the parameter values will be > 1 for the stressed condition and < 1 for the unstressed condition). [5]The multiplicative model provides goodness-of-fit results comparable to those obtained with substantially more complicated models in this class (refs).

$$DUR(p) = D_{mean}(p) \times S_1(c_1) \cdots \times S_n(c_n) \qquad (2)$$

[6]In a study by Smith et al. (1998), it was shown that an existing SoP model that was trained on speech from one professional speaker could be adapted to the duration patterns of another speaker using only a few sentences. [7]The duration model of the target speaker is called the multiplicative power model (Equation 3), which assigns weights to each factor that are speaker-dependent. [8]The results showed that a duration model based on 3.5 hours of recorded speech could be successfully adapted to a new speaker using only 6 sentences from this new speaker. [9]In order to adapt the generic duration module in the OGI Festival system to each of the target speakers that will be recorded in Aim 1, we will use an automatic adaptation approach that produces similar results to the multiplicative power model.

$$DUR(p;i) = D_{mean}(p) \times S_1(c_1)^{i1} \cdots \times S_n(c_n)^{in} \qquad (3)$$

Topic

Review

A: Conclusion

Review

A: Conclusion

C: Methods & Purpose

Case 4-7. Ending subsection to a Preliminary Studies Section. Bold = Topic, **A** = Analysis, **C** = Connection.

Summary. [1]The observations that **females** are more **susceptible to MS** and that fluctuations in **hormone levels** can alter the course of disease suggest an important **regulatory role** for **sex hormones** in this disorder. [2]Testosterone has immunosuppressive properties and has been shown to be an effective treatment that suppresses the development of experimental autoimmune disorders, including EAE. [3]Some studies have also shown that T therapy can be a viable treatment for human autoimmune disorders (54). [4]The data presented above from our lab, and those of other laboratories, strongly suggest that immune activation during the course of an autoimmune attack can inhibit the production of gonadal hormones. [5]This may have important implications for the pathogenesis of MS. [6]A reduction in circulating testosterone levels may be a consequence of an MS exacerbation and might interfere with the recovery from a disease attack by suppressing the hormonal regulation of the immune response (Fig. 6). [7]This reduction in circulating T levels may be one explanation for the high incidence of progressive MS in males.

[8]The overall goal of this proposal is to elucidate the mechanisms by which immune activation regulates testosterone production in EAE and to determine whether testosterone levels in male MS patients fluctuate with disease activity. [9]A complete understanding of the immune-endocrine axis will be important if testosterone replacement therapy is going to be considered a viable treatment for suppressing attacks of MS.

Topic, Review, & A: Speculation

Review

Significance & A: Speculation

A: Hypothesis

C: Purpose

Significance

Case 4-8. Preliminary Studies Section from an NIH R01 grant. Bold = Topic, **A** = Analysis, **C** = Connection.

Aim 1. Preliminary Results

[1]We determined the feasibility of *imaging the chick heart* in vivo using the **FDOCT system** developed and established in the lab shared by the PI and Dr. Smith (co-investigator)[11, 53-57]. [2]Using this system, we acquired images of the chick heart at HH13, HH14, HH16, and HH18 (see Fig. 1). [3]We found that we could image the whole heart at HH13 and HH14, but we could image only the OFT of HH16 and HH18 embryos due to limitations in OCT penetration depth (~1 mm), as the heart was growing larger. [4]Therefore, we will image the whole hearts of HH11 to HH14 chick embryos and the OFT of HH15 and HH16 embryos.
　　　[5]Since the heart beats faster at HH18 than at earlier developmental stages, analysis (and reconstruction) of heart motion is the most challenging at HH18. [6]Thus, in our preliminary studies, we focused on analyzing OCT image sequences of HH18 chick embryos. [7]We acquired sequences of 2D cross-sectional images (over several cardiac cycles) along the OFT of an HH18 chick embryo. [8]The longitudinal distance between cross-sectional planes (h) was 12.5 μm (see Fig. 3). [9]To reconstruct 4D images of the OFT, we then applied to the image sequences our developed (and implemented) synchronization procedure, which uses similarity of adjacent images to calculate the phase difference between two neighboring image sequences[1]. [10]In the proposed project, we will further develop our synchronization procedures to ensure proper reconstruction of a c-shaped heart and will apply these procedures to reconstruct 4D images of the chick heart at HH11 to HH14 and of the heart OFT at HH15 and HH16.
　　　[11]After 4D reconstruction of the chick OFT, to better visualize and quantify the morphology of the OFT, we segmented the reconstructed images; that is, we delineated the myocardium and lumen surfaces (with cardiac jelly in between). [12]To this end, we used a custom-made segmentation algorithm, based on a combination of optical flow techniques and active contour methods[58-60]. [13]Our preliminary segmentation procedures applied to 4D images of the heart OFT at HH18 resulted in segmentation of the myocardium and the lumen (Fig. 2) that were comparable to our manual segmentations. [14]In the proposed project, we will further develop our segmentation procedures to correctly segment the C-shape of the heart at HH11 to HH14, and we will apply it to reconstruct 4D images of the chick heart. [15]We will then use segmented images to visualize and to quantify motion and morphology of the chick heart at HH11 to HH16, and to develop subject-specific FEM models of the developing heart (Aim 3).

Topic

Review

A: Problem

C: Methods

Review

C: Purpose & Outcome

Review

C: Purpose, Outcome, & Methods

Case 4-9. Alternative heading outlines for a Progress Report Section. **(A)** Numbers in parentheses correspond to publications. **(B)** Aims and publications in parentheses. In **Alternative (B)**, the 4 aims from **Alternative (A)** have been expanded into 8 aims.

(A)
Progress Report
Introduction without a heading

AIM 1. ADAPTATIONS OF SPATIAL ORIENTATION FOR LOCOMOTION
1A. Vestibular-somatosensory Interaction for Locomotor Adaptation (1,2)
  • Unilateral vestibular loss.
  • Bilateral vestibular loss.
1B. Foot Nystagmus: A Feed-forward Mechanism for Locomotor Trajectory Control (3, 15–18)

AIM 2. ADAPTATIONS OF SPATIAL ORIENTATION FOR STANCE POSTURE
2A. After-effects of Stance on an Inclined Surface: An Adaptive Mechanism for Postural Vertical (13, 14)
2B. Differences in Preferred References Frames for Postural Orientation (2, 9)

AIM 3. CENTRAL MECHANISMS FOR LOCOMOTOR AND POSTURAL ADAPTATION
3A. Podokinetic After-rotation (4–7)
3B. Post-incline Lean (14)

AIM 4. MULTISENSORY CONTROL OF POSTURE (9, 10)
4A. Vestibular Information for Trunk Stabilization
4B. Haptic and Visual Substitution for Vestibular Loss

Manuscripts Published
Manuscripts Submitted

(B)
1. **Vestibular-somatosensory Interaction for Locomotor Adaptation** (Alternative I, Aim 1; Publications 1 and 2)
  • Unilateral vestibular loss.
  • Bilateral vestibular loss.

2. **Foot Nystagmus: A Feed-forward Mechanism for Locomotor Trajectory Control** (Alternative I, Aim 1; Publications 3 and 15–18)

3. **After-effects of Stance on an Inclined Surface: An Adaptive Mechanism for Postural Vertical** (Alternative I, Aim 2; Publications 13 and 14)

4. **Differences in Preferred References Frames for Postural Orientation** (Alternative I, Aim 2; Publications 12 and 19)

5. **Podokinetic After-rotation** (Alternative I, Aim 3; Publications 4–7)

6. **Post-incline Lean** (Alternative I, Aim 3; Publication 14)

7. **Vestibular Information for Trunk Stabilization** (Alternative I, Aim 4; Publications 9 and 10)

8. **Haptic and Visual Substitution for Vestibular Loss** (Alternative I, Aim 4; Publications 9 and 10)

Manuscripts Published
Manuscripts Submitted

Case 4-10. Example of sentence headings in a Progress Report Section.

4.1 Progress Report
Introduction without a heading

4.1.1 When patients were not matched to TBC or MSI treatment based on clinical features, long-term treatment outcomes were equivocal (Aim 1).

4.1.2 When patients with LBP were matched to a treatment based on clinical features, both pain and disability improved 7 weeks post-treatment initiation (Aim 1).

4.1.3 Neuromuscular responses were impaired in patients with LBP (Aim 2).

- Alterations in neuromuscular responses persisted up to 6 months in patients with LBP

- When patients with LBP were matched to a PT treatment based on shared clinical features, their neuromuscular responses improved with treatment.

- RUSI of the lumbar multifidus muscle helped identify subjects who were 'Eligible' for TBC-based stabilization exercises and who would likely respond favorably to this treatment.

4.1.4 Reliability was demonstrated in the TBC and MSI systems (Aims 1 and 2).

Case 4-11. Introductions from NIH Progress Report Sections. Bold = Topic, **A** = Analysis, **C** = Connection.

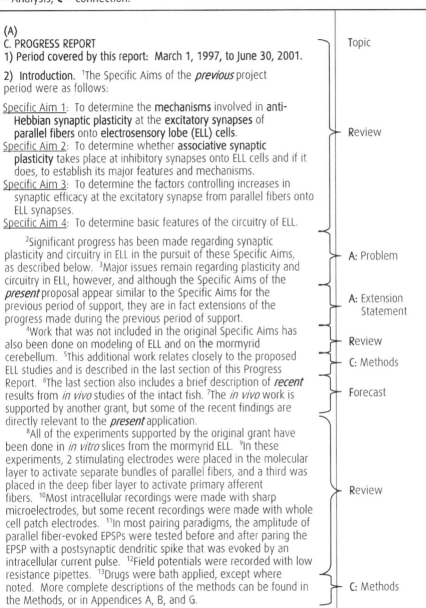

(A)
C. PROGRESS REPORT | Topic
1) Period covered by this report: March 1, 1997, to June 30, 2001.

2) Introduction. ¹The Specific Aims of the *previous* project period were as follows:

Specific Aim 1: To determine the **mechanisms** involved in **anti-Hebbian synaptic plasticity** at the **excitatory synapses** of **parallel fibers** onto **electrosensory lobe (ELL) cells.**
Specific Aim 2: To determine whether **associative synaptic plasticity** takes place at inhibitory synapses onto ELL cells and if it does, to establish its major features and mechanisms.
Specific Aim 3: To determine the factors controlling increases in synaptic efficacy at the excitatory synapse from parallel fibers onto ELL synapses.
Specific Aim 4: To determine basic features of the circuitry of ELL.

| Review

²Significant progress has been made regarding synaptic plasticity and circuitry in ELL in the pursuit of these Specific Aims, as described below. ³Major issues remain regarding plasticity and circuitry in ELL, however, and although the Specific Aims of the *present* proposal appear similar to the Specific Aims for the previous period of support, they are in fact extensions of the progress made during the previous period of support.

A: Problem

A: Extension Statement

⁴Work that was not included in the original Specific Aims has also been done on modeling of ELL and on the mormyrid cerebellum. ⁵This additional work relates closely to the proposed ELL studies and is described in the last section of this Progress Report. ⁶The last section also includes a brief description of *recent* results from *in vivo* studies of the intact fish. ⁷The *in vivo* work is supported by another grant, but some of the recent findings are directly relevant to the *present* application.

Review

C: Methods

Forecast

⁸All of the experiments supported by the original grant have been done in *in vitro* slices from the mormyrid ELL. ⁹In these experiments, 2 stimulating electrodes were placed in the molecular layer to activate separate bundles of parallel fibers, and a third was placed in the deep fiber layer to activate primary afferent fibers. ¹⁰Most intracellular recordings were made with sharp microelectrodes, but some recent recordings were made with whole cell patch electrodes. ¹¹In most pairing paradigms, the amplitude of parallel fiber-evoked EPSPs were tested before and after paring the EPSP with a postsynaptic dendritic spike that was evoked by an intracellular current pulse. ¹²Field potentials were recorded with low resistance pipettes. ¹³Drugs were bath applied, except where noted. More complete descriptions of the methods can be found in the Methods, or in Appendices A, B, and G.

Review

C: Methods

(B)
C. Progress Report

[1]Successful navigation in our daily environments requires the ability to adapt locomotion and posture in response to changes in surface conditions. [2]The objective of our previous grant cycle (dates) was to better understand how the kinesthetic and motor systems contribute to postural orientation during locomotion and stance. [3]This progress Report discusses experimental results from the previous grant cycle that focused on two adaptations: 1) locomotor trajectory adaptations in response to stepping on a circular treadmill and 2) postural adaptations in response to standing on a surface inclined in a pitch plane. [4]A list of our publications cited in the Progress Report (1-29) follows this section. [5]These previous studies form the basis for our new, proposed studies that seek to compare more directly the central and sensory mechanisms underlying adaptations to stepping on a rotating treadmill and stepping on an incline. [6]The proposed studies will test our central hypothesis, that common mechanisms control the adaptation of spatial orientation for both locomotion and stance posture. [7]New, preliminary, unpublished pilot studies for the new Specific Aims are in the rationale subsections for each experiment in the Methods Section of this proposal.

Topic

Review

Forecast

C: Purpose & Methods

Forecast

Case 4-12. Example of a Progress Report body subsection. Bold = Topic, **A** = Analysis, **C** = Connection.

1B. Foot Nystagmus: A Feed-forward Mechanism for Locomotor Trajectory Control (3, 15-18)
[1]Our previous studies of control subjects **stepping in place** on a rotating treadmill revealed a dynamic pattern of **foot rotation** relative to the trunk, a pattern that bears a close analogy to ocular **nystagmus** relative to the head. [2]We termed this pattern *angular-foot nystagmus* (AFN; 3). [3]In an extension of those studies, we examined forward locomotion around a curve and found a tight synchrony of AFN with the corresponding linear sequence of foot movement. [4]We termed this synchrony *linear foot nystagmus* (LFN; 17). [5]During the swing phase of locomotion, both AFN and LFN demonstrated stabilization of the foot in space just prior to ground contact, indicating an internally generated feed-forward mode of space-stabilizing motor control (16). [6]Our results supported the hypothesis that the amplitude ratio of AFN:LFN is expressed physiologically as a locomotor trajectory of a given curvature (17). [7]Preliminary results suggested that covert remodeling of AFN through stepping on a rotating treadmill or of LFN through stepping on a conventional linear treadmill produces predictable changes in the locomotor trajectory (18). [8]Based on our finding that linear treadmill exposure has long-lasting effects on locomotion characteristics (e.g., linear stride length), our proposed studies of postural after-effects of walking on an incline will be limited to walking in place on a stationary incline (Exp. 1, 3, 5-9). [9]Also, in Exp. 7, we will determine whether walking in place on a surface that is inclined laterally will produce long-lasting changes in locomotor trajectory and in the pattern of stepping.

Topic

Review

A: Speculation

Review

C: Methods & Purpose

Case 4-13. Example of a body subsection from a Progress Report Section. This example illustrates how analyses of reviews give rise to proposed hypotheses and how connection information can be expressed. **Bold** = Topic, **A** = Analysis, **C** = Connection.

Morphological Examination of ELL Cells

[1]**ELL cells** were recorded intracellularly and injected with biocytin for **morphological examination**. [2]The morphologies of the major cell types were described, including: MG cells, large ganglion cells, large fusiform cells, thick-smooth dendrite cells, small fusiform cells, horizontal cells, and a few granule cells. [3]Previous Golgi studies showed that the basilar dendrites of MG1 and MG2 cells are distributed differently within the layers of ELL (2), which led to our hypothesis that MG1 cells are inhibited by electrosensory stimuli ('off-center' cells), and that MG2 cells are excited by such stimuli ('on-center' cells). [4]We have now confirmed this functional hypothesis in *in vivo* studies (3). [5]In contrast to intracellular filling with biocytin, which labels axons very well, the Golgi method does not label axons well.

[6]We showed that the axons of MG1 and MG2 cells have different distributions within the layers of ELL. [7]Large ganglion and large fusiform cells are located in different layers of ELL. [8]The different distributions of MG1 and MG2 axons suggested the hypotheses that off-center MG1 cells inhibit on-center efferent cells (large fusiform) and that on-center MG2 cells inhibit off-center efferent cells (large ganglion; Fig. 3). [9]These hypothesized connections also make functional sense. [10]For example, inhibition of an on-center efferent cell by an off-center interneuron would enhance sensory responses of the efferent cell, whereas the opposite inhibition by an on-center interneuron would suppress sensory responses on the efferent cell.

[11]Both types of MG cells have axon terminals in the ganglion and plexiform layers where they can contact other MG cells (2). [12]Functional considerations, described in Significance, have suggested the hypothesis that MG1 cells selectively inhibit MG2 cells and vice versa. [13]However, there is, as yet, no morphological or physiological evidence for this hypothesis. [14]The hypothesized connections between these 2 types of MG cells, with each other and with efferent cells, have major consequences for ELL function, and experiments are planned in Aim 3 to determine whether they are present.

Topic

Review

A: Problem

Review

A: Speculation, Conclusion & Prediction

Review

A: Speculation, Hypothesis & Problem

Significance

C: Purpose

The Methods Section, Part 1

Most grant applications fail because of problems in the **Methods Section**.[1] You need to provide enough detailed, accurate, and clear methods information for reviewers:

- To understand your proposed research design and methods.

- To understand why you decided on particular features of your proposed research design and methods.

- To conclude that your proposed research design and methods will result in valid findings.

- To conclude that your proposed research design and methods are appropriate for you to achieve your proposed research objective and aims.

- To conclude that your proposed research design represents sound scientific methodology.

- To conclude that you, personally or through members of your research team, can successfully execute all of your proposed methods.

- To visualize your executing the proposed methods in a logical order.

- To view your credibility favorably.

---

[1] Unfortunately, there are few studies of why grant applications fail. The statement in the text is based on an informal analysis of NIH summary sheets and NIH-based research. Cuca, Janet & McLoughlin, William. Why clinical research grant applications fare poorly in review and how to recover. *Cancer Invest* 1987; **5**: 58.

In addition, you need to provide enough detailed, clear, and accurate information so that researchers – with your same or closely similar training and background – can (at least in theory) replicate your methods and come up with essentially the same data, results, and statistical analyses that you are proposing to achieve.

Chapter 5 covers basic organizational alternatives for the **Methods Section** and information to include when you are: (a) presenting your proposed research design and methods, (b) describing your **SOS** (_s_ubjects, _o_bjects of study, and _s_pecimens), and (c) describing your **MET** (_m_aterials, _e_quipment, and _t_ools). Chapter 6 focuses on how to draft procedures for collecting data and analyzing the data; and how to explain your potential methodological problems and proposed solutions. Chapter 6 includes guidelines on how to organize the **Methods Section** when some of the methodological features are the same or similar across aims or experiments – that is, when some of the methodological features are **shared** across the proposed aims or experiments. Chapter 6 also discusses the ending to the narrative.

Although the **Methods Section** needs detailed methods, you should not over-load the section with insignificant details and alternatives since they can cause reviewers to lose focus on the elegance and appropriateness of your research design. In addition, some agencies that fund grants for 1–3 years may require relatively short narratives, such as 3–5 pages, a length that does not allow much space for you to give too many details about your methods. Agencies granting large awards over 3–5 years, for example, may specify a longer narrative, such as 10–15 pages. In either case, you might not think you have enough space to adequately describe your proposed methods. Chapters 6.3.3 and 6.3.4 discuss ways to determine which methodological details to include, and Chapters 8.6.2 to 8.6.4 discuss strategies for shortening the text in order to make space for additional information.

Before writing, you may spend weeks, if not months, thinking about your proposed research design and methods. This forethought is good – and necessary – but given the complexity and importance of the **Methods Section**, excessive mental planning just shortens the time you have for drafting the section.

The sooner you begin drafting overview visuals and the text, the better. How soon is sooner? **_At least 2 to 3 months_** before a submission deadline for a scientific grant proposal of 10 or more pages.

Before drafting the text of the **Methods Section**, you should draft at least one, possibly 2 overview visuals of your proposed methods: (1) an overview table that identifies correspondences across your proposed aims, major methods, and timeline; and (2) an overview flow chart of major methodological tasks. Drafting overview visuals will force you to concretize your ideas about your research design and methods before you write. Also, drafting overview visuals is not a waste of time since you can include them in the narrative. See Chapter 5.2.2 and Cases 5-1 to 5-4, which discuss and give examples of overview visuals for the **Methods Section**.

5.1   Primary features of methods

You need to describe the *primary features* of your proposed research design and methods. Primary methodological features are those characteristics that define a scientific or technological approach or experiment, as recognized and determined by experts knowledgeable in currently accepted scientific, technical, and medical methodologies, practices, and theories.

> **Guideline 5.1** *When describing the design and details of your proposed methods, include primary methodological features.*

Although it is impossible to detail even a few primary features of methods in a book focused on writing, there are underlying principles of primary methodological features that can be identified here. What many primary methodological features share is their function: (1) individually or in combination, they serve *to control variables that may otherwise confound the data to be collected and interpreted*, and (2) they are designed to provide answers to your hypotheses or research questions. Thus, primary features include those that affect how you intend to control variables and that affect the reliability and validity of data that you collect.

For example, if you are describing a randomized clinical trial and you do not identify *objective* measures for inclusion criteria and exclusion criteria for subjects (see Chapter 5.7.4), you will have omitted a primary feature of your methodology. If you are conducting Western blot analysis and fail to include information on the specificity of the antibodies, this is a major omission of a primary feature. And if you are describing your use of analysis of co-variables (ANCOVA) to analyze data, but do not clarify what the dependent variables are, this omission is a critical error in primary content.

Before drafting the **Methods Section** – especially if you are a relatively new PI or a postdoctoral fellow – you might need to seek assistance to help you identify the primary features of your proposed methods. Otherwise, you might inadvertently omit necessary content.

> **Guideline 5.2** *Before writing the **Methods Section**, consider identifying the primary features of your proposed methods, using such aids as:*
> *(a) Technical books in your field.*[2]
> *(b) Online resources.*[3]
> *(c) Members of your research team.*
> *(d) A mentor.*
> *(e) A trusted colleague.*

---

[2] See, for example, Hulley, Stephen B. *et al. Designing Clinical Research: An Epidemiologic Approach*, 2nd Ed. Philadelphia, PA: Lippincott Williams & Wilkins; 2001.
[3] For example, see http://ccts.osu.edu/content/study-design (accessed November 28, 2014).

Your training, background, and experience should help you identify many, if not most, of the primary features of your proposed research design and methods. But for thoroughness, you should consider researching and double-checking this information with available resources. However, you will ultimately not know whether you have included adequate information to describe the primary features of your proposed research until the reviewers from your targeted funding agency tell you so, either by the score they assign to your grant proposal or by their written critiques.

5.2   Organization of the Methods Section

There is usually more than one way to organize subsections to a **Methods Section**. By drafting an overview visual(s) *before* drafting the text and by playing with different organizational alternatives using heading outlines *before* drafting the text, you can determine an arrangement that will help you describe your proposed research clearly and precisely within your funding agency's specifications about total page length of either the **Methods Section** or the narrative as a whole.

5.2.1   Basic organizational framework of the Methods Section

The organization and phrasing of headings in the **Methods Section** need to meet your funding agency's specifications yet, as much as possible, reflect your unique, proposed research (see Chapter 1.5). You need to divide the **Methods Section** into 3 main units – an introduction, a body, and an ending – and you need to divide the **Methods** body into body subsections. Each body subsection might be divided again into sub-subsections.

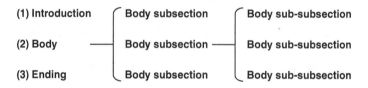

Unless otherwise indicated for clarity, the less cumbersome term *subsection* will be used for *sub-subsection*, sub-sub-section, and sub-sub-subsection.

The organizing principle for **Methods** subsections is the aims, as presented in the **Aims Section**. For example, if you are proposing 2 aims in the **Aims Section**, you will have at least 2 body subsections in the **Methods Section**: the first **Methods** body subsection is for the proposed methods to achieve Aim 1, and the second **Methods** body subsection is for the proposed methods to achieve Aim 2. Similarly, if you are proposing 3 aims, you will have at least 3 **Methods** body subsections.

Guideline 5.3   *Organize* **Methods** *body subsections to correspond to the number of aims in the* **Aims Section** *and their sequence in the* **Aims Section**.

The left-hand side of Figure 5-1 presents a generic organization of a **Methods Section**. Each **Methods** body subsection presents the methods to one of the proposed aims. Also as shown on the left-hand side of Figure 5–1 (see asterisks), you will need to vary this generic organization if some of your proposed aims *share* methods. Variations to this generic organization in order to accommodate shared methods across aims are discussed in Chapter 6.11.

5.2.2 Overview visuals

A strategy to help you organize the **Methods Section** and also to help reviewers quickly understand your proposed research design is to draft an overview visual(s) (Chapter 1.11.2). While drafting the **Methods Section**, you can use the overview visual as a guide, and you can also include it in the **Methods Section**. An overview visual identifies key features of your proposed research design.

> **Guideline 5.4** *Draft an overview visual(s) before drafting a heading outline of the* **Methods Section** *and before drafting any* **Methods** *text.*

An overview visual is usually placed in the introduction to the **Methods Section**, but it can also be used in any **Methods** body subsection to help reviewers better understand your proposed research design for an individual aim.

> **Guideline 5.5**
> (a) *Locate an overview visual of the entire project in the introduction to a* **Methods** *Section.*
> (b) *Consider locating a* **Methods** *overview visual for a particular aim in the introduction to the relevant* **Methods** *body subsection.*

Two common types of overview visuals are the overview table and the overview flow chart. Cases 5-1 to 5-5 present overview tables, and Case 5-6 presents an overview flow chart from a **Methods** introduction.

● **Overview table**. An overview table presents key features of the proposed research in columns and rows. With only a few eye movements across this table, reviewers can learn about the relationships of key features of your proposed methods to each other and to the proposed aims, and about your proposed schedule.

Guideline 5.6 suggests the design of an overview table for an introduction to a **Methods Section**.

> **Guideline 5.6** *Consider including an overview table in the* **Methods** *introduction.*
> (a) *Use column 1 for the proposed aims.*
> (b) *Use the middle column(s) to describe key features of the methods.*
> (c) *Use the far-right column for the time line.*

The overview tables in Cases 5-1 to 5-3 illustrate the typical design of an overview table described in Guideline 5.6. The aims are listed in the first (far-left) column and are briefly explained in key phrases from the aims in the **Aims Section**. The aims are not usually presented verbatim from the **Aims Section** in overview visuals due to space constraints. The middle column(s) characterizes significant features of the proposed methods, such as the number of experiments or tasks (Case 5-1), a list of the proposed procedures to achieve each aim (Cases 5-1 to 5-3), the numbers or types of **SOS** for each aim, especially when the proposed research involves human subjects (Cases 5-1 and 5-3), and the names of the team members or sites responsible for particular procedures (Cases 5-2 and 5-3). The last (far-right) column presents the time schedule of the activities in months or years.

In Case 5-4A, the overview table with a quarterly time schedule consumes over half the table. If submission requirements do not specify such a quarterly breakdown of procedures, you can reduce the quarterly columns to one column, as shown in Case 5-4B, which will free up columnar space for other information. Case 5-4B presents such a revision to Case 5-4A, with the middle column ready to receive specific methodological information.

In examining Cases 5-1 to 5-3, you may have noticed that they are a revision of the overview worksheet (see Chapter 1.11.2 and Table 1-6). With just a few changes, the overview worksheet can be revised into an overview table.

An overview table in a **Methods** body subsection is slightly different from an overview table in a **Methods** introduction. In a **Methods** body subsection that addresses only one aim, an overview table identifies relationships among methodological features for that particular aim. As a result, the overview visual in a **Methods** body subsection does not identify the aim in the first column since the entire table focuses on one aim, as should be indicated in the title of the visual. However, chronological details of activities for the particular aim are often provided, as shown in Case 5-5.

● **An overview figure.** Complex spatial, chronological, or causal relationships among key features of your research design or procedures may be difficult for you to write about or may be difficult for reviewers to initially, quickly conceptualize. In these cases, an overview figure is helpful.

> **Guideline 5.7** *Consider using an overview figure to capture complex spatial, chronological, or causal relationships among key features of your proposed research design or procedures.*

Common overview figures include overview maps and photographs, structural designs, and flow charts. An overview flow chart is particularly useful in providing a chronological perspective on a clinical trial.

Case 5-6 presents an overview flow chart of activities for a randomized, controlled trial that is being proposed for Aim 3.[4] The activities have been divided into 3 phases, indicated on the left side of the chart. If the overview flow chart characterizes a trial

---

[4] An overview flow chart as described here is a variation of a CONSORT flow diagram for clinical research papers (www.consort-statement.org/consort-statement/flow-diagram0; accessed November 28, 2014).

that covers all of the proposed aims, then instead of the phases listed on the left-hand side of the chart, you indicate the aims by their number as enumerated in the **Aims Section**. You should use the same key terms (and abbreviations) in the overview visual that you use in the **Aims Section**. Also, you need to introduce the overview visual before the reviewers encounter it, using terms and abbreviations that you already use in the **Aims Section** (see Chapter 7.7.3).

Guideline 5.8

(*a*) *Use key terms and abbreviations in the **Methods** overview visual that match those in the **Aims Section**.*

(*b*) *In your references to the overview visual in the text, use key terms and abbreviations that match those in the **Methods** overview visual and in the **Aims Section**.*

5.2.3 Heading outlines for Methods body subsections

Once you draft an overview visual(s) and before drafting the text, you should begin organizing each **Methods** subsection and further dividing each body subsection into subsections. As shown in Figure 5-1, the **Methods** introduction and ending are not usually divided into subsections. In other words, they are usually presented in paragraphs.

The heading outline to a **Methods Section** functions as your roadmap for drafting the text. You will use its headings to **Methods** body subsections and sub-subsections in the text.

Guideline 5.9 *In creating a heading outline:*

(*a*) *Do not divide the **Methods** introduction and ending into subsections.*

(*b*) *Do divide the **Methods** body subsections into sub-subsections.*

(*c*) *Use an itemization or enumeration scheme for **Methods** body subsections and sub-subsections.*

(*d*) *In the heading to each **Methods** subsection, indicate the number of the corresponding aim and repeat the corresponding aim.*

(*e*) *In the heading to each **Methods** subsection, use key terms and phrasing that evoke the aims as phrased in the **Aims Section**.*

(*f*) *Draft grammatically parallel headings to **Methods** subsections.*

The heading outlines in Cases 5-7 to 5-9 reflect features of headings identified in Guideline 5.9.[5,6] You should record the itemization scheme in the heading outline that you will use in the text. By recording the itemization scheme in the heading outline, you can keep track of your overall organization while drafting individual **Methods** sub-sections. Also, by deciding on the itemization scheme *before* drafting, you can draft the text faster and with fewer revisions. In addition, the itemization scheme can help your

---

[5] These heading outlines will also be used in Chapter 6.

[6] These heading outlines use indentation to make each subsection more readily apparent; however, when using the heading outlines to draft the narrative, you would not use such indentation in the text.

reviewers keep track of the hierarchical relationship across **Methods** subsections. Unless you are otherwise instructed by submission requirements, you will need to choose the itemization scheme that best helps the reviewers understand the organizational hierarchy of the subsections and sub-subsections for each of your proposed aims.

Using heading outlines, you can experiment with different itemization schemes without substantial effort. Notice that there are different itemization schemes in Cases 5-7 to 5-9 (see Chapter 1.11.3). Case 5-7 shows an itemization scheme that is numeric-alpha, and Case 5-8 shows one that is alpha-numeric and that uses sequentially numbered experiments for the sub-subsections. The itemization scheme in Case 5-9 is entirely numeric.

Another useful feature of heading outlines is illustrated in Cases 5-7 to 5-9: the headings to the **Methods** subsections identify, by number, the corresponding aim from the **Aims Section**, either at the beginning or at the end of the heading. In Cases 5-7 and 5-8, each heading to a **Methods** body subsection begins with the relevant proposed aim identified by number, and in Case 5-9, each heading to a **Methods** body subsection ends with the relevant proposed aim in parentheses.[7]

5.3   The content of Methods subsections and sub-subsections

The creation of a **Methods** heading outline and the drafting of **Methods** text presuppose a familiarity with types of content that need to be covered in a **Methods Section**. Funding agencies sometimes require different types of **Methods** content. However, there is considerable similarity in **Methods** content across funding agencies, even though they sometimes describe their required **Methods** content differently. **Methods** content described in Chapters 5 and 6 of this book draws on similarities across funding agencies' submission requirements. However, you should not rely only on the **Methods** content described here. You need to review and then to double-check the content that your funding agency requires for its **Methods Section**.

In this book, the **Methods Section** is divided into 22 categories of **Methods** content, as shown in the circles of Figure 5-1. The **Methods** content numbered 1–9, in various combinations, comprise the **Methods** introduction, and the **Methods** content numbered 10–16 are included in a **Methods** subsection. The ending subsection is often not specified in submission requirements, but in this book, the ending is considered a required, important **Methods** subsection, and its content is identified in Figure 5-1, items 17–22. Methods content for the ending is discussed in Chapter 6.12.

5.4   Methods introduction

You will find many **Methods Sections** that do not have introductions. *This is a mistake.* A **Methods** introduction can focus the reviewers' attention on the research

---

[7] Case 5-9 presents a heading outline from a narrative submitted to NSF, which typically uses the term *objective* for *aim*.

purpose, the research design and time line, and on any particular characteristic of the **Methods** that you want the reviewers to notice, such as expected outcomes.

5.4.1   Layout and brevity of Methods introductions

The **Methods** introduction is typically not given a heading in order to save space. However, if you use a heading, it is usually in run-in style (Chapters 1.11.3 and 8.6.2). A few choices for the heading to a **Methods** introduction are: **Introduction, Overall Approach, Overall Strategy**, and **Overview**. The heading outline in Case 5-7 indicates that the **Methods** introduction is untitled, which means that it is a paragraph without a heading (an *informal introduction*).

Each of the **Methods** introductions in Cases 5-8 and 5-9 uses a formal overview of the proposed methods in **D.1 Overview** and **D.1 Project Overview**, respectively. When there is a formal overview, the **Methods** introduction typically presents the research objective from the **Aims Section** and identifies any particular information that you want reviewers to notice.

The **Methods** introduction needs to be brief in order to move reviewers quickly into the proposed methods.

> **Guideline 5.10**   *Keep the introduction to less than 15% of the total length of the Methods Section.*

It is difficult to incorporate all possible types of introductory information listed in Figure 5-1 into a **Methods** introduction, and still maintain its brevity. Guideline 5.11 offers a few strategies to help you achieve a short **Methods** introduction:

> **Guideline 5.11**   *To keep the **Methods** introduction brief:*
> *(a) Place some of the introductory information into a **Methods** ending.*
> *(b) Integrate information about the research team into the body subsections that describe the proposed procedures.*

Even though one **Methods** introduction does not typically cover all types of information listed in Items 1–9 in Figure 5-1, a **Methods** introduction needs at least 2 types of information, as noted in Guideline 5.12.

> **Guideline 5.12**   *Include the following types of information in a **Methods** introduction:*
> *(a) Research objective (Item #1 in Figure 5-1).*
> *(b) Key design features of the proposed methods (Item #2 in Figure 5-1).*

Case 5-10 presents a **Methods** introduction with an overview table, and Case 5-11 presents 2 **Methods** introductions without overview visuals. Each introduction identifies the research purpose and the primary features of the proposed research design and methods (indicated as connection information in Case 5-11). Even though the introduction in Case 5-11B is much shorter than that in Case 5-11A, the shorter introduction still provides a good overview to orient reviewers to the methods.

5.4.2   Research team

Information about your research team (and their sites) may be necessary to support the proposed budget and to indicate the responsibilities of the research team (and research sites) in terms of the proposed methods. The **Preliminary Studies Section** addresses the research team in terms of its preparedness to successfully undertake the proposed research. However, the **Methods Section** is also where you explain the proposed division of labor. Information about the research team may also be necessary to explain why investigators at different sites are involved. The overview table in Case 5-2 lists information on the division of labor by research site, and the overview table in Case 5-3 lists information on the division of labor by each research team member.

   If you include information on an overview table that indicates different sites, you will likely also need an explanation in the introduction that justifies your use of multiple sites.

5.4.3   Forecast

A forecast is a phrase or sentence that identifies the content and organization of an upcoming subsection. A forecast is frequently used in introductions to help reviewers notice particular information so that they can better understand the organization of the section. The **Methods** introduction in Case 5-10 includes forecasts in sentences 4 and 5: sentence 4 identifies where shared methods are explained, and sentence 5 indicates where to find the justification for the sheep model.

   The forecast does not usually communicate substantive content, so you can omit it if you need space for other information (Chapter 8.6.4). However, if you intend to include a forecast, for efficiency you should consider writing it *after* the text and the organization of the **Methods Section** are stable in order not to revise it.

> **Guideline 5.13**   *To create space for more substantial **Methods** content, consider omitting the forecast (but not an overview visual).*

5.5   Drafting of Methods body subsections

As earlier mentioned (Chapter 5.2.1; Figure 5-1), the body of a **Methods Section** is divided into subsections: one body subsection for each proposed aim, plus a possible subsection for shared methods (see Chapter 6.11 on shared methods).

> **Guideline 5.14**
>
> (a) Divide the **Methods** body into subsections, with each subsection dedicated to describing the methods for a proposed aim.
> (b) Sequence the **Methods** body subsections in the order in which you sequence and number the proposed aims in the **Aims Section**.
> (c) Further divide the **Methods** subsections into sub-subsections.
> (d) Consider sequencing the **Methods** subsections and sub-subsections in the sequence shown in Figure 5-1.
> (e) Use organizational parallelism across headings to **Methods** body subsections.
> (f) Avoid using more than 3 levels of subsections to describe the proposed methods for each aim.

The heading outline in Case 5-7 illustrates how a **Methods Section** (called a **Research Design and Methods Section** in this example) can be divided into subsections and then further subdivided. Notice how the rationale subsection has not been further subdivided, but the experimental design subsection has been subdivided into 3 subsections under Aims 1 and 2, whereas the experimental design in Aim 3 has only 2 subsections. Due to readability issues, you should not subdivide a **Methods** unit into sub-sub-subsubsections. In other words, you should have no more than 3 levels of **Methods** subsections describing the methods to each aim.

As noted in Guideline 5.14(*e*), **Methods** body subsections need to reflect *organizational parallelism* as much as possible. Organizational parallelism is a term that covers 2 particular features of a text: (1) the headings of subsections at the *same* level in the organizational hierarchy are grammatically parallel to each other and (2) the headings of subsections at 2 *different* levels in the organizational hierarchy do not need to be grammatically parallel to each other, but they often are.

For example, in Case 5-7, all of the headings for the 3 **Methods** body subsections (**1**, **2**, and **3**) are at the same level in the organizational hierarchy and are grammatically parallel. Each heading begins with **Specific Aim**, followed by the number of the aim and a period; then each aim is phrased as a noun followed by modification. The first specific aim begins with **Definition**, which is a type of noun called a *nominalization*. The second aim begins with **Characterization**, also a nominalization, as is **Identification**, which begins the third aim.

For another example of organizational parallelism, notice that all 3 **Methods** body subsections in Case 5-8 (itemized with **D.3**, **D.4**, and **D.5**) have the same 2 types of subsections: **Hypothesis** and **Experiment N**. Also, each experiment is named, with the name beginning with an infinitive verb. Using verbs to identify experiments helps reviewers focus on the activity or task that you will perform. In Case 5-9, all of the 4 body subsections at the same level (**4.2**, **4.3**, **4.4**, and **4.5**) begin with a verb ending in + **ing**, and all of the headings to the sub-subsections under each of these 4 body subsections begin with infinitive verbs. For example, under the subsection **4.2**, the 3 sub-subsections begin with the infinitive verbs **Design**, **Construct**, and **Prepare**.

The remaining sections of this chapter describe the first 3 types of content that may be found in **Methods** body subsections, as shown in Figure 5–1.

- Rationale (Chapter 5.6).

- Subjects, objects of study, specimens (SOS) (Chapter 5.7).

- Materials, equipment, and tools (MET) (Chapter 5.8).

5.6   Rationale

In the context of a scientific grant proposal, the rationale is background information that explains or justifies *why* you have decided on a particular methodological feature(s). A rationale can help convince reviewers that your research design and your methods are not arbitrary – that you deliberately selected them in order to achieve your proposed research objective and aims. In most cases, a particular methodological feature needs justification if it is *not* standard in the field, if it is a variation of a standard procedure, or if it is brand new.

A **rationale** subsection is typically located as the first subsection in a **Methods** body subsection. A **rationale** subsection can also be given the heading **Hypothesis** or **Rationale and Hypothesis**, if it includes a hypothesis, or it can bear a unique heading. The heading outlines in Cases 5-7 and 5-8 show the organizational plans for including rationale subsections in **Methods** subsections, as indicated by the headings **Rationale** in Case 5-7 and **Hypothesis** in Case 5-8. Cases 5-12 and 5-13 present examples of **Rationale** subsections.

5.6.1   What constitutes "rationale" information?

Rationale is a type of background information, so you explain it in much the same way that you explain background information in a **Background Section** or a **Preliminary Studies Section**. That is, you explain rationale by including content that comprises a scientific argument, as noted in Guideline 5.15:

> **Guideline 5.15**   *To justify a feature(s) of your proposed methods, create a rationale subsection with content that comprises a scientific argument: topic, review, analysis, connection, and significance (TRACS).*

The right-hand columns of Cases 5-12 and 5-13 identify the types of **TRACS** content comprising the rationale subsections.

The topic, review, and connection information of a scientific argument are particularly noteworthy in a rationale subsection. The topic is the *specific methodological feature* that you are justifying, the review gives important background on this feature that serves to justify it, and the connection is the relationship between that feature and your proposed methods. Case 5-12A justifies the experimental focus on identifying *performance criteria*, Case 5-12B justifies the experimental focus on the *suppression of testosterone production*, Case 5-13A justifies the focus on *Bp strains*, and Case 5-13B justifies the use of *a chick model of heart development* to better understand human heart development. Each of these rationale subsections also provides the connection between the justified topic and the proposed methods.

In a rationale subsection, you typically review and analyze previous research, but you can also review the research team's skills and experience by reviewing information from non-narrative sections of the grant proposal, such as a biographical sketch. A review and analysis of your skills and experience in a rationale subsection can bolster your credibility.

Example 1 illustrates a passage from a rationale subsection that reviews the skills and experience of the PI and Co-I (also notice the citation) and that draws positive conclusions about these skills and experience, which serve to bolster the investigators' credibility:

Example 1
[1]Our successful purification of SarA in a form capable of binding diverse DNA targets (e.g., *cis* elements upstream of the *agr* and *cna* promoters), together with the established expertise of the Co-I in the use of PCE-assisted binding site selection (PCE-ABS) (ref), places us in a unique position to identify additional SarA targets within the *S. aureus* genome. [2]Finally, the molecular expertise of the PI in the molecular biology of *S. aureus* and the successful production of appropriate *sar* and *agr* mutants will allow us to confirm the results of our PVR-ABS experiments by directly testing the *sar*-mediated regulation of the genes *cis* to putative SarA-binding sites.

If you are justifying a **new** (or modified) methodological feature or methodology, such as a procedure or piece of equipment, you should address its significance. As used here, a **new** method is a method that has been primarily, previously used only in the preliminary research (whether or not it was published).

> **Guideline 5.16** *If you plan to use or to develop a new methodological feature or methodology, include significance in a rationale subsection.*

5.6.2   Where is rationale information located?

Where you locate the rationale, especially when it includes preliminary research, indicates its scope.

> **Guideline 5.17** *Consider including rationale information in any of these locations:*
> *(a) In the **Methods** introduction.*[8]
> *(b) In the first **Methods** body subsection.*
> *(c) Within the description of a proposed procedure.*
> *(d) In the **Preliminary Studies Section.***

---

[8] Since rationale can include preliminary research, this may be one reason why NIH specifies that the **Preliminary Studies Section** is placed under the **Approach** heading, which locates the rationale before any in the body subsections in the **Methods Section** (www.nih.gov/grants/funding/phs398/phs398.html; I-45; accessed November 28, 2014).

Rationale information justifying a methodological feature that is relevant to all of the proposed aims can be included in the **Methods** introduction or as an initial **Methods** body subsection, before any of the methods are described. The heading outline in Case 5-8 indicates such a rationale subsection (itemized as, for example, the heading **D.2 Hypothesis**) that is located before the methods to any of the proposed aims. In Case 5-13B, the justification of a model that relates to more than one aim is also located before any descriptions of the methods.

Rationale information that justifies a methodological feature of only one aim can be included as the first sub-subsection in the relevant **Methods** body subsection. For example, in Case 5-8, the first subsections in **D.3**, **D.4**, and **D.5** are the rationale, entitled **Hypothesis**. This location indicates the narrow scope of the justification, which covers only the methods in the **Methods** body subsection.

Another common location for rationale information that relates to the methods of one aim is within the description of proposed procedures. Such *integrated rationale information* is useful when only a brief justification is needed. Example 2 illustrates a procedural passage with rationale information (in bold italics) from a **Methods** subsection.

Example 2

[1]Intracellular recordings will be made with sharp microelectrodes (150–200 MΩ) filled with 2% biocytin in 2M KMeSO4. [2]***Recording with sharp electrodes has been successful in this laboratory in the past [e.g., 42, 44, 67] and has an advantage for plasticity studies over patch recording in that the cell is not dialyzed, and 'run down' is not a problem.*** [3]***Patch recording may well prove useful for some purposes, however.*** [4]An experimental set up will be established in this laboratory for visualized patch recording from cells of ELL. [5]It will be possible to use this same set up for selected experiments on this proposed project on the cerebellum.

Rationale information can also be included in the **Preliminary Studies Section** when the same rationale is applicable to both the proposed methods and the preliminary research.

Locating rationale in the **Preliminary Studies Section** can free-up space in the **Methods Section** for other information. If you locate rationale in the **Preliminary Studies Section**, then in the **Methods Section**, at relevant points, you need to cross-reference *back* to the rationale in the **Preliminary Studies Section**. However, doing so might lower the readability of the **Methods** subsection. It is preferable to include rationale in the **Methods Section** if you have the space there, to create a self-contained **Methods Section**. If you include rationale in the **Methods Section**, then in the **Preliminary Studies Section**, you can cross-reference *forward* to the relevant rationale information in the **Methods**.

Guideline 5.18   *If rationale information relates to both preliminary and proposed methods, include the rationale in the **Methods Section**, and then in the **Preliminary Studies Section**, include a cross-reference to the relevant **Methods** subsection.*

5.7 Subjects, objects of study, and specimens

Subjects, objects of study, and specimens (**SOS**) are what you will manipulate under controlled experimental conditions to collect data. The **SOS** might include humans, mice, brain tissue, leaves, or cells. Cases 5-14 to 5-16 present subsections that describe **SOS**. We describe 8 types of content that descriptions of **SOS** can potentially include:

(1) Source/recruitment of SOS.

(2) Number of SOS and how you determined the number.

(3) Inclusion and exclusion criteria.

(4) Maintenance and retention of SOS.

(5) Informed consent.

(6) Confidentiality (for subjects).

(7) Safety and risk issues (for subjects, vertebrate animals, and research staff).

(8) Rationale/justification.

Depending on submission requirements from your funding agency, you will not necessarily include all 8 types of content in an **SOS** subsection. For proposed research that involves humans, NIH requires items 6, 7, and 8 to be described in a section called **Protection of Human Subjects**, but NSF does not require such a section. You need to become familiar with your funding agency's submission requirements regarding **SOS**.

5.7.1 Where in the Methods Section to locate descriptions of subjects, objects, and specimens

To determine where to include descriptions of proposed **SOS**, you need to consider: (a) whether the proposed **SOS** relate to the methods of only one aim, (b) whether they relate to the methods of more than one proposed method (termed *shared SOS*), or (c) whether the proposed **SOS** are the same or closely similar to those used in the preliminary research.

Guideline 5.19

(a) *For proposed SOS that are used in the methods for only one of the proposed aims, describe the SOS in the **Methods** body subsection of the relevant proposed aim.*

(b) *For proposed SOS that are used in the methods for multiple proposed aims, consider describing the shared SOS in a shared methods subsection.*

(c) *For proposed SOS that are the same or closely similar to those used in the preliminary research, consider describing these SOS in the **Methods** body subsection or in the relevant **Preliminary Studies** subsection, but not in both.*

The most useful and desirable location for a description of the proposed **SOS** is in a **Methods** body subsection (Guideline 5.19a). However, for **SOS** that are the same across proposed methods, you can include the description of the shared **SOS** in a shared methods subsection. Also, for proposed **SOS** that you also used in the preliminary research, you can describe them either in the relevant **Preliminary Studies** subsection or in the relevant **Methods** subsection.

SOS

Preliminary Studies ◄─────────────► Methods

For shared **SOS**, you insert a pair of cross-references: one in the relevant **Methods** subsection, to cross-reference back to the **SOS** description in the **Preliminary Studies** subsection; and one in the **Preliminary Studies** subsection, to cross-reference forward to the relevant **Methods** subsection. These pairs of cross-references help reviewers efficiently track the relationship between your preliminary and proposed methods, in addition to learning about your proposed **SOS**.

5.7.2   Source and recruitment of subjects, objects, and specimens

The *source* of **SOS** is from where you intend to procure the **SOS**.

● **Recruitment of human populations**. When the **SOS** are subjects (sometimes called *participants*), source information involves recruitment information. You need to identify from where you intend to recruit them and how, such as by advertising in local newspapers and on websites and radios, or simply through word-of-mouth. In Case 5-14, sentence 3 indicates that control subjects will be recruited from advertisements in a medical school newspaper and from spouses and friends of other subjects, and sentence 6 identifies the source of *UVL* (unilateral vestibular loss) *patients*, who will be referred by *local otolaryngologists*.

● **Specific person as a source of referral for SOS**. If the source of the **SOS** is a specific person (e.g., word-of-mouth), you need to identify the person by name and the person's affiliation. If this is the first time you have mentioned this person by name in the narrative, you also identify the first name, such as in Example 3. Also, rather than using the ambiguous *Dr.*, you should identify the person's degree (i.e., *Ph.D.* or *M.D.*). In clinical studies, this person is often the study recruiter or study coordinator.

Example 3
Zebra finches will be readily available at the Brazilian site from a breeding colony initiated by Jose Smith, PhD, and from local bird breeders. (Modified from sentence 2 in Case 5-16.)

If someone will be gifting the **SOS**, you should identify the person and mention this gift in the text since it will affect your proposed budget.

● **Company as a source for SOS.** If the source of the **SOS** is a company, you identify both its name and location in parentheses. If the source is not any one company but many, you can mention them generally, such as *local bird breeders* in sentence 2 of Case 5-16 (Example 3 above).

5.7.3   Number of subjects, objects, and specimens

When describing **SOS**, you need to identify their numbers and explain *how* you determined the numbers. The *how* is important to show that your research design is not arbitrary and that you are deliberately seeking to control all relevant factors that may affect data analysis and statistical significance. *How* information is particularly important when you have subjects, in which case your explanation should include a power analysis derived from either your own preliminary data or a previously published report that uses similar techniques, a similar type(s) of subject, and a similar research objective. Example 4 illustrates a sentence that explains how the numbers of subjects was determined.

Example 4
[1]The numbers of subjects are dictated by 2 factors: 1) the number of subjects who can be enrolled and retained in the study based on preliminary research, and 2) the power calculations for the statistical significance tests to be used for data analysis. [2]We will test at least 50 subjects who use tamoxifen and 50 control subjects who do not. [3]Power calculations will be provided in the analysis portion of this **Methods Section**. (Case 5-15, Sentences 1–3.)

5.7.4   Inclusion and exclusion criteria

You need to identify the significant features that are guiding your decision to include or exclude **SOS** in your proposed research. These significant features are your inclusion and exclusion criteria. They reflect the quality of your research design and your control of variables that may otherwise confound your results and ultimately affect the validity of data and your interpretation. You may need to justify any or all of your inclusion and exclusion criteria in order to help reviewers conclude that your research design is not arbitrary and that your proposed methods exhibit sound features of scientific methodology.

● **Groups.** If you are proposing to categorize individual **SOS** according to a particular feature, you should consider describing the **SOS** in terms of groups. Also, you should not just assign each group a number; you should assign each group a descriptive name that reflects the most important, distinctive feature of the group. Descriptive group names can help your reviewers more readily recall the defining feature of each group.

Guideline 5.20   *Assign descriptive names to each group of SOS.*

You also need to describe your control **SOS** with the term *control* in order to indicate your understanding of scientific methodology and to help your reviewers evaluate the research design. Example 5 identifies the control subjects with the term *control*.

Example 5
Thirty control subjects and 30 subjects with chronic unilateral vestibular loss will be studied in each of the 5 proposed experiments. (Case 5-14, sentence 1.)

The groups of **SOS** can be the organizing principle of an **SOS** subsection in that each group is discussed in its own subsection or paragraph. For example in Case 5-14, the first paragraph (entitled **Control Subjects, Exps. 1–6**) discusses the control subjects, and the second paragraph, the test subjects. Case 5-15 presents an alternative organization for **SOS**, illustrating an **SOS** subsection organized by types of **SOS** content. The first subsection in Case 5-15 focuses on recruitment and numbers of subjects, and the second, on eligibility (inclusion and exclusion criteria). Even though the organization for the description of **SOS** is different in Cases 5-14 and 5-15, a good characterization of each **SOS** still emerges.

● **Description of SOS.** You need to describe features of your **SOS** that are relevant to your proposed research. If you provide no description and only name the **SOS**, reviewers might question why you selected them. Example 6 presents a passage describing *spHCN* as the **SOS**. Sentence 1 names the **SOS**, sentence 2 describes the significant feature of the spHCN that is relevant to the proposed research, and sentence 3 explains how this significant feature is relevant to the proposed research. Notice that the terms *inclusion criteria*, *exclusion criteria*, and *eligibility* are not used since those terms are usually reserved for humans and possibly non-human primates and other vertebrates.

Example 6
[1]For the initial experiments, we will use spHCN, which is a hyperpolarized-activated ion channel from sea urchin that is homologous to the mammalian HCN channels. [2]spHCN activates at less negative voltages than the mammalian HCN channels. [3]The less hyperpolarized activation range of spHCN channels will allow us to test for S4 movement without having to clamp the oocytes to extreme negative potentials, which will activate endogenous channels in the *Xenopus* oocytes and will contaminate our current measurements.

Example 7a describes significant features of subjects who will be studied. If these features are criteria for their selection, then rather than just describing them, you can include their features in a list, such as in Example 7b, which is a revision of Example 7a. The use of lists makes the criteria more obvious.

Example 7
a. C.2. *Experimental Subjects* – [1]In Aim 3, we will enroll 3 experimental subjects (age 18–75) to obtain sufficient pilot data to perform a power analysis for a future study.

[2]Each subject will have had an ischemic or hemorrhagic stroke >1 year prior to enrollment. [3]Each will also have a severe impairment in the affected arm, with baseline upper extremity Fugl-Meyer Assessment score 8–24, and elbow and shoulder spasticity (Ashworth score ≤3); they will have shoulder abduction–elbow flexion dyssynergia. [4]Exclusion criteria are exercise intolerance; co-morbidities limiting movement, including orthopedic problems; cognitive dysfunction preventing compliance with instructions; participation in other studies; and plans to initiate or discontinue any physical/occupational therapy during the period of enrollment.

b. C.2. *Experimental Subjects* – [1]In Aim 3, we will enroll 3 subjects to obtain sufficient pilot data in order to perform a power analysis for a future study. [2]Inclusion criteria are that each subject will: (1) be between 18 and 75 years old, (2) have had an ischemic or hemorrhagic stroke >1 year prior to enrollment, (3) have a severe impairment in the affected arm (baseline upper extremity Fugl–Meyer Assessment score 8–24), (4) have elbow and shoulder spasticity (Ashworth score ≤3), and (5) will exhibit shoulder abduction–elbow flexion dyssynergia. [3]Exclusion criteria are: exercise intolerance, co-morbidities limiting movement (e.g., orthopedic problems), cognitive dysfunction preventing compliance with instructions, participation in other research studies, and plans to initiate or discontinue any physical/occupational therapy during the period of enrollment.

- **Identification and formatting of SOS criteria**. For **SOS** that are humans, and possibly for non-human primates, you need to identify the criteria that you will use to admit or exclude **SOS** for your study. These criteria are necessary to show that your research design is not arbitrary, that you are seeking to control variables, and that you are identifying, in advance, those **SOS** that are most likely to respond to your procedures or intervention. If you have few inclusion and exclusion criteria, you might be tempted to mention them descriptively, such as in Example 7a. This handling of criteria is a problem since you are not actually presenting criteria; you are only describing features (which may or may not be eligibility criteria). If the features that you are describing are the eligibility criteria for the **SOS**, you should use such terms as *inclusion criteria*, *exclusion criteria*, and *eligibility* to help reviewers understand that your acquisition of **SOS** will not be arbitrary.

As shown in Cases 5-14 and 5-15, it is useful to present inclusion and exclusion criteria in itemized, horizontal lists. Vertical lists are usually avoided because they take too much textual space. Notice that sentences 5 and 9 in Case 5-14 and sentence 11 in Case 5-15 use the terms *inclusion criteria* and *eligibility criteria* in the lead-ins to the lists, clarifying the type of content. The use of a list for eligibility criteria implies a controlled, deliberate research design, which can enhance your credibility.

Guideline 5.21 *Consider using itemized lists to present inclusion and exclusion criteria, and consider using the terms **inclusion criteria**, **exclusion criteria**, and **eligibility**.*

- **Detail and objectivity**. You need to describe the inclusion and exclusion criteria and descriptive features of your **SOS** in detail and as objectively as possible in order to reflect sound scientific methodology.

> **Guideline 5.22** *To describe eligibility criteria as objectively as possible:*
> *(a) Avoid subjective terms as much as possible.*
> *(b) For subjective terms that you do include, define and/or quantify them.*

Such terms as *normal, healthy,* and *able-bodied* are subjective and can be interpreted differently, depending on the context and on the experience and training of a reviewer. If you use subjective terms, you need to define them. For example, in sentence 5 of Case 5-14, the subjective criterion *normal* occurs in items 3, 9, and 10 of the inclusion criteria, and in each case, *normal* is objectively defined: in item 3, *normal* will be objectively determined by the scoring system used in the Manual Muscle test; in item 9, *normal* will be determined by objective scores on horizontal rotation tests; and in item 10, *normal* will be defined by the ability of subjects to achieve a particular, quantifiable level of performance. For another example, sentence 12 of Case 5-15 does not mention that the subjects will be required to have *very good vision*, which is a subjective characterization. Instead, the acuity threshold is quantified: *20/20 or better.*

5.7.5   Maintenance, retention, and termination procedures

- **Maintenance procedures**. These procedures involve how you will keep **SOS** that have no or little control over their environment, from when they are selected or accepted into the study, until you collect data from them. The explanation of how you intend to maintain **SOS** for study should be crafted carefully since maintenance procedures can introduce variables that could affect the data that you will collect, their interpretation, and statistical analysis.

For example, sentence 3 in Case 5-16 explains that the birds will be maintained in an acoustically isolated environment for 12–14 hours before experimentation, and sentence 4 justifies this maintenance procedure by explaining its purpose. Also, if the **SOS** are animals, maintenance needs to reflect humane handling.

- **Retention procedures**. You need to describe retention procedures for subjects. These procedures involve how you will keep subjects adequately participating in the study until you no longer need them. The rationale for your retention procedures can also be included in the discussion. As shown in Example 8, retention procedures usually involve the payment of subjects.

Example 8
[1]To facilitate retention of subjects given their possible participation over several days of testing, they will sign an agreement to attend all sessions, and they will receive their payment for all sessions that they complete. [2]Healthy subjects will mainly be

drawn from employees who work close to the laboratory. [3]Based on our previous studies, we do not expect attrition problems with our subjects.

- **Termination procedures.** These procedures cover what you will do with **SOS** that are primarily vertebrate animals, when you no longer need them for your study. Describing the termination of vertebrate animals may be difficult because of public controversy surrounding animal use in scientific, medical, and technical research. Such terms as *euthanized, sacrificed, terminated,* and *life humanely ended* are sometimes used, to control connotations and to reflect the PI's humane terminal procedures. However, if the type of termination affects the collected data, you need to be more specific. For example, sentence 5 in Case 5-16 indicates that the birds will be decapitated. This specific information is needed because the research analyzes mRNAs, which rapidly degrade after death, so there is a need to rapidly kill, dissect, and freeze the tissue for further analysis, and other termination methods are not compatible with these needs.

If experiments do not result in termination, you should say so, and you should explain what will happen with the **SOS**.

Guideline 5.23 *When describing proposed maintenance, retention, and termination procedures, specify those that affect the collected data.*

5.7.6   Research ethics

For studies that use human and vertebrate animal **SOS**, you need to express your awareness of research ethics, as indicated by submission requirements from your funding agency.

This expression might involve, for example, your agreement to follow the 1974 Helsinki Declaration, that you will obtain Internal Review Board (IRB) approval for your study design and study documents before experimentation begins, and that all study materials, including materials relating to informed consent, will be developed and maintained so as to protect the confidentiality of enrolled subjects. Also, explaining exactly how you will seek informed consent will help clarify to reviewers that you are aware of legal requirements involving human subjects. Your explanation of procedures for obtaining informed consent can be presented in a single sentence, illustrated in Example 9, or in a short subsection, shown in Example 10.

Example 9
Subjects who pass our initial screening criteria will be informed of the experimental procedures, risks, benefits, and their rights as experimental subjects, and subjects agreeing to participate will read and sign a consent form.

Example 10
Informed Consent and Medical Release. [1]Once prospective subjects arrive for testing, they will be given the opportunity to ask questions before signing the consent form.

[2]If they elect to sign, they then will be asked to sign a medical release form allowing us to obtain relevant information from their providers (including oncologists and eye doctors). [3]All subjects so far have signed this optional release.

5.7.7   Blinding and subject confidentiality

If your research involves blinding, you need to explain: (1) who will be blinded, (2) at what point in the procedures that blinding will occur, (3) how data records will be maintained to ensure successful blinding, and (4) the point in the study at which blinding will be lifted.

Even if blinding is not part of your research design, you need to describe your efforts to ensure subject confidentiality – efforts that may need more description than a brief mention that you will follow confidentiality requirements, such as HIPPA (the United States Health Insurance Portability and Accountability Act).

Guideline 5.24   *When dealing with human SOS, describe how you will maintain subject confidentiality.*

As illustrated in Example 11, you may need to include how a particular subject's data will be handled anonymously, how the data will be stored, who will have access to the data, and who will have authority to retrieve it.

Example 11
Confidentiality: [1]Every effort will be made to protect subject confidentiality. [2]Identifiable subject data will not be published, and identifiable data will not be seen by non-study personnel without the subject's written approval. [3]Spreadsheets that contain visual and ophthalmic data, and breast cancer treatment histories will be linked to numerical identifiers. [4]The spreadsheets containing experimental or medical data will not contain subject names, addresses, or telephone numbers. [5]The numerical identifiers that match subject names will be kept in separate spreadsheets and will only be accessed by study personnel (i.e., PI, Co-I, or approved personnel under their supervision).

5.7.8   Safety issues

You need to address the ways in which the safety of your subjects, vertebrate animals, and research staff may be at risk, given your proposed procedures.

Guideline 5.25
(a) *Describe safety and risk issues that your subjects, vertebrate animals, and research personnel may be subjected to, and what you intend to do to eliminate, mitigate, reduce, or prevent threats to safety.*
(b) *The more invasive the procedures for humans and vertebrate animals, the more you need to describe safety, risk, and care issues.*

Some funding agencies require safety and risk information to be described in special supplementary sections for human subjects and vertebrate animals.[9] Case 5-17 gives a **Vertebrate Animal Section** that identifies safety information in paragraphs 3 and 4. If your funding agency does not mention safety concerns, you need to address them briefly when describing your **SOS**. Types of safety information include:

- Medical staff or veterinary supervision.

- Care and prevention procedures.

- Possible safety incidents that subjects and vertebrate animals may be subjected to, including risks, harms, and minor discomfort.

- Possible safety incidents that research staff may be subjected to, including risks, harms, and minor discomfort.

- Plans to prevent or to mitigate risks, and how to handle any safety incidents that may arise.

- Degree of (un)likelihood of a safety incident.

- Substantiation of the degree of (un)likelihood that the risk will occur.

Examples 12 and 13 present 2 brief subsections from narratives describing procedures that involve risks to human subjects. Example 12 addresses safety issues involving the research staff's handling of blood and the discomfort that subjects may experience when giving blood, a risk identified in sentence 3. Even less- or non-invasive procedures need to be described in terms of safety, as illustrated in Example 13, which identifies risks associated with the use of bright light.

Example 12
Safety: [1]The proposed research involves the use of potentially infectious human blood. [2]Appropriate precautions will be taken in the PI's laboratory, to prevent the transmission of blood-borne diseases to laboratory personnel. [3]Safeguards will include the following positive procedures: (1) hepatitis shots will be mandatory for all personnel engaged in the proposed research; (2) protective gloves and laboratory coats will be worn by all personnel; (3) blood spills will be cleaned promptly with a disinfectant solution; and (4) sharps (e.g., needles) will be disposed of, in appropriate puncture-resistant containers and in accord with university policies for hazardous materials.

Example 13
[1]Before we begin the visual function tests on each subject, the examiner will check to make sure the proper filters are in place. [2]In addition, the examiner will never knowingly leave the machine without filters in place. [3]Before testing, subjects will be instructed not to look through the eyepiece until the examiner tells them that it is safe to do so. [4]The light levels for fundus photography are brighter than physiologic light

---

[9] For example, NIH requires a **Protection of Human Subjects Section** and a **Vertebrate Animal Section** (www.nih.gov/grants/funding/phs398/phs398.html; I-48 and I-49; accessed November 28, 2014).

levels, but the light levels for photography are used routinely for clinical testing without any known hazards to the subjects.

Because safety regulations, recommendations, and policies periodically change – especially those about human safety and animal care – you need to review submission requirements before drafting **SOS** subsections and supplemental **SOS** sections. Funding agencies in the United States determine information for these supplemental sections to a certain extent based on federal regulations, such as Protection of Human Subjects (45 CFR Part 46) and the Animal Welfare Act (7 USC 2131 et seq.); recommendations from special interest groups, such as the Panel on Euthanasia of the American Veterinary Medical Association; and policies from funding agencies.

5.7.9   Justification of subjects, objects, and specimens

Types of **SOS** and significant features of **SOS** are typically described and justified in order to help reviewers understand the carefully designed, non-arbitrary character of the proposed research. **SOS** are usually justified in a rationale subsection, illustrated in Case 5-13B, sentences 1 and 2.

> **Guideline 5.26** *Justify any important aspect of your* **SOS** *that reviewers might not find obvious.*

5.8   Materials, equipment, and tools

Materials, equipment, and tools (**MET**) are the tangible and intangible objects you will manipulate to observe the **SOS** and to collect and analyze the data. Types of **MET** content include:

(1) Name of the **MET**.

(2) Type of the **MET**.

(3) Source of the **MET**.

(4) Function or purpose of the **MET**.

(5) Source of energy for the **MET** (optional).

(6) Advantages of the **MET** over other **MET** (optional).

(7) Potential problems with the **MET** and solutions (optional).

(8) Safety issues associated with the **MET** (optional).

(9) Justification (optional).

(10) Significant physical and non-tangible components[10] of the **MET** (optional).

(11) Function or purpose of each significant component of the **MET** (optional).

(12) Relationship of the **MET** to the user, procedures, and the **SOS** (optional).

---

[10] Such as the composition of particular materials.

Two common questions that PIs have about describing a **MET** are:

What kind of details do I include?
How many details do I include?

The details that you need to include are the name, type, source of the **MET**, and the purpose for which you are using it (items 1–4 above; see Chapter 5.8.1), and any other information that is relevant to the proposed data-collection procedures and research design (items 5–12 above). The extent to which you describe the **MET** depends, in large part, on the extent that your **MET** are standard or customized (see Table 5-1B).

5.8.1   Descriptions of materials, equipment, and tools

You need to identify 4 characteristics of the **MET**:

(1) Name of **MET**.

(2) Type of **MET**.

(3) Source of **MET**.

(4) Function or purpose of **MET**.

The name of the **MET** is integrated into the sentence, and the type and source are placed in parentheses, after the name. The type of **MET** can be described in terms of a model or version (if relevant). The source is the manufacturer of the **MET**, which can also include the place of manufacture. If you mention both the manufacturer and the place of manufacture, then when later describing another **MET** from the same manufacturer, you need only mention the manufacturer's name, not the place of manufacture. In Example 14, a confocal microscope is named, and its type is identified by model number.

Example 14
Sectioned tissues from the GI tract, mesenteric LN, and spleen will be examined using a laser scanning confocal microscope (Zeiss LSM 510 META, 4-laser microscope) available in the laboratory of Dr. John Smith (Co-Investigator).

5.8.2   Degree of standardization and customization

In addition to identifying the name, type, and source of a **MET**, you can describe the **MET** in terms of any additional features (such as in items 5–12 above) that are particularly relevant to your proposed procedures and research design. The extent that you describe these features depends on the degree to which the **MET** is standard or customized.

Guideline 5.27
(a) *The more the **MET** is standard, the less descriptive detail you give.*
(b) *The less the **MET** is standard, the more descriptive detail you give.*
(c) *For standard **MET** that are described elsewhere, refer readers to the publication and briefly describe it.*
(d) *Consider including a visual of any nonstandard **MET**, such as a photograph or a sketch.*

Table 5-1A provides a visual to help explain Guideline 5.27. Standardization can range from *novel* (fully customized) to *standard*. Application of Guideline 5.27 is illustrated in Case 5-18. In this **MET** subsection, the standard equipment **Humphrey Field Analyzer, Model 745**, is described in just a short paragraph of 2 sentences. In contrast, the **4-channel Maxwellian View Device** that the PI will customize and build is described in 5 paragraphs, covering 26 sentences.

● **Novel materials, equipment and tools (fully customized; completely non-standard; uniquely designed)**. A uniquely designed **MET** needs detailed description – even if a description has already been published. A visual of the fully customized **MET**, if it is a major **MET**, should also be included to help reviewers better conceptualize it.

Case 5-18 presents a description of a uniquely designed piece of equipment: the **Balance Test Device**. It is named and its purpose identified in the introductory paragraph (sentence 3). Sentences 3–8 (comprising Section 4.2 in Case 5-18) describe its physical components, its source of power (hydraulic drive), and its customization and function, with details on when the construction will occur, the types of customization, and their purposes.

● **Semi-standard materials, equipment, and tools**. When proposing to use a standard **MET** that you will partially customize – or that is already partially customized – you identify:

(1) Name, type, and source of the **MET**.

(2) Distinctive feature(s) of the customization.

(3) Purpose and/or advantage of the customization.

(4) Relationship of the **MET** to the user, procedures, and the **SOS**.

In addition, a photograph or sketch of the partially customized **MET** can be useful to help reviewers visualize how it is different from the standard.

In the 6-paragraph description of the partially customized **4-channel Maxwellian View Device** in Case 5-19, sentence 8 names the device and identifies its purpose. In sentence 9, the PI explicitly tells the reviewers that only the modification (that is, the unique features) will be described. The PI's experience in modifying the standard is explained in sentence 10. Sentences 11–15 identify major components of the **MET** and their respective functions, with the advantages of the customization given in sentence 15. The remaining subsection continues with the discussion of the modifications, the functions of each modification, and the advantages of each. Notice that the discussion of the advantages, especially in paragraphs 5 and 6, point to design features that reduce or eliminate confounds. Paragraph 3 explains the relationship of the device to its users.

Case 5-20 presents another description of a partially customized **MET**. In this example, the OMAG is not a standard piece of equipment, but OMAG has been previously published and described. Therefore, the current system is first described for reviewers to better understand a description of the upcoming partial customization that follows in the second paragraph. Also, the purpose and advantage of each customized feature are described.

- **Standard materials, equipment, and tools.** In addition to identifying the **MET** in terms of basic information (name, type, source), you also need to describe the function or purpose unless the function or purpose is inherent in its name. For example, the name *DNA sequencing kit* (Example 12) includes its function, so you would provide no additional statement about its function.[11] For a proposed standard **MET** that is a particular model, you can provide details about the model beyond the basic identifying information of name, type, and source in order for the reviewers to be able to replicate (at least in theory) what you are doing with the MET. Such additional information can include:

  (1) Significant physical and non-tangible components of the **MET**.
  (2) Function or purpose of the model in comparison to other models.

In Case 5-19, the *Humphrey Field Analyzer II, Model 745* is briefly described in sentences 6 and 7 in terms of its model number, function, and advantage over other models.

5.8.3   Where to locate descriptions of materials, equipment, and tools

The obvious and most typical location for a **MET** description is in the **Methods Section**. However, a **MET** description can also be located in the **Preliminary Studies Section**, if the same equipment was used in the preliminary research.

$$\text{Preliminary Studies} \xleftarrow{\quad \text{MET} \quad} \text{Methods}$$

In most cases, you will describe the **MET** that was also used in preliminary research, in the **Methods Section**, to create a self-contained **Methods Section** that does not require the reviewers flipping pages to find the **MET** information. In addition, you include a cross-reference in the **Preliminary Studies Section**, to point reviewers to the **MET** description in the **Methods Section**. An alternative is to include the description of the **MET** in the **Preliminary Studies Section**, in which you would include a pair of cross-references: one in the relevant **Preliminary Studies** subsection to identify the relevant **Methods** subsection for reviewers, and one in the **Methods Section** to let reviewers know where the **MET** is described in the **Preliminary Studies Section**.

---

[11] However, you also need to explain what gene or genes you are looking for, with the kit, which could be considered a more specific purpose statement.

In terms of where to locate descriptions of **MET** within a **Methods Section**, standard **MET** can be located in a brief paragraph after the **SOS** or can be integrated into the first relevant procedural sentence since a standard **MET** does not need extended discussion. However, a novel and a semi-standard **MET** can be described in a subsection immediately after the description of a standard **MET**.

Guideline 5.28

(a) *For proposed **MET** that are the same as or closely similar to preliminary **MET**, locate descriptions of these **MET** in a **Methods** subsection or in a **Preliminary Studies** subsection.*

(b) *For proposed **MET** that are neither the same as, nor closely similar to preliminary **MET**, locate descriptions of these **MET** in the **Methods Section**, either immediately after the **SOS** or integrated within procedural descriptions.*

Chapter 6 continues the discussion of the **Methods Section**, focusing on data-collection, data-sharing, and data-analysis procedures, potential problems and possible solutions, a shared **Methods** subsection, and the ending to the narrative.

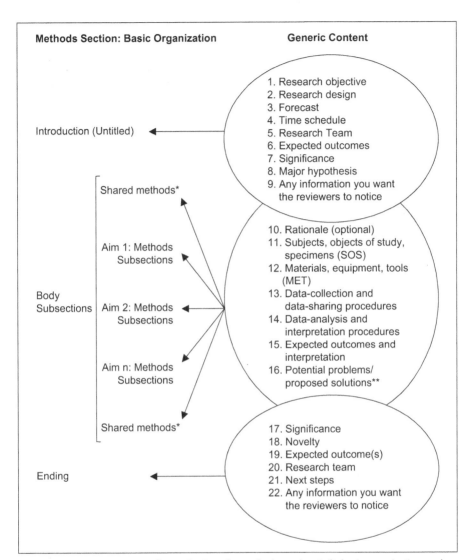

Methods Section: Basic Organization   **Generic Content**

Introduction (Untitled) ◄———

1. Research objective
2. Research design
3. Forecast
4. Time schedule
5. Research Team
6. Expected outcomes
7. Significance
8. Major hypothesis
9. Any information you want
   the reviewers to notice

Body
Subsections

Shared methods*

Aim 1: Methods
Subsections

Aim 2: Methods
Subsections

Aim n: Methods
Subsections

Shared methods*

10. Rationale (optional)
11. Subjects, objects of study,
    specimens (SOS)
12. Materials, equipment, tools
    (MET)
13. Data-collection and
    data-sharing procedures
14. Data-analysis and
    interpretation procedures
15. Expected outcomes and
    interpretation
16. Potential problems/
    proposed solutions**

Ending ◄———

17. Significance
18. Novelty
19. Expected outcome(s)
20. Research team
21. Next steps
22. Any information you want
    the reviewers to notice

Figure 5-1. Generic content and organization of a Methods Section. Left: basic organization, and right: generic content.

*A shared **Methods** subsection can occur immediately after the **Methods** introduction, immediately before the first body subsection; or immediately after the last body subsection and before the ending. Shared **Methods** subsections are discussed in Chapter 6.11.

For shared **Potential Problems and Proposed Solutions, this subsection is not located in a shared **Methods** subsection; rather, it is presented after the last relevant **Methods** subsection.

Table 5-1. Descriptions of materials, equipment, and tools (MET).

A. Extent of MET description is dependent on degree of MET standardization

The more standard the MET, the less descriptive the detail.

The less standard the MET, the more descriptive the detail.

For standard MET described elsewhere, refer reviewers to the publication and briefly describe it.

Novel   Semi-standard   Standard

B.

Type of MET	Typical Content	Location of MET Description
Novel	Name, function or purpose, distinctive and non-distinctive physical and operational features, advantage of the customized MET over another MET, any associated hazards. Consider including a photo of a new MET.	Describe MET in a dedicated MET subsection or a separate MET paragraph, before the procedures.
Semi-standard	Name, source, function, or purpose of the MET and of the partial customization, distinctive physical and operational features of the customization, advantages of the customized features, any associated hazards. Consider including a photo showing its modification.	Describe MET in a dedicated MET subsection or in a separate MET paragraph, before the procedures.
Standard	Name (common or brand name), source (manufacturer and location of the manufacture), function or purpose (optional), any associated hazards.	Integrate MET: (a) within the procedures or (b) in a separate MET subsection or paragraph, before the procedures.

Case 5-1. Overview table from a Methods introduction.

Table 1: Summary of Experiments

Aims	Exp	Description	# Subjects		Time
			Normal	BVL	
1. Sensorimotor and biomechanical mechanisms	1	Identify mechanisms contributing to AP postural control based on response timing	10	10	Yr 3
	2	Identify mechanisms that compensate for feedback time delays	10	10	Yr 4
2. Sensorimotor integration of uncorrelated sensory inputs	3	Determine whether the IC model accounts for responses evoked by uncorrelated sensory-orientation cues	10	10	Yr 1
	4	Apply optimization methods to identify rules that govern sensorimotor integration	0	0	Yr 1–5
3. Sensorimotor control of ML sway	5	Characterize sensorimotor integration strategies for controlling ML sway	10	10	Yr 2
	6	Determine the effect of stance width on ML sensorimotor integration	12	0	Yr 5

Case 5-2. Overview table showing responsibilities of research sites.

Table 3: Collaboration between State U and State Tech

Aims	Methods and Primary Responsibilities		Yr
	State U	**State Tech**	
Methods common to Specific Aims 1–4	Immunocytochemistry Coagulation potential Molecular biology Cell attachment & durability	Immunocytochemistry Coagulation potential Molecular biology Cell attachment & durability	1–5
1. Characterization of EPCs *in vitro*	Isolation and passaging	Isolation and passaging	1–2
2. Hemodynamic preconditioning of EPCs	*In vitro* flow preconditioning (ECM-coated grafts) Baboon AV shunt	*In vitro* flow preconditioning (Tissue-engineered constructs) Baboon AV shunt	1–5
3. Surface matrix protein preconditioning of EPCs	*In vitro* surface preconditioning (ECM-coated grafts) Baboon AV shunt	*In vitro* surface preconditioning (Tissue engineered constructs) Baboon AV shunt	1–5
4. *In vivo* evaluation of endothelialized grafts	Surgical graft implants Baboon AV shunt (ECM-coated grafts)	Surgical graft implants Baboon AV shunt (Tissue-engineered constructs)	1–5

Case 5-3. Overview table showing researchers' responsibilities.

Summary of Proposed Research

Aims	Methods	Researchers	Embryo Groups	Timeline
1. Characterize OFT motion, morphology, and growth	*In vivo* OCT and Doppler OCT imaging	Dr. Smith	Normal (n = 12) OTB (n = 12 + 12 sham) VVL (n = 12 + 12 sham)	Years: 0–3 Years: 0–3 Years: 2–4
	4D image reconstruction & segmentation	Dr. Jones	Normal OTB VVL	Years: 0–3 Years: 1–4 Years: 2–4
2. Determine hemodynamic forces and gene expression patterns on the OFT walls	Blood pressure measurements	Dr. Jones	Normal (n = 12) OTB (n = 12 + 12 sham) VVL (n = 12 + 12 sham)	Years: 0–3 Years: 0–3 Years: 2–4
	FEM modeling	Dr. Jones	Normal OTB VVL	Years: 1–3 Years: 2–4 Years: 3–4
	Gene expression	Dr. Smythe	Normal (n = 12) OTB (n = 12 + 12 sham) VVL (n = 12 + 12 sham)	Years: 0–2 Years: 0–2 Years: 1–3

Case 5-4. Introduction to a Methods Section with an overview table.
(A) Introduction with a research design table showing a quarterly time schedule.
(B) Research design table revised to reduce space for the time schedule and to create space for a middle column where tasks can be described.

(A) Time Line Indicated by Quarters

5.1 Project Overview

[1]The plan of research for this project is divided into 4 major objectives, each involving 2–3 specific tasks (see Table 1). [2]Most tasks address one or more hypotheses (identified by bold in the sections that follow), some of which will help explain key effects that have been noted in previous work, while other hypotheses are more exploratory in that they address effects that are anticipated but have not been studied to date. [3]Objectives 1–3 will be performed by Smith, Jones, and a Ph.D. student from our university, and Objective 4 will be carried out by Smythe and one or more of his undergraduate students from the university.

Table 1. Timeline for the major objectives and tasks

Activity	Year I	Year II	Year III
Objective 1. Preparing Physical Model			
Task 1.1 Design physical model			
Task 1.2 Construct physical model			
Prepare analytical methods			
Objective 2. Studying Horizontal Effects			
Task 2.1 Characterize steady-state			
Task 2.2 Apply sequential treatments			
Task 2.3 Model horizontal effects			
Objective 3. Studying Vertical Effects			
Task 3.1 Characterize stead-state			
Task 3.2 Apply sequential treatments			
Task 3.3 Model vertical effects			
Objective 4. Characterizing Microbiology			
Task 4.1 Map community structure			
Task 4.2 Determine role of DIRBs			

(B) Revised Overview Table from (A): Time Line Reduced and a New Middle Column Included

Table 1. Overview of the proposed objectives and tasks

Activity	Description	Time
Objective 1. Preparing Physical Model		
Task 1.1 Design physical model		Yr 1
Task 1.2 Construct physical model		Yr 1
Task 1.3 Prepare analytical methods		Yr 1
Objective 2. Studying Horizontal Effects	*Reorganization of the time line in Case 5-4A frees up this middle space for a description of the proposed procedures or study personnel.*	
Task 2.1 Characterize steady-state		Yrs 1–2
Task 2.2 Apply sequential treatments		Yr 2
Task 2.3 Model horizontal effects		Yrs 1–2
Objective 3. Studying Vertical Effects		
Task 3.1 Characterize steady-state		Yrs 2–3
Task 3.2 Apply sequential treatments		Yrs 2–3
Task 3.3 Model vertical effects		Yrs 2–3
Objective 4. Characterizing Microbiology		
Task 4.1 Map community structure		Yrs 1–3
Task 4.2 Determine role of DIRBs		Yrs 1–3

Case 5-5. Example of an overview table of assessments for a proposed aim.

Table 2. Schedule of Assessments for Aim 3

Consent and Assessments	Initial Screening for Enrollment	Baseline Assessment	Visit 1- Treatment Initiation 0-week	Visit 2- 2-Week Assessment	Visit 3- Treatment Completion; 6-week Assessment	Visit 4- 12-week Assessment	Visit 5- 24-week Assessment
Trial Periods	< 48 hours of AON dx	<1 week of screening	Within 7 days of initial screening	2 weeks	6 weeks	12 weeks	24 weeks
Informed Consent	X						
Medical Hx	X						
Physical Exam		X				X	X
OCT Imaging of the Eyes		X				X	X
MRI of Brain*		X					X
Serum Pharmacokinetics			X		X		

Assessment							
Serum Safety Labs	X		X	X		X	X
Serum/PB MC Cytokine Labs			X		X		X
Serum/PB MC MMP-9/sICAM-1			X		X		X
PBMC CAMP			X		X		X
EDSS	X	X				X	X
25' Walk/9 HPT	X	X				X	X
Adverse Events		X	X	X		X	X

* MRI at 24 weeks performed on only those subjects with clinically isolated syndrome or MS.

Case 5-6. An overview flow chart covering the methods of one aim.

Figure 1. Overview of the Pilot Study (Aim 3)

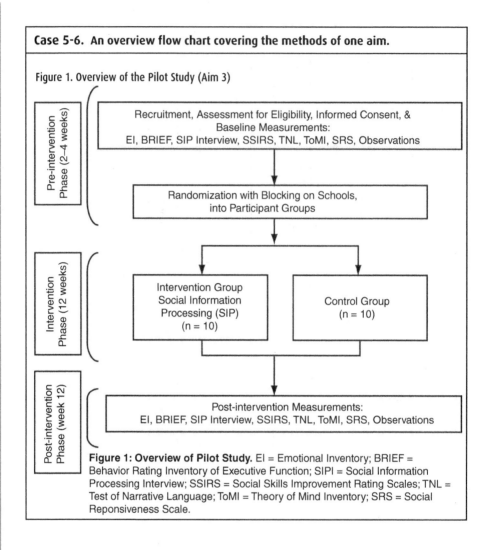

Figure 1: Overview of Pilot Study. EI = Emotional Inventory; BRIEF = Behavior Rating Inventory of Executive Function; SIPI = Social Information Processing Interview; SSIRS = Social Skills Improvement Rating Scales; TNL = Test of Narrative Language; ToMI = Theory of Mind Inventory; SRS = Social Reponsiveness Scale.

Case 5-7. Heading outline of a Methods Section.

D. RESEARCH DESIGN AND METHODS
Introduction (without a heading)

1. Specific Aim #1: Defining the relationship between SarA transcription and the production of SarA.
   A. Rationale.
   B. Experimental design.
      a. Determination of the temporal accumulation of SarA by Western blot.
         • Methods.
      b. Determination of the temporal accumulation of SarA by EMSA.
         • Methods.
      c. Determination of the temporal accumulation of SarA by transcriptional activation.
         • Methods.

   } Body Subsection

2. Specific Aim #2. Characterization of the mechanism by which SarA regulates expression of the S. aureus collagen adhesin gene.
   A. Rationale.
   B. Experimental design.
      a. Complementation of the defect in *cna* transcription observed in SarA mutants.
         • Methods.
      b. Characterization of the SarA DNA-binding site upstream of *cna*.
         • Methods #1: Localization of the SarA-binding site upstream of *cna*.
         • Methods #2: Mapping of the SarA-binding site upstream of *cna*.
         • Methods #3: Defining the sequence characteristics of the SarA-binding site upstream of *cna*.
      c. Correlation of SarA binding with the regulation of collagen adhesin gene transcription.
         • Methods.

   } Body Subsection

3. Specific Aim #3. Identification of *S. aureus* virulence factors under the direct control of SarA.
   A. Rationale.
   B. Experimental design.
      a. Methods #1:
      b. Methods #2:

   } Body Subsection

Case 5-8. Headings from a Methods Section. The headings to the body subsections are the actual aims from the **Aims Section**. The shared methods section in this **Methods Section** is at the end of the section.

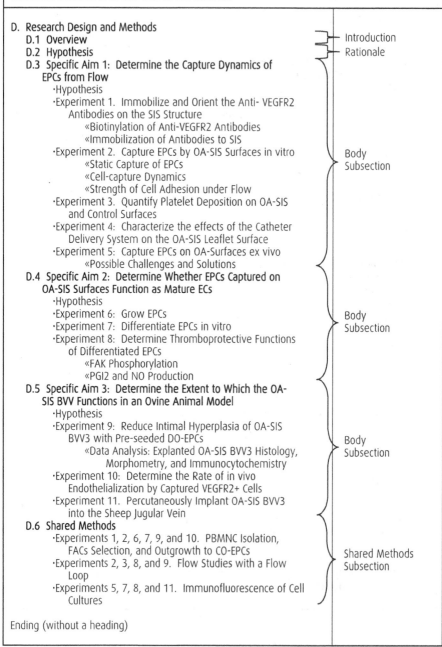

D. Research Design and Methods
  D.1 Overview
  D.2 Hypothesis
  D.3 Specific Aim 1: Determine the Capture Dynamics of EPCs from Flow
    ·Hypothesis
    ·Experiment 1. Immobilize and Orient the Anti- VEGFR2 Antibodies on the SIS Structure
      «Biotinylation of Anti-VEGFR2 Antibodies
      «Immobilization of Antibodies to SIS
    ·Experiment 2. Capture EPCs by OA-SIS Surfaces in vitro
      «Static Capture of EPCs
      «Cell-capture Dynamics
      «Strength of Cell Adhesion under Flow
    ·Experiment 3. Quantify Platelet Deposition on OA-SIS and Control Surfaces
    ·Experiment 4: Characterize the effects of the Catheter Delivery System on the OA-SIS Leaflet Surface
    ·Experiment 5: Capture EPCs on OA-Surfaces ex vivo
      «Possible Challenges and Solutions
  D.4 Specific Aim 2: Determine Whether EPCs Captured on OA-SIS Surfaces Function as Mature ECs
    ·Hypothesis
    ·Experiment 6: Grow EPCs
    ·Experiment 7: Differentiate EPCs in vitro
    ·Experiment 8: Determine Thromboprotective Functions of Differentiated EPCs
      «FAK Phosphorylation
      «PGI2 and NO Production
  D.5 Specific Aim 3: Determine the Extent to Which the OA-SIS BVV Functions in an Ovine Animal Model
    ·Hypothesis
    ·Experiment 9: Reduce Intimal Hyperplasia of OA-SIS BVV3 with Pre-seeded DO-EPCs
      «Data Analysis: Explanted OA-SIS BVV3 Histology, Morphometry, and Immunocytochemistry
    ·Experiment 10: Determine the Rate of in vivo Endothelialization by Captured VEGFR2+ Cells
    ·Experiment 11. Percutaneously Implant OA-SIS BVV3 into the Sheep Jugular Vein
  D.6 Shared Methods
    ·Experiments 1, 2, 6, 7, 9, and 10. PBMNC Isolation, FACs Selection, and Outgrowth to CO-EPCs
    ·Experiments 2, 3, 8, and 9. Flow Studies with a Flow Loop
    ·Experiments 5, 7, 8, and 11. Immunofluorescence of Cell Cultures

Ending (without a heading)

Annotations (brackets at right):
- Introduction
- Rationale
- Body Subsection
- Body Subsection
- Body Subsection
- Shared Methods Subsection

Case 5-9. Heading Outline of a Methods Section. In this example, the term *objective* is used for *aim*.

4. PLAN OF RESEARCH	
4.1 Project Overview	Introduction
4.2 Preparing the Physical Model System (Objective 1) 　4.2.1 Design the 2D physical model (Task 1.1) 　　　　Selection of materials 　　　　Preliminary reactive-transport modeling 　4.2.2 Construct physical model system (Task 1.2) 　4.2.3 Prepare analytical methods for Objectives 2 and 3 (Task 1.3)	Body Subsection
4.3 Studying Horizontal Effects (Objective 2) 　4.3.1 Establish and characterize horizontal steady-state (Task 2.1) 　4.3.2 Apply sequential treatments (Task 2.2) 　4.3.3 Model horizontal effects (Task 2.3)	Body Subsection
4.4 Studying Vertical Effects (Objective 3) 　4.4.1 Establish and characterize vertical steady-state (Task 3.1) 　4.4.2 Apply Sequential Treatments (Task 3.2) 　4.4.3 Model vertical effects (Task 3.3)	Body Subsection
4.5 Characterizing Microbiological Processes (Objective 4) 　4.5.1 Map microbial community structure (Task 4.1) 　4.5.2 Determine role of dissimilatory iron reducing bacteria (Task 4.2)	Body Subsection

Case 5-10. Methods introduction with an overview table. Bold = Topic, **A** = Analysis, **C** = Connection.

D. Research Approach

D.1 Hypothesis and Overview

[1]The proposed research will address important questions regarding the ability of **OA-SIS leaflets to capture EPCs** and **to support the differentiation** of these **EPCs** into fully **functional ECs** that are capable of **preventing thrombus and intimal hyperplasia in BVVs.** [2]Our central hypothesis is that ECM-protein SIS leaflets that are coated with EPC-directed antibodies capture EPCs from flow and allow for the subsequent spreading, proliferation, and normal functioning of ECs. [3]Table 3 presents an overview of the proposed research design. [4]Methods for Aims 1-3 are described in Sections D.2, D.3, and D.4, respectively; methods common to the experiments for Aims 1-3 are described in D.5. [5]See Section F for a justification of the sheep model and the number of sheep to be used in Exps. 5, 10, 11, and 12.

> Topic
> C: Purpose
>
> A: Hypothesis
>
> Forecast

Table 3: Summary of Proposed Research

Aims	Description	Time
1. Determine if the OA-SIS surface is non-thrombogenic and determine the best method to. capture EPCs	**1. Immobilize and orient** anti-VEGFR2 antibodies on the SIS surface **2. Capture** EPCs in vitro **3. Quantify** platelet deposition on the SIS surface **4. Characterize** effects of the catheter delivery system on antibody-coated SIS leaflets **5. Capture** EPCs *ex vivo*	Yr 1-3
2. Determine the function of captured EPCs	**6. Grow** EPCs **7. Differentiate** EPCs in vitro **8. Differentiate** EPCs in vitro under flow **9. Determine** thrombotic functions of EPCs	Yr 2-4
3. Determine the ability of ECs to reduce thrombus and intimal hyperplasia on the OA-SIS BVV in an ovine model	**10. Quantify** intimal hyperplasia on BVVs with CO-EPCs; 6 sheep, each implanted with 2 BVV3s: one OA-SIS BVV and one uncoated control BVV **11. Determine** the rate of in vivo endothelialization by captured EPCs; 12 sheep, each implanted with 3 suspended leaflets: OA-SIS leaflet, anti-IgG1-streptavidin-coated SIS leaflet, and uncoated control SIS leaflet **12. Implant** vales and evaluate at 3 months; 8 sheep, each implanted with 2 BVV3: either OA-SIS BVV, anti-IgG1-streptavidin-coated SIS leaflet, or uncoated BVV	Yr 3-5

> Methods, Overview & Table

Case 5-11. Introductions to 2 Methods Sections (A and B) with no research design tables. Bold = Topic, **A** = Analysis, **C** = Connection.

(A)

D. RESEARCH DESIGN AND METHODS
[1]The 2 HCN channels that we will use in the proposed research are **sphCN**, from the sea urchin, and the mammalian **HCN1**, from the mouse. [2]All experiments will be based on cysteine-specific reagents. [3]Cysteines will be introduced in the HCN channels using site-directed mutagenesis. [4]The cysteine-substituted channels will be expressed in Xenopus oocytes and recorded with a 2-electrode voltage clamp (TEVC), a cut-open oocyte voltage clamp, or macro-patches using the patch-clamp technique. [5]To determine the mechanism of voltage activation in the hyperpolarization-activated cyclic nucleotide-gated ion channel family (HCN channels), we will characterize the conformational changes in HCN channels during activation in order to identify the voltage sensor in HCN channels. [6]Our hypothesis is that the fourth transmembrane domain S4 is the voltage sensor in HCN channels.
    [7]We will start by looking for conformational changes of S4. [8]However, other transmembrane domains will also be tested. [9]We will use 2 different methods to determine whether S4 in HCN channels moves as a voltage sensor: cysteine accessibility (Specific Aim 1) and fluorescence measurements (Specific Aim 2). [10]In addition, we will use the cut-open technique to measure the gating currents in HCN channels in order to correlate the movement of the gating charges to the fluorescence signals and to estimate the gating charge contribution of the different S4 charges (Specific Aim 3). [11]All the data from the different Specific Aims will be used in the construction of a biophysical model of voltage-activation in HCN channels (See Analysis section).

Topic
C: Methods

C: Purpose

A: Hypothesis

C: Purpose & Methods

C: Methods & Forecast

(B)

D. Methods

[1]The objectives of this proposal will be accomplished through 14 experiments distributed among 3 aims. [2]The experiments in Aims 1 and 2 are designed to examine the **regulation of the HPG axis in the SJL mouse model of EAE**. [3]The experiments in Aim 3 will focus on the regulation of the HPG axis in male patients who are at high risk for developing clinically definite MS. [4]Each experiment will have its own subject and control group. [5]The methods of statistical evaluation for Aims 1 and 2 are presented at the end of this Methods section.

Topic
C: Purpose & Methods

Forecast

Case 5-12. Two examples of rationale subsections (A and B). Bold = Topic, **A** = Analysis, **C** = Connection.

(A)

Rationale. [1]Results from our previous experiments using single-input perturbations [102] and those from preliminary studies using dual-input perturbations (Fig. 15) indicate that sensory reweighting has a dominant role in postural regulation. } Review

[2]A secondary feature of postural regulation is the active control of stiffness and damping of the system [102]. [3]We hypothesize that the postural control system uses a particular combination of stiffness, damping, and sensory-weighting factors to optimize a **performance criterion**. } A: Hypothesis Topic

[4]A performance criterion defines a physical value or a combination of values associated with the postural control system. [5]Examples include energy expenditure, root-mean-square (rms) values of kinematic variables such as body sway and sway velocity, and rms values of kinetic variables such as torque and rate-of-change of torque. } Review

[6]In a given environmental and stimulus condition, the performance criterion varies as a function of postural control parameters (stiffness, damping, sensory-channel weights). [7]If the postural control system optimizes its performance by selecting control parameters that minimize a performance criterion and if this performance criterion is known, we can predict how the postural control system will perform in an arbitrary environment and stimulus condition. } A: Prediction

[8]In Exp. 4, we will identify candidate performance criteria and then will test these criteria by comparing predicted responses to actual responses obtained in previous experiments [102], that presented single-input stimuli, and responses that will be obtained in Exp. 3 using dual-input stimuli. } Forecast & C: Methods

(B)

Rationale: [1]Although the **production of testosterone (T)** is inhibited during the course of EAE, the **mechanisms governing this inhibition** have as yet to be defined. } Topic & Review A: Problem

[2]The production of T is primarily regulated via interactions among the hypothalamus, anterior pituitary, and the gonads. [3]Inflammatory mediators released into the blood during infection or inflammation have been shown to suppress T production. } Review

[4]This inhibition is thought to occur by: (1) interfering with the secretion of gonadotropins by the anterior pituitary, or (2) directly inhibiting the synthesis of T by the gonads, or (3) a combination of (1) and (2). } A: Speculation

[5]The following experiments are designed to determine the extent to which these 2 pathways are regulated during the course of EAE. } Forecast & C: Purpose

Case 5-13. Two examples of rationale subsections (A and B). Bold = Topic, **A** = Analysis, **C** = Connection.

(A)

Rationale and Hypothesis. [1]We have observed that the **1026b strain of** *Bp* **and at least 3 clinical** *Bp* **isolates efficiently infect the GI tract of adult mice** following **low-dose oral inoculation** and then **disseminate** to multiple different organs over a several-month period. [2]However, it is not known whether all strains of *Bp* can cause enteric infection or whether certain strains are particularly virulent following oral inoculation. [3]Nor is it known whether certain *in vitro* properties, such as cell invasion and replication, can be corrected with *in vivo* virulence, which is an essential step in developing assays for identification of enteric virulence determinants. [4]To address these questions, a low-dose oral challenge model in BALB/c mice will be used to screen a panel of 30 *Bp* isolates for enteric tropism and for virulence. [5]Selected high and low virulence strains of *Bp* will then be further evaluated *in vitro* to assess their ability to invade and replicate in intestinal epithelial cells and macrophages. [6]We hypothesize that most or all *Bp* strains can cause enteric infection and that enteric virulence will correlate with increased invasion and replication in intestinal epithelial cells and macrophages

Topic & Review

A: Problem

C: Methods & Purpose

A: Hypothesis

(B)

C.1 Early Development of the Embryonic Heart. [1]The proposed research uses a **chick model of heart development** because this model has been highly successful; results often apply to humans because developmental processes are highly conserved among vertebrate species. [2]Further, the chick embryo is easy to access; it is transparent and can be easily observed and manipulated without affecting cardiac growth; and our research team has extensive experience working with chick embryos and is involved in developing technologies to image the beating chick heart *in vivo*.[28-34] [3]In early development stages, human and chick hearts begin as straight, tube-like structures that then loop and twist, transforming into "s-shaped" tubular hearts. [4]After HH24, formation of cardiac valves and chambers begins.[24,22] [5]At HH24, the chick heart is still tubular-shaped and consists of: atrium, ventricle, and OFT. [6]The OFT connects the primitive ventricle to the arterial system via the aortic sac, and during early stages of development, the OFT is very sensitive to changes in normal developmental conditions (including hyperglycemia), leading to cardiac defects. [7]Further, the OFT provides a specific cardiac segment to study, from which ECs can be extracted for *in vitro* experiments. [8]The ability of glucose to alter EC gene and protein expression in the chick OFT is the focus of our proposal.

Topic & C: Methods

Significance

A: Conclusion

Review

C: Purpose

Case 5-14. Example of a subject subsection organized by study groups.

Subjects. [1]Thirty control subjects and 30 subjects with chronic UVL will be studied in each of the 5 proposed experiments. [2]Some subjects will participate in more than one experiment.

Control Subjects, Exps. 1–6. [3]Control subjects will be recruited from advertising in the State U. Medical School newspaper and from spouses and friends of subjects with chronic UVL. [4]Control subjects will be matched to chronic UVL subjects by age, sex, and body height and weight. [5]All control subjects will be required to meet stringent inclusion criteria: they must (1) undergo a standard neurological examination by Dr. Jones to assess criteria (2)–(10); (2) have no history or evidence of orthopedic, muscular, or neurological disability; (3) have normal strength in both legs as scored by the Manual Muscle Test; (4) be able to stand on toes and on heels, and have normal ankle-, knee-, and hip-joint range of motion; (5) not be on any medication that might affect balance; (6) have 20/40 or better vision with or without correction; (7) be between 21 and 69 years old; (8) be able to stand independently for at least 20 min; (9) have normal vestibular function as defined by normal horizontal vestibulo-ocular reflex (VOR) gain and phase for 0.05, 0.2, and 0.8 Hz horizontal rotation testing, which will be performed in Dr. Kelly's laboratory (refs); and (10) have normal somatosensory function in the feet as defined by detection of 125 Hz vibration at ankles and toes, detection of < 1° great toe and ankle joint motion (kinesthesia thresholds), and normal threshold of light-touch thresholds as assessed by Semmes–Weinstein hairs on the plantar surface (refs).

Subjects with Chronic Unilateral Vestibular Loss, Exps. 1–5. [6]Local otolaryngologists will refer chronic UVL patients, and Dr. Smith already has 10 chronic UVL subjects, aged 28–40, in his database whom we can recruit. [7]Subject selection will focus on patients who have lost vestibular function on one side as a result of a labyrinthectomy, vestibular nerve section, and/or acoustic neuroma removal at least 1 year before testing. [8]Chronic UVL subjects will be required to meet all of the inclusion criteria for control subjects except for criterion (9), their vestibular loss. [9]In addition, they will be required to meet the following inclusion criteria for UVL (refs): (1) no evidence of vestibular abnormality on the intact side; (2) absent or severely depressed caloric responses in the involved ear; (3) VOR gains within normal limits and phase leads exceeding the 95th percentile of the normative values (refs) for 0.05 and 0.2 Hz rotations; (4) horizontal VOR gain within clinic normal limits and 4–6 s time constants in response to pseudorandom rotations with frequencies between 0.01 and 1.5 Hz; and (5) for patients with acoustic neuroma, no signs of CNS (cerebellar or vestibular) damage. [10]In addition, all chronic UVL subjects will be well-compensated by conventional clinical criteria: no spontaneous nystagmus, no asymmetrical postural alignment in stance on a firm surface, and the ability to stand on an anterior-posterior, sway-referenced surface with eyes closed (Neurocom Equitest). [11]In addition, the relative amount of functional compensation will be evaluated with 3 standardized questionnaires about the effects of current symptoms on function: the Dizziness Handicapped Inventory (DHI, refs), the Activities of Balance Confidence (ABC, refs), and the Vestibular ADL Scale (refs).

Case 5-15. Example of a subject subsection organized by scientific methodology.

C. Subjects:

C.1 Subject Recruitment and Numbers of Subjects. [1]The numbers of subjects are dictated by 2 factors: 1) the number of subjects who can be enrolled and retained in the study based on preliminary research, and 2) the power calculations for the statistical tests to be used for data analysis. [2]We will test at least 50 subjects who use tamoxifen and 50 control subjects who do not use tamoxifen. [3]Power calculations will be provided in the analysis portions of this Methods Section. [4]When the P.I. conducted longitudinal studies of over 200 AMD and control subjects, the 18-month return rate exceeded 90%. [5]Techniques for maximizing subject retention are specified in the Human Subjects Section of this proposal. [6]We expect to recruit about at least 20 new eligible short-term (10-12 months) tamoxifen users per year for each of study-years 2 and 3, and at least 10 new eligible short-term tamoxifen users during study-year 1. [7]Information regarding subject demographics is given in the Human Subjects Section.

C.2 Eligibility. [8]Two groups of subjects will be tested: 1) women who use 20 mg tamoxifen daily on a continuous basis (the tamoxifen subjects), and 2) women who have never used tamoxifen (the control subjects). [9]All subjects will be 40-69 years old at study entry. [10]Tamoxifen subjects will have used tamoxifen for 10 months to 1 year at the time of study entry. [11]Subjects will meet the same eligibility criteria used for our preliminary studies of tamoxifen and for many of our other clinical studies. [12]Subjects will be required to have: a) 20/20 or better acuity in the test eye, and 20/25 or better acuity in the fellow eye, b) IOP < 22 mm Hg in each eye, c) a maximum of 5 diopters myopia in each eye, d) normal color vision for each eye, e) medical histories that are negative for eye disease, diabetes, ocular surgery, and ocular hypertension, and negative for the use of any medications (other than tamoxifen or oral contraceptives) that are known to affect vision, and f) every tamoxifen subject must have an ECOG score of 0, meaning that every tamoxifen subject must be fully active and able to carry on all pre-cancer activities without restriction.

[13]Women will be excluded from the study if, after testing, they are determined on masked evaluation of stereoscopic color optic-nerve-head and fundus photographs to have pathologic or suspicious ocular changes that are likely to be unrelated to tamoxifen use. [14]The presence of small, non-confluent macular drusen or a comparable level of macular pigmentary change will not be grounds for exclusion, nor will RNFL thickening or associated changes that are detected only via the HRT or the OCT. [15]Use of blood pressure medication will be an exclusion factor because, as stated in the Preliminary Results, high blood pressure or its medications may affect visual adaptation.

Case 5-16. Example of a subsection describing a vertebrate object of study.

D.2 Birds. [1]The experiments will be performed in zebra finches, the species for which most data are available on the anatomy of the song system and central auditory pathways, as well as on the gene expression and electrophysiological responses to song in NCM (reviewed in refs). [2]Zebra finches will be readily available at the Brazilian site from a breeding colony initiated by Dr. John Smith and from local bird breeders. [3]Adult zebra finches will be acoustically isolated overnight (12–14 h before the experiments) in a sound-proof chamber. [4]This condition minimizes auditory activity and associated gene expression in NCM (refs) and will provide a baseline condition for all experiments. [5]The birds will then be decapitated without experiencing any auditory stimulation (unstimulated controls) or after hearing conspecific song (as detailed in the context of specific experiments), and brain specimen taken (see below). [6]Females will be used because they do not sing, so that self-stimulation by singing can be avoided when comparing control and song-stimulated birds. [7]The responses in males will be assessed in selected experiments.

Case 5-17. Vertebrate Animal Section.

VERTEBRATE ANIMALS:
[1]The animal care facilities and programs of the Kelly Institute meet the requirements of the law and NIH regulations. [2]The College of Veterinary Medicine at State University, of which the Kelly Institute is a part, is accredited by the American Association of Laboratory Animal Care. [3]The animals are subjected to regular veterinary care on a routine basis. [4]An Animal Care Protocol has been approved for these studies: Protocol #1000-0001, approved 10/5/09 for a three-year period.

1. Description of animal use; species and numbers of animals to be used. [5]Rabbits will be used to prepare polyclonal sera against peptides and expressed proteins. [6]Mice of the BALB/c strain will be used for hybridoma production for preparing structure-specific monoclonal antibodies.

> [7]Rabbits: Maximum number of animals used each year 1-5: 2 Total: 10
> [8]Mice: Maximum number of animals used each year 1-5: 10 Total: 50

2. Justification for use of animals. [9]Rabbits will be required for preparing polyclonal sera against synthetic peptides and expressed proteins. [10]Rabbits are the standard species for preparation of such sera. [11]All rabbit handling and immunizations will be conducted by the Department of Laboratory Animal Medicine at State University on an at-cost basis, or from commercial vendors.

[12]Mice of the BALB/c strain will be used for hybridoma production as those are compatible with the myeloma cells to be used for fusions. [13]Mice are the standard and best characterized species for preparing specific monoclonal antibodies. [14]MAb will be produced in tissue culture; no production of ascitic fluids in mice will be carried out.

3. Veterinary Care. [15]The services of the Center for Research Animal Resources are available if any other veterinary care or consultation is required, and they are available on-call 24 hours per day. [16]Any animals suffering clinical disease will be examined and treated by a veterinarian. If the condition is untreatable the animal will be euthanized.

4. Discomfort, Distress and Injury. [17]The experiments in this research will not cause undue discomfort or distress to the animals being used. [18]However, when needed, veterinary care will be provided; if the condition is untreatable the animal will be euthanized by approved methods.

5. Euthanasia. [19]Euthanasia will be performed by methods specified and approved by the AVMA Panel on Euthanasia.

> [20]Mice: Euthanized by CO_2 anesthesia.
> [21]Rabbits: Barbital overdose - I.V. administration.

Case 5-18. Subsection for a fully customized piece of equipment.

---

4. Test equipment. [1]Four test devices will be used for subject screening and for the proposed experiments: a vertical-axis rotation chair, a 2-axis rotation device, a balance test device, and a high-bandwidth rotating platform.

4.1 Vertical-axis and 2-axis Rotation Devices (Figs. E and F). [2]Both of these devices will be used for VOR screening. They are described in detail, in Resources.

4.2 Balance Test Device (Figs. A and B). [3]The balance test device is for screening tests and experimental tests (Exp. 3 and 5). [4]This custom-designed device includes: a) instrumentation to measure body sway, sway velocity, and COP, and to collect EMG data, b) a motor-driven visual surround with the rotation axis collinear with the subject's ankle-joint axis for experiments evoking AP sway, c) a motor-driven support surface with rotation axis collinear with the test subject's ankle-joint axis for AP sway experiments, d) a backboard assembly to restrain the subject and to allow AP body sway only about the ankle joints, and e) an overhead motor drive to deliver servo-controlled torque perturbations to the backboard assembly. [5]Sideways stance on this device provides surface rotation that evokes ML body sway and an additional backboard assembly restrains the test subject and allows ML body sway about an axis between the feet at ankle-joint level (Fig. B, p. 12). [6]These 2 backboard assemblies and 2 additional backboard assemblies will allow us to investigate stance-width changes on ML sway (see 4.4 below and Constrained body motion, p. 32). [7]Modifications will be made to the visual surround in the first year to improve its dynamic characteristics. [8]The visual surround will be re-oriented to provide a visual stimulus to evoke ML body sway for Exp. 5.

4.3 High-bandwidth Rotating Platform (Fig. C). [9]This device will be used for the experimental tests (Exp. 1, 2, 6). [10]A rotating platform will be constructed in Year 1 of the grant. [11]Its construction and placement will not interfere with the use of the balance test device and the rotation equipment used for VOR tests. [12]The platform will be hydraulically driven to provide a high-bandwidth rotation of the support surface. [13]The rotation axis will be collinear with the subject's ankle joints for experiments evoking AP sway, or will be between the feet at ankle-joint level for experiments evoking ML sway. [14]The platform will have a similar design and capabilities as the device used by Kearney and colleagues for the study of ankle-joint reflexes and biomechanics [53]. [15]The high bandwidth (about a 20-Hz, small-signal bandwidth) is necessary to improve the resolution in order to detect the time-of-onset of different passive and active mechanisms that contribute to the generation of corrective torque. [16]This device will be used for experiments testing the timing of different postural mechanisms (Exp. 1), compensation for time delays (Exp. 2), and the control of ML sway (Exp. 6).

4.4 Backboards (Fig. C). [17]The proposed custom-made devices will use a version of a backboard support frame to constrain the subject's body-sway motion. [18]The purpose of the backboard is to simplify the biomechanics of the body. [19]This simplification of biomechanics will allow us to focus on understanding sensorimotor integration rather than characterizing multi-segment control. [20]Our previous results, comparing backboard versus freestanding conditions, have shown no major differences in sensorimotor integration strategies [102]. [21]However, in the proposed experiments, we will continue our previous practice of repeating tests in a subset of subjects in free-standing or less-constrained conditions to insure that we detect any conditions where the backboard might influence the experimental results.

[22]Two backboards will be used. [23]One will constrain the body to sway as a single-link inverted pendulum, with sway occurring only in the AP direction about the ankle joint axis (Exps. 1, 2, 3). [24]A second will constrain the body to sway as a single-link inverted pendulum with sway occurring only in the ML direction about an axis passing between narrowly placed feet at ankle level (Exp. 5).

Case 5-19. Subsection describing standard and customized equipment and tools.

D.2.3 INSTRUMENTATION. [1]Visual function data will be obtained using 2 apparatuses: 1) a Humphrey Field Analyzer II, Model 745, and 2) a custom-built, 4-channel Maxwellian View device. [2]Each will be housed in the PI's laboratory and will be dedicated to the research. [3]Visual function data to be used mainly for screening purposes will be obtained with several other pieces of equipment. [4]Visual acuity will be assessed using a Humphrey Instruments model 599 autorefractor, color vision will be assessed using the D-15 test and a Nagel anonaloscope, and spatial contrast sensitivity will be assessed with a Pelli-Robson chart. [5]The autorefractor is in general clinical use; all other devices are housed in the PI's laboratory and will be dedicated to the research.

(1) Humphrey Field Analyzer II, Model 745. [6]This model of the Humphrey Field Analyzer measures SWAP visual fields and white-on-white automated static threshold 24–2 visual fields, the latter using the Swedish Interactive Threshold Algorithm Standard program. [7]SITA Standard white-on-white visual fields can be measured in about half the time required for the Full Threshold algorithm originally used for white-on-white fields

(2) 4-channel Maxwellian View Device [8]This device will be built specifically for the proposed research, to measure increment-threshold sensitivities in the fovea and in the visual periphery; most features of this device will be those of a conventional Maxwellian View system. [9]These features are not described. [10]The PI has built and now operates 2 Maxwellian View devices, one mainly for testing clinical populations and the other mainly for conducting basic psychophysical studies of visual adaptation. [11]The proposed device will combine some of the features from each of these 2 devices and will include several unique features to allow threshold to be measured reliably in the visual periphery.

[12]The new device will include 2 pairs of conventional 2-channel Maxwellian View systems, each with 1 mm diameter exit pupils. [13]One of these pairs (the central pair) will project test and background lights through the pupil and onto the retina along the direction of fixation. [14]Another pair (the peripheral pair) will project test and background lights through the pupil and onto the retina at a 20° angle from fixation. [15]By presenting test/background stimulus combinations through 2 separate pairs of channels, we can ensure that all test stimuli are spatially uniform and that all background stimuli are spatially uniform in the immediate region of the test stimuli.

[16]Another feature unique to the proposed device will be its means for facilitating subject alignment. [17]Proper alignment is more crucial for testing in the visual field periphery than in the fovea alone because the peripheral stimulus beams diverge from the foveal stimulus beams. [18]Because use of a bite bar reduces subject compliance for a longitudinal study and precludes some people from being tested at all, we will use a flat chinrest and rubber eye guard, as we do for our existing clinical Maxwellian View device. [19]However, in our new device, we will add 2 unique features, which, together, will take the place of a bite bar assembly. [20]One feature is the adjustability to be incorporated into the eye guard assembly for translation in the x (horizontal), y (vertical) and z (in and

out) directions. [21]Once positioned, the eye guard assembly will be locked in place. [22]The other feature is a non-elastic headband with a velcro patch to be attached to a complementary velcro patch connected to a translatable and lockable projection from the eye guard assembly. [23]These additional features will allow us to align subjects and to maintain their alignment in essentially the same manner as we do for our basic psychophysical experiments. [24]The eye guard will have 2 additional features. [25]Trial lenses will be insertable to correct for a subject's refractive errors (spherical equivalent), and the eye guard will be firmer than the one we now use, to prevent compression and consequent translation of eye position towards the device.

[26]The proposed device will include an infrared video camera for monitoring subject alignment. [27]One infrared light will be presented in Maxwellian View along with the peripheral pair of visible lights, and another infrared light will be presented in Maxwellian View along with the central pair of visible lights. [28]Thus, this Maxwellian View system will be a 6-channel device, although only 4 of the channels will contain visible light. [29]By comparing the positions of the 2 infrared Maxwellian View beams at the surface of the cornea, we will monitor the stability of subject alignment in the z direction. [30]To view the pupil, in order to monitor alignment in the x and y directions, an additional infrared light, not in Maxwellian View, will illuminate the subject's eye.

Case 5-20. Subsection describing a partially customized tool.

D.1.1 Modification of the existing OMAG system

[1]The current OMAG system (see Fig. 1) operates at 842 nm wavelength in the PI's lab. [2]It has the following features: 1) optic-fiber implementation similar to a spectral domain OCT system (refs), thus providing OMAG with portability, 2) spatial resolution of $16 \times 16 \times 8 \mu m$ (x-y-z), 3) optical ranging capability of 6.4 mm in air (of which 3.2 mm is used for structural imaging and 3.2 mm, for blood perfusion imaging), 4) an imaging speed of 10 frames per second (fps), with each frame consisting of 1000 A scans, i.e. axial scans), 5) volumetric imaging time of 50 s in a Labview environment with a view field of $2.2 \times 2.2 \times 1.7$ mm^3. [3]We will modify this system for use in Aims 3 and 4 until an optimal OMAG system is constructed (see below).

[4]We will first modify the OMAG system, to solve its current problems, and then we will use the modified OMAG system to perform the imaging tasks that we will describe in Aims 3 and 4. [5]A problem is that, in preliminary studies, when we used OMAG to image cerebral blood perfusion in mice, we had to manually position the mouse's head on a stereotaxic stage, which in turn was mounted on a 2-dimensional (x-y) manual translational stage, which required the mounting stage to be wheeled away from the probe beam for surgical procedures. [6]These efforts were time consuming, requiring ~15 minutes just to place the head in the precise spot for imaging, indicating problems with the ability of OMAG to be used efficiently and accurately in a clinical and research settings. [7]Manual positioning was also problematic since we need to ensure that the head of the mouse (which we used across experiments) was consistently positioned under the probe, at the same location. [8]To make the OMAG system more efficient and user-friendly, and to ensure consistency in positioning of the mouse's head, we will purchase a precision stereotaxic stage and will incorporate it into the OMAG system. [9]We will also purchase and incorporate a computer controlled x-y linear translation stage to automate sample positioning under the probe. [10]These modifications will not only stabilize the sample but will also reduce the time needed to place the sample under the imaging probe. [11]This work is important because we need to be able to observe blood flow alterations at the same imaging spot in the cortex area over time, in order to understand how local cerebrovascular perfusion is altered under the various experimental conditions. [12]Continuing the imaging studies using our existing system will allow us to identify potential problems in system development so that the optimized system can be constructed in a faster time frame. [13]Further, continuing the imaging studies in parallel to building the new system will be more cost-effective than building the new system from the outset.

The Methods Section, Part 2

Chapter 6 continues the description of the **Methods Section** begun in Chapter 5 and focuses on:

Data-collection procedures
Data-analysis procedures
Data interpretation and expected outcomes
Potential problems and proposed solutions
Shared methods
Ending

In Chapter 5, Figure 5-1 gives the generic content and organization of a **Methods Section**. For convenience, this figure is reproduced and renumbered in Chapter 6 as Figure 6-1. Also for convenience, Case 5-9, showing a heading outline of a **Methods Section**, is also reproduced in this chapter as Case 6-1. Chapter 6 gives generic heading outlines in Tables 6-1 and 6-2. The term *subsection* will continue to be used in Chapter 6 to refer to a **Methods** subsection or sub-subsection, unless phrased otherwise for clarity.

6.1　Data-collection and data-analysis subsections and procedures

Proposed procedures are the actions and tasks that you intend to execute in order to acquire data that you will then analyze, interpret, and ultimately disseminate in professional journals and at professional meetings.

In the narrative, the specific actions that you and members of your research team intend to perform in order to achieve your proposed research objective and aims are variously termed *procedures, protocols, tasks, methods, experiments,* and *methodological activities.* These terms are not synonymous even though they are often used synonymously. However, because their features are similar, the term **procedure** will be used here for procedures, protocols, tasks, methods, experiments, and methodological activities.

As shown in Figure 6-1, the rationale, **SOS**, and **MET** precede the data-collection and -analysis procedures in each body subsection that gives procedures. However, if you have no rationale, **SOS**, or **MET** subsections, or if they are described elsewhere in the narrative, such as in the **Preliminary Studies Section**, then the data-collection and data-analysis procedures directly follow the **Methods** introduction.

Procedures are described in chronological order, in the order in which you will begin them. As a result, you will describe data-collection procedures before data-analysis procedures.

> **Guideline 6.1**　*Describe data-collection procedures before data-analysis procedures.*

To better understand how Guideline 6.1 and Figure 6-1 interrelate and can be applied, consider a research design for proposed research that has: 3 aims, a rationale subsection that covers the methods for the 3 aims, 2 experiments proposed for Aim 1, 3 for Aim 2, and one for Aim 3 (for potential problems and proposed solutions, see Chapter 6.10). Table 6-1 shows different ways that a narrative for this hypothetical proposed research can be organized.

● Table 6-1A shows an organizational alternative in which each experiment has its data-collection and data-analysis procedures described separately from the other experiments. This alternative is useful when the data-collection and data-analysis procedures for each experiment are different.

● Table 6-1B shows an organizational alternative in which the data-analysis procedures for the experiments proposed for each aim are the same. In this case, each procedural body subsection has only one data-analysis passage that covers the experiments for a particular aim.

● Table 6-1C presents an organizational alternative in which all of the proposed experiments are different, but the data-analysis procedures across the experiments are the same. In this third case, there is only one data-analysis set of procedures in subsection **C.5**. Following Guideline 6.1, data-collection procedures are described separately from data-analysis procedures, and the data-analysis procedures follow their relevant data-collection procedures.

Two other organizational alternatives are shown in Table 6-2. In Table 6-2, consider a research design that has: 3 aims, a rationale subsection that covers the methods for all 3 aims, 2 different procedures that will be run for data collection, and 4 different sets of data-analysis procedures (one data-analysis set for Aim 1, 2 data-analysis sets for Aim 2, and one data-analysis set for Aim 3).

● As shown in Table 6-2A, the 2 data-collection procedures are described before the data-analysis procedures: one data-analysis set of data-analysis procedures for Aim 1, 2 data-analysis sets for Aim 2, and 1 data-analysis set for Aim 3.

● Table 6-2B addresses the same scenario as in Table 6-2A, except that in Table 6-2B, the research design is for a clinical study involving subjects, an intervention that involves a piece of equipment, and a set of procedures that involve subjects interacting with the equipment. As shown in Table 6-2B, subsection C.2 has the subjects and equipment described before the data-collection procedures and the data-analysis procedures are described, following Guideline 6.1.

You should try different heading outlines in advance of drafting the text in order to identify an organization that allows you to sequence procedures that reflect your research design and that arrange data-collection procedures before data-analysis procedures.

Case 6-1 (reproduced from Case 5-9) presents a heading outline of an actual **Methods Section** that shows 4 **Methods** procedural body subsections (4.2, 4.3, 4.4, and 4.5) – one for the proposed methods to each aim (termed *objective* in the heading outline). These 4 procedural body subsections are further divided into sub-subsections of procedures. Similarly, Case 6-2 presents a heading outline of a **Methods Section** with 4 **Methods** procedural body subsections (C.1, C.2, C.3, and C.4). The heading **Protocol** in each procedural body subsection indicates where the data-collection procedures are located, and the heading **Analysis** indicates the subsections where the data-analysis procedures are located.

6.2   Phrasing of headings to procedural body subsections

The major headings to procedural body subsections indicate their corresponding aims from the **Aims Section**. There are 4 types of headings from which to choose, for procedural body subsections: generic, semi-generic, standard, and non-generic.

● **Generic** or **standard headings** are phrased in standard methodological terms. The headings **Rationale and hypotheses**, **Protocol**, **Analysis**, **Interpretation**, and **Potential complications** in Case 6-2 are generic. The absence of grammatical parallelism is allowable for headings that name generic or standard procedures.

● **Semi-generic procedural headings** include a non-generic term and at least one generic term indicating the type of action or task. A semi-generic procedural heading can also name a generic or standard procedure (see below). The heading **Aorto-iliac graft procedure** in Case 6-3 is a semi-generic procedural heading since the term **procedure** is a generic term, and **Aorto-iliac graft** is a non-generic term that refers to the proposed research.

- **Non-generic headings** comprise substantive terms that directly refer to key terms in the proposed research. They identify subsections of **Methods** procedures without using standard terminology. Non-generic headings usually lead with a verb-related term that identifies the procedural action or task described in the subsection. The procedural headings **Preparing the Physical Model System** in Case 6-1, **Preparing slices** in Case 6-4, and **Localization of the SarA-binding site upstream of cna** in Case 6-5 all begin with a verb-related term (**Preparing** is from the verb *prepare* and **Localization** is from the verb *localize*).

The non-generic or semi-generic headings to procedural subsections at the same level in the organizational hierarchy need to be *grammatically parallel*. Verb-related terms that lead the headings to procedural subsections can be an infinitive verb, a participial phrase, or a nominalization or gerund. For example, in Case 6-1, the procedural headings at the first level in the organizational hierarchy are all in the **verb + ing** (participial) form, but they all could have been consistently phrased with other verb-related grammar:

Participial Phrase:	Preparing the Physical Model System
	Studying Horizontal Effects
	Studying Vertical Effects
	Characterizing Microbiological Processes
Infinitive Verb:	Prepare the Physical Model System
	Study Horizontal Effects
	Study Vertical Effects
	Characterize Microbiological Processes
Nominalization or Gerund:	Preparation of the Physical Model System
	Studying of Horizontal Effects
	Studying of Vertical Effects
	Characterization of Microbiological Processes

Guideline 6.2

(*a*) *Phrase non-generic and semi-generic procedural headings at any one level in the organizational hierarchy, in parallel grammar.*

(*b*) *Include at least one verb-related term in a procedural heading.*

(*c*) *Phrase the verb-related term in one of 4 grammatical forms: an infinitive verb, an infinitive verb phrase, a participial phrase, or a noun in a nominalization or gerund form.*

6.3   Organization and phrasing of procedures

You need to describe your proposed data-collection and data-analysis procedures precisely and clearly so that reviewers can: (a) readily understand them, (b) determine whether or not the procedures are accurate, (c) mentally visualize your performing the procedures, (d) readily determine how they reflect principles of sound scientific methodology, (e) determine whether the research design and procedures eliminate confounding variables, and (f) make an educated guess as to whether you will achieve your proposed research objective and aims.

If procedures are inaccurate due to clarity problems, reviewers will not likely assume that you know the correct procedures and you "just" have problems explaining them. They will likely assume that you do not understand the procedures. Also, if procedures are not clear, reviewers will not likely use their own knowledge to reconstruct what you mean. And if your procedures are not complete in significant ways, reviewers will likely not use their own knowledge to fill in critical gaps of procedural content.

6.3.1 Chronological order for procedural descriptions

You need to sequence procedural subsections in chronological order, in the order in which you will begin the procedures identified in a subsection heading. In Table 6-2A, the procedures described in **C.2** will be begun before those in **C.3**, simply because the procedures in **C.2** are described before those in **C.3**.

Not only are procedural subsections sequenced chronologically in the order in which you will begin them, but within the procedural subsections, the procedures are also sequenced in chronological order, in the order in which you will begin them. In the procedural descriptions in Examples 1 and 2, both passages illustrate sentences with procedural information sequenced chronologically in terms of when the actions will begin.

Example 1
[1]In brief, ePTFE (5–7 cm long, 4 mm i.d.) will be implanted bilaterally as aorto-iliac grafts (Fig. 5). [2]After a midline incision, the aorta and iliac arteries will be dissected free from surrounding tissues. [3]Following anti-coagulation with heparin, the aorta will be ligated at the bifurcation of the iliac arteries. (Case 6-3, sentences 3–5.)

Example 2
[1]Fish will be deeply anesthetized in cold anesthetic (MS-222, 1:10,000), the brain will be rapidly exposed, and the cranial nerves of the medulla will be cut. (Case 6-4, sentence 1.)

The 3 sentences in Example 1 describe the procedures for performing 5 verb actions, with the procedures described in the order in which the PI will begin them: first, ePTFE will be implanted bilaterally; then a midline incision will be made; next, the aorta and iliac arteries will be dissected free from surrounding tissues; then the aorta will undergo anti-coagulation with heparin; and last, the aorta will be ligated. Example 2 describes the procedures for performing 3 verb actions, with each action mentioned in the order in which it will be begun: first, the fish will be anesthetized; then the brain will be exposed; and last, the cranial nerves will be cut.

> **Guideline 6.3** *For clarity, describe all proposed procedures in chronological order, in the order in which the PI or another research team member will begin executing them.*

Readers assume that procedural sentences are described in chronological order, as explained in Reader Assumption 1:

Reader Assumption 1: *Readers assume that the first procedure that is described will be started before the procedures that are described later, unless otherwise explicitly indicated.*

In order to counter (or to reinforce) Reader Assumption 1, or to clarify the chronology, you can add chronological terms, such as *after, at the same time, before, following, while, next,* and *then,* to procedural sentences, as explained in Guideline 6.4:

> **Guideline 6.4**   *Use chronological terms in procedural sentences:*
> (a) *To counter Reader Assumption 1 in order to indicate that the procedures described first will not be performed first, or*
> (b) *For clarity, to reinforce or to emphasize Reader Assumption 1, that the procedures will be executed in the order described.*

6.3.2   The grammar of a procedure

A procedure is phrased with at least one action verb that begins with **will** (popularly called the *future verb*), whether the verb is phrased in active or passive grammar.

will + infinitive verb   grammar of an action verb in an active clause
will + be + past participle   grammar of an action verb in a passive clause

The following examples of action verbs all begin with **will** and are followed by an infinitive verb in active or passive grammar.

Active Procedural Verbs	**Passive Procedural Verbs**
will write	*will be written*
will process	*will be processed*
will determine	*will be determined*
will calculate	*will be calculated*
will reconstruct	*will be reconstructed*

In Example 3, the first sentence is identifiable as a passive procedural sentence since its verb is phrased *will be removed.* In contrast, the second sentence, which gives background, is not a procedure because its verb – *has averaged* – does not begin with **will**.

Example 3
[1]The aorta and graft segments **will be removed** en bloc, dissected free of surrounding tissue, blocked, sectioned, and stained. [2]Overall graft patency **has averaged** about 85%. (Case 6-3, sentences 12 and 13.)

Instead of **will**, you might be tempted to use the term *intend to* or *plan to* before the infinitive verb. However, these terms address your mental state, so they suggest less commitment or less certainty that you will perform the action verbs. As a result, you should not often use *intend to* or *plan to* in procedural descriptions.

> **Guideline 6.5** *For every procedure:*
> (a) *Use at least one action verb that begins with **will**.*
> (b) *Avoid using **intend to** or **plan to** when describing procedures.*

You might begin executing your proposed procedures before writing the narrative or submitting the grant proposal. If so, you still need to describe the procedures using **will**, as if you have not yet begun them. Without **will**, you do not have a procedural description.

6.3.3 Core procedural information

At a minimum, procedures need to convey 3 types of *core* procedural information:

(1) *What* – **the verb action**: the verb action comprising the procedure.

(2) *Which* **SOS and MET**: the objects that will be manipulated or affected by the verb action.

(3) *How*: the manner in which the verb action will be executed in order to affect or to manipulate the **SOS** or **MET**.

The first sentence in Example 1 above is analyzed here in terms of its core procedural information:

What **verb action:** *will be implanted*
Which **SOS and MET**: *ePTFE*
How implantation will be achieved: *bilaterally*

(1) **What – the verb action.** A procedure needs an action verb, which is the task or activity that someone or something will perform. The form of the action verb begins with **will** (see Chapter 6.3.2).

Example 4a presents a procedural sentence in active grammar that can be used in a **Methods Section**. Example 4b shows the same action verb but in passive grammar. Examples 4c and 4d are more specific alternatives to the procedures in Examples 4a and 4b, specifications that are achieved by the addition of a prepositional phrase that includes additional *what* information – 2 action verbs in a prepositional phrase: *by synthesizing and annealing*. For readers to interpret verbs in a prepositional phrase as procedural verbs, the prepositional phrase needs to be associated with an action verb that begins with **will**, such as *will be generated* in Example 4c and *will generate* in Example 4d.

Example 4
a. We ***will generate*** appropriate fragments.
b. Appropriate fragments ***will be generated***.
c. Appropriate fragments ***will be generated by synthesizing and annealing*** complementary pairs of oligonucleotides. (Case 6-5, modified from sentence 3.)
d. We ***will generate*** appropriate fragments ***by synthesizing and annealing*** complementary pairs of oligonucleotides.

Note that if an additional action verb in a passive procedural sentence is *using*, the preposition *by* is not typically used, as illustrated in Example 5.

Example 5
End-to-side anastomoses ***will be constructed ~~by~~ using*** proline sutures. (Case 6-3, sentence 7.)

(2) Which SOS and MET. A procedure needs more than an action verb; it needs an object – something that is affected or manipulated by the action verb. For example, in Example 4, the *appropriate fragments* are being affected by the action verb since they are what will be *generated*, and in Example 5, the *anastomoses* are being affected by the action verb since they will undergo *construction*.

(3) ***How*** **information.** Another type of core procedural information explains ***how*** the action verb will be executed. ***How*** information is sometimes termed ***adverbial information***. In Example 4c, ***how*** information is *by synthesizing and annealing complementary pairs of oligonucleotides*. In Example 5, the ***how*** information is *using proline sutures*.

As shown in Examples 1 and 2, the core procedural information (***what***, ***which***, and ***how***) can be – and often is – presented in one sentence. In addition, ***how*** information can also be presented in a list or in a separate procedural sentence to improve the readability of the text or to clarify the procedure. Example 6a is a procedural sentence with the action verb *will be generated*. This verb is specified with ***how*** information after the colon, in the enumerated list of procedures. This sentence can also be rephrased into 2 shorter sentences, shown in Example 6b, where the second sentence conveys ***how*** information.

Example 6
a. A family of synthetic stimuli that interpolate between the natural whistles and the synthetic tones ***will be generated by***: 1) determining extremal points of the sound wave by a parabolic fit, 2) calculating the instantaneous frequency and amplitude, 3) reconstructing the sounds by a sinusoidal interpolation of the extremal points, and . . .
b. [1]A family of synthetic stimuli that interpolate between the natural whistles and the synthetic tones will be generated. [2]*To generate these synthetic stimuli, we will*: 1) determine extremal points of the sound wave by a parabolic fit, 2) calculate the instantaneous frequency and amplitude, 3) reconstruct the sounds by a sinusoidal interpolation of the extremal points, and . . .

As shown in Example 6b, *To generate these synthetic stimuli* leads the second sentence. If you present ***how*** information in a separate sentence, the original procedural verb (in this case, *generate*) or a noun derivative (such as *generation*) needs to be used early in the separate sentence to associate the ***which*** and the ***how*** information in the second sentence (which also serves as a transition between the sentences; see Chapter 8.4 on given and new information).

> **Guideline 6.6** *When explaining **how** information:*
> (a) *Consider using a **by** prepositional phrase.*
>   ***Exception**: If the verb action is the verb **use**, in its participial form **using**, then omit the preposition **by**.*
> (b) *For readability, consider presenting **how** information in an itemized list or in a sentence separate from the **what** information.*
> (c) *For **how** information presented in a separate sentence, use the original procedural verb or a noun derivative early in the **how** sentence.*

6.3.4 Additional details about procedural descriptions

Procedures can be elaborated with non-core procedural information, which is supplemental information that clarifies or provides details about the core procedures. Four types of non-core procedural information are:

(1) *Who (or **what**)* will execute the procedures: the agent.

(2) *When* you intend to perform the procedures, relative to other procedures.

(3) *Where* you will perform the procedures and the spatial relationships among the **SOS** and **MET** while you execute the procedures.

(4) *Why* you selected particular features of your proposed actions or tasks; that is, their methodological purpose, benefit, and/or advantage.

We address each of these types of non-core procedural information:

(1) *Who*: **the agent**. *Who* information identifies the person or thing executing the verb action. This person or thing is called the ***agent***. There are 3 ways to handle *who* information in a procedural description: (a) by not identifying the agent, (b) by identifying the agent when the agent *is **not*** the PI, and (c) by identifying the agent when the agent *is* the PI.

● **The unidentified agent**. In a **Methods Section** of a narrative, readers assume that the agent who will be executing the proposed procedures will be the PI (or someone whom the PI directly supervises).

> ***Reader Assumption 2:*** *In procedural sentences, reviewers assume that the PI will perform the procedures if no other agent is identified.*

A result of this assumption is that procedural descriptions often do not include *who* information because the PI often times – if not most of the time – is the agent. The entire procedural subsection **Preparing slices** in Case 6-4 does not mention who will execute the procedures; the assumption is that the PI will execute them.

● **The identified agent who is not the PI**. When Reader Assumption 2 does not apply – that is, when the agent is not the PI – you need to identify who the agent is in order to negate the assumption.

Guideline 6.7   *If the agent to a procedure is not the PI, identify the agent.*

The procedural sentences in Examples 7 and 8 include *who* information. In Example 7, the agents will perform the balance exercises while using the ABF device. In Example 8, Dr. Jane Smith will train the 3 physical therapists, who will then provide the rehabilitation protocols and will teach the subjects.

Example 7
The control group will perform balance exercises with a physical therapist, and *the experimental group* will perform the same exercises with a physical therapist while using the ABF device. (Case 6-6, sentence 2.)

Example 8
Three physical therapists will be trained by *Dr. Jane Smith*, who will provide these rehabilitation protocols and will teach subjects how to use the ABF device. (Case 6-6, sentence 7.)

The agent is often a human, which is why this procedural content is called *who* information. However, the agent can also be a non-human, such as a **MET** or a process, as shown in Examples 9 and 10 (in bold), respectively:

Example 9
The sodium-free ACSF will reduce depolarization and its associated cell damage during slice preparation. (Case 6-4, sentence 10.)

Example 10
The movement of the reference mirror towards the incident beam will image blood flow in one direction, away from the direction of the incident beam. (Case 6-7, sentence 2.)

In a narrative of a center or network grant application, you need to identify *who* information – the procedures that each site will be responsible for performing. In addition, in a narrative to a project that involves team members, you need *who* information to clarify the division of labor. The identification of *who* information for sites and for research team members is sometimes also included in an overview table. (See Cases 5.2 and 5.3.)

• **The identified agent who is the PI.** Two common writing questions about an agent who is the PI are: (1) whether to identify the PI-agent, given Reader Assumption 2 and, if so, how? and (2) whether to use *we*, *I*, or *the PI* in order to refer to the PI-agent.

The short answer to the first question is that even though readers assume that the PI is the agent in procedural sentences, you can still identify the PI as the agent (especially if the verb action is mental activity, such as the verb *hypothesize*). However, the pronoun *we* is usually used when referring to the PI, not the pronoun *I*. This usage could reflect the requirement that the supporting institution's sign off is necessary for the submission, which essentially makes not only the PI but also the supporting institution responsible for the execution of the proposed research. Thus, the plural *we* is accurate and is preferably used.

Also, if you identify the PI in a first-person pronoun, such as *we*, active phrasing is acceptable, but passive is not. Example 11a with the first-person pronoun *we* as the grammatical subject is more appropriate than Example 11b, where the first-person pronoun *us* is integrated into the passive sentence.

Example 11

a. To further localize the SarA corresponding to progressively smaller regions of the DNA upstream of cna, **we** will synthesize DNA fragments corresponding to progressively smaller regions of the DNA upstream of cna and use these fragments in EMSA experiments utilizing purified SarA. (Case 6-5, sentence 2.)

b. ‡To further localize the SarA corresponding to progressively smaller regions of the DNA upstream of cna, DNA fragments corresponding to progressively smaller regions of the DNA upstream of cna will be synthesized **by us**, and these fragments will be used in EMSA experiments utilizing purified SarA.

Guideline 6.8

(*a*) *Refer to the PI as **we** or **the PI** if the sentence is active.*

(*b*) *If referring to the PI as **we** in the narrative, do not use **by us** in a passive sentence to refer to the PI.*

(*c*) *Consider mentioning the PI as the agent to emphasize or to clarify that the PI will perform the action or task, rather than another member of the research team.*

(*d*) *Consider mentioning the PI as the agent for a grammatical reason, such as to establish grammatical parallelism, to avoid a dangling modifier, or to sequence content from given to new information.*

Chapter 8.5 further discusses the agent in relation to active and passive grammar of procedural sentences and also discusses agentless procedural sentences.

(2) **When** information. *When* information is sometimes called *chronological* or *time* information. *When* information includes chronological terms that indicate: (a) the sequence in which you intend to perform the verb actions and tasks, relative to other actions and tasks, (b) the frequency of the actions and tasks, and (c) their duration.

Examples of chronological terms that indicate sequence are: *after, at noon, before, first, next, prior to, second, subsequently, for 60 s, then,* and *while.* Chronological terms indicating frequency include *frequently, seldom, never, always, rarely, weekly,* and *twice.* Some terms can help you clarify the duration of procedures, such as *for as long as* and *for 2 minutes.* The following procedural sentence identifies **when** information in bold:

Example 12

Subjects from both groups will receive balance training ***once a day for 45 minutes per session, 5 days a week for 2 weeks, while*** they are inpatients. (Case 6-6, sentence 5.)

When using chronological terms, time units can be abbreviated to save space, but their longer versions are not first introduced in parentheses after the term since they are so common. The following are common abbreviations of time units in scientific English:

> s   *second*
> ms   *millisecond*
> h   *hour*

There are different conventions for abbreviating time units. The units listed above are from the *AMA Manual of Style* (11th Edition). Another alternative for *hour* is *hr*. When using abbreviations for time units, you should use abbreviations from one recognized authority in writing, such as the *AMA Manual of Style*, which has lists of standard abbreviations for chronological terms and chronological units. Note that you do not add + **s** for plural if the time unit is abbreviated, but you do add + **s** for plural when the time unit is not abbreviated, as illustrated by these examples:

Abbreviated singular time unit:	**Non-abbreviated plural time unit:**
3 min	*3 minutes*
6 ms	*6 milliseconds*
2 hr	*2 hours*

(3) ***Where* information**. Procedural sentences can also identify spatial information, termed ***where*** information. ***Where*** information can include: (a) the source of **SOS** and **MET**, (b) the spatial proximity of **SOS, MET**, and investigators while procedures are being performed, and (c) when there is more than one laboratory site, the location where particular procedures will be performed. ***Where*** information includes not only geographic locations but also anatomical sites, such as the anatomical sites identified in Example 13.

Example 13
[1]Following anti-coagulation with heparin, the aorta will be ligated *at the bifurcation of the iliac arteries*. [2]Graft implants will be anastamosed *to the aorta, proximal to the aortic ligation and to each common iliac artery just proximal to the takeoff of the internal iliac arteries*. (Case 6-3, sentences 5 and 6.)

(4) ***Why* information**. *Why* information explains the methodological purpose for a procedure – that is, what you intend to achieve by executing a procedure. Conditions in which you will likely need to explain the purpose of a proposed action are given in Guideline 6.9:

Guideline 6.9   *Provide **why** information:*
(a) *To clarify the purpose of particular procedures.*
(b) *To indicate that a particular action was not chosen arbitrarily but rather for a sound methodological reason.*
(c) *To clarify the benefits or advantages of a procedure.*

Methodological purpose is usually phrased in these grammatical structures:

> *(in order) to*
> *(in order) for*
> the verbs *allow* and *provide*
> *so that*

A common way to phrase the methodological purpose is to use *in order to* or its short form *to*, followed by an infinitive verb. When using *(in order)to*, you can include the phrase either before or after the procedural verb of the sentence. The sentences in Example 14 include expressions of methodological purpose phrased with to + infinitive verb.

Example 14
a. We will determine whether two, 3D volumetric images are needed **to image the direction of blood flow**. (Case 6-7, modified from sentence 6.)
b. **To image the direction of blood flow**, we will determine whether two, 3D volumetric images are needed.

For clarity, methodological purpose is expressed in the long form *in order to* when there is another infinitive verb (to + infinitive verb) in the sentence. In this case, the higher-level purpose is phrased with *in order to*, and the lower-level purpose is phrased with *to*. In Examples 15a and 15b, the long form *in order to correlate* indicates that the action verb *correlate* is the higher-level purpose of *to measure*. In other words, the purpose of measuring is *to correlate the movement of the gating charges*.

Example 15
a. We will use the cut-open technique **to measure** the gating currents in HCN channels **in order to correlate** the movement of the gating charges to the fluorescence signals.
b. **In order to correlate** the movement of the gating charges to the fluorescence signals, we will use the cut-open technique **to measure** the gating currents in HCN channels.

The long phrase *in order for* also means methodological purpose; however, its shorter form *for* has various meanings, only one of which is methodological purpose.

In addition to expressing methodological purpose, **why** information provides justification for procedures. Justification is usually explained as the benefit or advantage of a particular procedure. For example, sentence 3 in Example 16 addresses the benefit of *the interface method*.

Example 16
[1]All initial experiments will be done in an interface chamber with the top surface of the slice exposed to a moistened mixture of 95% O_2 and 5% CO_2, and with the bottom surface exposed to a steady flow of normal ACSF at a rate of 1–1.5 ml per minute. [2]Both the interface and submerged methods have been used in the past. [3]*The interface method has the advantages of better oxygenation of the tissue, smaller shock artifacts following electrical stimulation of fiber bundles, and greater recording stability*. (Case 6-4, sentences 14–16.)

(5) Procedural background information. Whether a narrative has a **Background Section** or not, background information can occur in procedural descriptions to help reviewers understand the research context for particular procedures. Types of background information typically found in procedural descriptions include:

- Preliminary and previous studies.

- Definitions of terms.

- Explanations of equations and assumptions.

- Assumptions underlying proposed procedures.

- Extent to which the procedures are standard, established, novel, or modifications of standard or established procedures.

- Characteristics of the procedures.

- Known consequences of procedures.

As seen in sentences 2 and 3 of Example 16 above, verbs in sentences that present background information are not phrased with *will*. In Example 17, the first sentence uses the procedural verb *will perform*. However, the verb in the second sentence is *are*, a verb form that signals that this second sentence is not procedural.

Example 17
[1]The control group **will perform** balance exercises with a physical therapist, and the experimental group **will perform** the same exercises with a physical therapist while using the ABF device. [2]*The balance exercises are a modification of Frenkel's exercises (1902), widely used in rehabilitation to improve balance in patients with somatosensory ataxia* (ref). (Case 6-6, sentences 2–3.)

Many, if not most, of the sentences in subsections that describe procedures in computer science, mathematical analysis, and modeling are not procedural but are background explanations. Background information in procedures of computer science, mathematical analysis, and modeling is discussed further in Chapter 6.6.

6.4   How many procedural details to include

It can be difficult determining how much detail to include beyond core procedural details, especially when your reviewers are likely already familiar with them and especially when you have length limits imposed on the narrative that preclude extensive discussion of the methods. Guideline 6.10 gives strategies for determining how much procedural detail to include:

Guideline 6.10

(a) *The less standard a procedure, the more you describe it in detail.*
(b) *The less established a procedure in your laboratory, the more you describe it in detail.*
(c) *If a procedure has been described in a previous subsection, cross-reference back to the previous description, by subsection number.*

● **Length of the narrative.** The longer the narrative, the more textual space is at your disposal in the **Methods Section** for details. In short narratives (5 or fewer pages) and even in some longer ones, you may have difficulty adequately describing your procedures and still not go beyond the maximum prescribed page length. In such a case, you need to include core *what*, *which*, and *how* procedural information and reduce non-core procedural information.

● **Standard procedures.** Just because you will be performing standard procedures, you cannot assume that you do not need to describe them since your technical reviewers will already know them. You need to describe standard procedures in enough detail for reviewers to evaluate your understanding of them, your credibility, and whether they are appropriate for your proposed research. Guideline 6.11 gives strategies for describing standard procedures.

> **Guideline 6.11** *For standard procedures:*
> (a) *Identify them by their standard name.*
> (b) *Include a citation to a seminal article that describes in detail the standard procedure that you will execute.*
> (c) *Briefly and specifically summarize the standard procedures if you have not already done so earlier in the narrative. If you have already done so, cross-reference to that earlier summary by its subsection number.*
> (d) *Consider describing the advantage or benefit of using the standard procedure.*
> *Exception: For minor standard procedures, consider only identifying them by their standard name and including a citation to an article that describes them.*

Example 18 shows a passage from a **Methods Section** that describes standard procedures. Notice that the term *standard* is used, a citation is provided in sentence 1 to support the conclusion that the procedures are standard, and a brief explanation is included. Also, as shown in sentence 2 of Example 18, the description of the standard procedure is prefaced with *briefly* for reviewers to understand that you are deliberately being brief.

Example 18
[1]EMSA experiments will be done using *standard* procedures (5). [2]*Briefly*, 32P-labeled DNA fragments will be mixed with purified SarA diluted to concentrations ranging from 1 to 100 nM (this concentration range was chosen based on preliminary experiments demonstrating a band shift with 55 nM SarA). [3]After an appropriate incubation period, SarA-DNA complexes will be resolved by native gel electrophoresis, as illustrated in Preliminary Results. (Case 6-5, sentences 5–7.)

● **Novel procedures.** You need to describe your novel procedures, even if you have already published them.

> **Guideline 6.12** *For novel procedures:*
> (a) *Consider assigning the novel procedures a name and consistently use it. If you have published these procedures, use the same name as in the publication.*

(b) *When first mentioning your novel procedures in the narrative, regardless of the section, use the term* **novel** *or* **unique** *to describe them.*

(c) *Whether or not your novel procedures have been published, detail and identify their purpose, and their benefit or advantage.*

(d) *If you have already described the novel procedures in an earlier subsection, cross-reference back to the earlier description by the subsection number.*

For example, Case 6-7 identifies the *digital frequency modulation approach* as *novel* (e.g., sentence 17).

● **Established laboratory procedures.** Established procedures in your laboratory might be standard, novel, or modifications of procedures in your discipline. Guideline 6.13 gives strategies for describing your established procedures:

Guideline 6.13 *For procedures that are established in your laboratory:*

(a) *Consider giving the established procedures a name and use this name consistently. If the novel procedures have been published, use the same name as in the publication.*

(b) *Mention that the procedures are established in your laboratory.*

(c) *Cite to one of your publications, if any, that has a full description of the established procedures.*

(d) *Whether or not your established procedures have been published, describe them, and identify their purpose and/or advantage over other procedures.*

(e) *If you have already described your established procedures in an earlier subsection, refer back to the earlier description by subsection number.*

Sentence 2 of Case 6-3 provides an illustration of a procedural section that is explicitly described as *well-established* in the PI's laboratory. Even though the procedures have already been described in the cited publication, they are described briefly in sentences 3–7. The advantages of using these established procedures are described in sentence 14.

● **Modifications of standard or established laboratory procedures.** Guideline 6.14 gives strategies for how to describe your proposed procedures that are essentially modifications of standard or established procedures in your laboratory:

Guideline 6.14 *For proposed procedures that are modifications of standard or established procedures in your laboratory:*

(a) *Identify the procedures by their standard or established name, and cite a source that describes in detail the standard or established procedures.*

(b) *Mention that your proposed procedures are a modification of a standard or an established procedure.*

(c) *Identify the purpose of the modification.*

(d) *Identify the benefit or advantage of the modification.*

(e) *Briefly detail the modification.*

(f) *If you have already described the modification in an earlier subsection of the narrative, cross-reference to the earlier description by subsection number.*

Sentences 1–5 in Case 6-6 illustrate a description of a data-collection procedure that is a modification of a standard procedure, as noted in sentence 3, where it is characterized as a modification of *Frenkel's exercises*. The nature of the modification is explained in sentence 4, which is the addition of *standing and walking exercises, with and without vision, under challenging surface conditions*.

6.5 Reducing subjectivity in descriptions of procedures

The more you reduce subjectivity in procedural descriptions, the clearer will be your procedures and the more you will enhance your credibility as an investigator. Guideline 6.15 identifies different ways to reduce subjectivity:

> **Guideline 6.15** *To reduce the subjectivity of your methodological descriptions:*
> (a) *Objectify subjective terms by quantifying or defining them.*
> (b) *Provide criteria to back up your decisions about your research design.*
> (c) *Substantiate your analytical statements (e.g., problem statements, hypotheses, speculations, and conclusions; see Chapter 2.2.3).*
> (d) *Provide background information to explain why you have chosen particular methods to achieve your research objective and aims.*

● **Objectifying subjective terms.** Many terms in descriptions of procedures are subjective in the sense that they can mean different things to different people or their meanings change depending on the context. Such subjective terms include *a few, some, many, healthy, ill, normal, abnormal, small,* and *large*. To help reviewers interpret your procedures as you intend them, you need to avoid subjective terms or at least quantify them or define them. Example 19 includes the subjective terms *small* and *large*, both of which are quantified.

Example 19
The size of the acoustic inventories required for this approach will range from **small** (< 2000 units, such as used in diphone synthesis) to **large** (> 50,000 units, such as used in unit selection synthesis).

● **Providing criteria for procedures.** Another way to achieve a less subjective procedural description and to indicate that your procedures are not arbitrary is to provide criteria for methodological decisions. Example 20 provides criteria for determining which DNA fragments will be used first, and Example 21, criteria for determining the number of subjects to study.

Example 20
The experiments described below will be done starting with the DNA fragments *that exhibit the highest affinity for SarA (i.e., that exhibit a gel shift with the lowest concentration of SarA and are not competitively inhibited in the presence of other, unlabeled DNA fragments)*. (Case 6-5, sentence 11.)

Example 21
Based on previous experience and power analysis (*C.2*), we will aim for a minimum of 12 specimens that we will be able to evaluate from each dose group.

Example 20 gives 2 criteria in parentheses, for determining *the highest affinity for SarA*, and Example 21 leads with 2 criteria that the PI will use to determine the number of specimens to study.

• **Substantiating conclusions about your procedures.** Reviewers will not necessarily agree with your conclusions about your proposed procedures just because you say so. You need to explain your procedures with details aimed at substantiating any conclusions that you include. Example 22 gives the conclusion that the preliminary research *will greatly facilitate* the proposed research. Details about this preliminary work, shown in sentence 2, help support the conclusion. If you are not prepared to substantiate a conclusion, you should omit it.

Example 22
[1]Appropriate fragments will be generated either by synthesizing and annealing complementary pairs of oligonucleotides or by PCR. [2]The fact that we have already sequenced the region extending 930 pb upstream of the cna transcriptional start site (19) *will greatly facilitate* the synthesis of the appropriate DNA targets. (Case 6-5, sentences 3 and 4.)

6.6 Procedural descriptions in computer science, mathematical analysis, and modeling

A few words are needed about procedural descriptions in computer science, mathematical analysis, and modeling, given their exceptional character. In these fields, it is common to find procedural passages with very few procedural sentences but with extensive background information to explain equations and assumptions. This background explanation does not replace procedural sentences; the background supplements the procedures. You still need to include at least one procedural sentence with a procedural verb – one that uses **will** – for reviewers to understand how the background information relates to your proposed research.

> Guideline 6.16   *When describing proposed procedures to achieve research aims in computer science, mathematical analysis, or modeling, include at least one procedural sentence that identifies how the background information relates to the proposed procedures.*

An example illustrating this guideline is in Case 6-7. This procedural subsection consists of 18 sentences, 14 of which are background and only 3 of which are procedural (the fourth non-background sentence [sentence 9] is imperative [assume]). The 3 procedural sentences are reproduced below as Examples 23, 24, and 25. The procedural description in Case 6-7 would be incomplete with only background information. Each of these 3 procedural sentences is distinguished from the sentences that give background information through their procedural verbs that use *will*.

Example 23
In our proposed research, *we will determine* whether two, 3D volumetric images are needed to image the direction of blood flow. (Case 6-7, sentence 6.)

Example 24
We *will incorporate* this novel digital frequency modulation approach into our optimal OMAG system. (Case 6-7, sentence 17.)

Example 25
As a result, we *will use* the following approach: the analysis in Section C.1.2 *will give* one image of blood flow direction when the mirror moves towards the incident beam, and the analysis in this section *will give* an image of blood flow in the opposite direction, as if the mirror is moving away from the incident beam. (Case 6-7, sentence 18.)

To make procedural sentences in computer science, mathematical analysis, and modeling narratives easier for reviewers to notice, you need to locate the procedural sentences in focus positions of paragraphs (Chapter 1.8.1).

Guideline 6.17   *In a **Methods** subsection to a computer science, mathematical analysis, or modeling research project, or in a **Methods** subsection that includes mathematical analysis, locate procedural sentences in a focus position of a paragraph – either at the beginning or at the end of a paragraph.*

This guideline is also illustrated in Case 6-7, where the initial sentence in the second paragraph is procedural, with its verb *will determine* (Example 23 above), and the procedural sentences in the last paragraph (Examples 24 and 25 above) constitute the entire short, last paragraph of the subsection.

6.7 Data-sharing and data-analysis subsections and procedures

Most of the discussion about procedural sentences thus far has presented features of procedures that are common to data-collecting, data-sharing, and data-analysis procedures. However, there are also features specific to data-sharing and data-analysis procedures.

(1) **Data-sharing procedures**. Data-sharing procedures are those actions that you execute in order to exchange your data with other members of your research team. Data-sharing procedures are necessary, for example, in center or network grants, where members of the research team are at different sites. Types of information typically found in data-sharing procedural descriptions are:

- Means of communication, such as email, meetings, websites, and videoconferences.

- Procedures for creating a research atmosphere that encourages data- and information-sharing.

- Ongoing collaborations across the different sites.

Case 6-8A presents a subsection with data-sharing procedures from the narrative of a center grant application that is proposing a research team of over 30 members. Such a large research team warrants the inclusion of a subsection for data-sharing plans. However, in most instances, data-sharing will be among collaborators within the same institution or across only a few institutions. Sometimes collaborators might be in different countries. When you do not have an unusually large number of research team members that might require extensive discussion of data-sharing plans, the explanation of data-sharing can be integrated into a data-collection subsection. This situation is shown in Case 6-8B where sentences 5 and 6 address data-sharing.

When data-sharing procedures are included, you need to make sure that the budget includes expenses related to data sharing, such as travel expenses for collaborators at different sites to visit each other for professional purposes.

(2) **Data-analysis subsection and procedures**. Data-analysis procedures involve the actions and tasks that will be carried out to identify, derive, interpret, and compare and contrast patterns in the collected data. Data-analysis subsections are variously entitled **Analysis, Data Analysis, Statistical Analysis**, or **Statistics**. Typical content of a data-analysis subsection can include:

- Type or form of the data to be analyzed.

- Type of data-analysis procedure.

- Description of the data-analysis procedure.

- Rationale or purpose of the data-analysis procedure.

- Assumptions relevant to the data-analysis procedure.

- Relationship of the analytical results to the proposed hypothesis.

- Independent and dependent variables.

- Number of **SOS** needed for analyses.

- Statistical significance.

- Comparisons of collected data.

Data-analysis procedures can be included in dedicated data-collection subsections, located at the end of each **Methods** subsection that gives the proposed methods to an aim. Tables 6-1 and 6-2 show different locations where you can include data-analysis subsections. In addition, the 4 analysis sub-subsections in C.1, C.2, C.3, and C.4 of Case 6-2 are also data-analysis subsections. However, data-analysis procedures can be located in data-collection subsections, especially when the data-analysis procedures are short, such as just one or 2 sentences.

6.8   Expected outcomes and interpretation of data

Expected outcomes are the major results that you anticipate as a consequence of executing methods for your proposed aims. Your interpretation of data indicates what your data could mean. Content comprising interpretation of data can include:

- Speculations, conclusions, predictions, and hypotheses about what potential findings and analytical results might mean.

- Relationship between expected outcomes and the proposed hypothesis.

- Relationship of expected outcomes to each other.

- Comparisons of expected outcomes to previous outcomes from other research.

- Background information to clarify interpretations.

- Preliminary studies to support speculations.

Expected outcomes and interpretation of data can be in a dedicated subsection with a heading, such as **Expected Outcomes**, **Interpretation of Data**, **Outcomes and Interpretation of Data**, or **Interpretation of Experimental Results**. Expected outcomes and interpretation of data can also be located within data-collection subsections, especially when the interpretation relates to an individual procedural action or task.

   Procedures that will be executed in order to achieve your anticipated results or to address anticipated problems are *ancillary methods*. As illustrated in sentence 10 of Case 6-5, ancillary methods are often explained alongside expected outcomes and interpretation of data.

   When the discussion of interpretation is extensive and relates to results that you expect from an entire group of proposed procedures, you can include an interpretation subsection. For example, in Case 6-2, each procedural body subsection (**C.1**, **C.2**, **C.3**, and **C.4**) includes an interpretation subsection after the data-analysis procedures.

6.9   When proposed procedures are preliminary procedures

Your proposed procedures might be those that you have already used in preliminary studies. In this situation, you need to decide where to locate the procedural descriptions: in the **Methods Section** or in the **Preliminary Studies Section**.

Procedures
Preliminary Studies ◄───────────► Methods

The recommendation here is to locate the core procedures in the **Methods Section** to achieve a **Methods Section** that is self-contained. A self-contained **Methods Section** will help reviewers focus on the procedures.

> **Guideline 6.18**   *Write **Methods** subsections that are as self-contained in procedural content as possible.*

What is a self-contained **Methods** subsection? It is one that has enough details about your proposed methods that reviewers do not need to flip to another section in the narrative, such as the **Preliminary Studies Section**, for a description of methods. If you draft **Methods** subsections with only cross-references to procedures in other sections, you are inviting reviewers to lose focus on your research design, which may interfere with their understanding of your proposed research. In addition, even though you may cross-reference to procedural descriptions in other sections of the narrative, reviewers might not seek out and read those other descriptions.

Thus, when describing your proposed procedures, containment of procedures in the **Methods Section** is a goal.

6.10   Potential problems and proposed solutions

For reviewers to assess your research skills, your problem-solving skills in scientific methodology, and your research design, they need to learn about problems that might occur while you perform your proposed procedures and what you intend to do if the problems occur.

You may find it difficult writing about potential problems since you might be revealing weaknesses in your research design. However, it is far better to be proactive than reactive. Your proposed solutions to potential problems – including alternative strategies, workarounds, and fall-back methods – are critical when trying to convince reviewers that: (a) your proposed solutions will keep your research moving forward, (b) your proposed solutions will allow you to achieve your proposed research objective and aims, and (c) your research results will be significant, novel, reliable, and valid, even if the potential problems occur.

6.10.1   Content for a potential problems/solutions subsection

A wide variety of problem-related content can be described in a potential problems/ solutions subsection:

- A potential problem.

- A proposed solution or work-around to the problem.

- (Un)Likelihood of the problem occurring.

- Support for the (un)likelihood of the problem occurring.

- Source of the problem.

- Consequences of the problem.

- Importance or minimization of the problem.

- Support for the minimization.

- Likelihood that the proposed solution(s) will solve the problem or will help you achieve your stated research purpose, aims, and anticipated outcomes.

- Benefits and drawbacks of the solution.

- Alternative solutions.

- Effects of the problem on data collection or data interpretation.

6.10.2   Locations for potential problems/solutions information and subsections

There are 3 locations where you can include potential problems and your proposed solutions in a narrative.

(1) **Within descriptions of data-collection procedures**. If you discuss problems and solutions relatively briefly, you can integrate the information within data-collection subsections. For example, Case 6-5 presents a data-collection subsection with problem-solving information integrated in sentences 9 and 10.

(2) **In dedicated potential problems/solutions subsections**. You can create one or multiple dedicated potential problems/solutions subsections, in which case you include one of these subsections at the end of each **Methods** procedural body subsection. These subsections discuss the potential problems/solutions that may occur in the procedures for each individual aim. Case 6-2 shows 4 subsections entitled **Potential complications** – each positioned at the end of each **Methods** body subsection.

If you anticipate problems in the methods for one or 2, but not all, of your proposed aims, you can include just one potential problem/solution subsection in the relevant **Methods** body subsections. Thus, for example, in Case 6-2, if you anticipate no complications for Experiment 3, you can omit the subsection entitled **Potential complications for Exp. 3**, even though the other subsections have a corresponding subsection.

Yet another alternative is to include a potential problem/solution subsection and to explain briefly that no problem is anticipated and why. The purpose of identifying your expectation for no problems is for reviewers to understand that you have not inadvertently failed to discuss potential problems and solutions for a particular aim (however, see next item).

If you create only one potential problem/solution subsection, you can locate it at the end of the **Methods** body subsection (see Item #16 in Figure 6-1). This sole subsection is useful when you anticipate similar potential problems and solutions across procedures for the aims. However, a problem with creating this subsection is that your **Methods Section** ends on a negative note, even if you offer solutions to the problem. To avoid a negative ending, you need to add an ending subsection to the narrative (Chapter 6.12).[1]

(3) **With outcomes information.** Another location for information about potential problems and proposed solutions is within a subsection that gives outcomes information. This type of subsection is shown in Case 6-10.

In Case 6-10, sentence 1 identifies the research purpose of the methods (*characterization*) and the ultimate, proposed outcome *of this first section*, and sentences 2 and 3 justify this part of the study by explaining that variants in the capsid structures are already known to exist. Sentence 4 further explains methods. The problem discussion begins at sentence 5, which is essentially a topic sentence introducing the problem-solution discussion that runs from sentences 6 through 13. The actual problem is identified in sentence 6: the inability to predict some subtle, functional effects of structural changes in the capsids. Background information on functional effects is given in sentence 7, and sentences 8–13 discuss solutions.

What is particularly noteworthy in Case 6-10 is the focus on capsids early in the paragraph and the focus on capsids at the end of the paragraph, implying that the problem identified in sentence 6 will not hinder the PI in achieving the stated research purpose and anticipated outcomes, given the possible solutions explained in sentences 8, 9, 11, and 12. Thus, as illustrated by the progression of information in Case 6-10, potential problem/solution information that is integrated into an outcomes paragraph can be useful to show that the problem will not prevent you from achieving your proposed research objective and aims, given your proposed methodological solutions.

Guideline 6.19

(a) *Explain potential problems and proposed solutions by: (i) integrating them into a procedural subsection, (ii) locating them in a dedicated subsection, or (iii) locating them with outcomes information.*

(b) *Add an ending section after the **Methods Section** to avoid ending the narrative with a discussion of problems.*

(4) **No methodological problems.** If you do not expect any problems with your procedures, you say so and then justify *why* this is the case. This justification is presented in the **Methods** introduction or in a potential problems/proposed solutions subsection at the end of **Methods** body subsection. Reasons that might help you justify why you do not expect methodological problems include:

---

[1] Even if you do not have a subsection dedicated to potential problems and proposed solutions at the end of the **Methods** body, we still suggest creating an ending subsection.

- The PI's extensive, successful experience with the procedures – an explanation that can also enhance the PI's credibility.

- Preliminary studies that do not suggest problems.

- Established routines.

- The transparently simple nature of the procedures.

Examples 26 and 27 show how you can explain that no problems are expected due to the extensive experience of the investigators.

Example 26
The procedures are well-established in our laboratory, and we do not anticipate any problems.

Example 27
Co-investigator Smith has extensive experience in transforming the spectral characteristics of a source speaker to those of a target speaker, and based on his previous work, we do not expect any complications.

6.11   Shared methods

The organization of the **Methods Section** is dependent not only on the funding agency's submission requirements, but also on the number and types of methods that your proposed aims share. Methods that are common across your aims are called *shared* or *common methods*. Case 6-2 shows a heading outline to a narrative with an initial shared methods subsection (see the heading **Common Methods** at C.0), and Case 6-11 presents an initial shared methods section, termed **Shared Methods**.

Guideline 6.20   *Determine whether to include a shared methods subsection based on the extent that the methods across your proposed aims are the same or are closely similar.*

You need to describe the shared methods in essentially the same phrasing and grammar, and you need to cross-reference to sections that describe them.

Shared methods can be described in one of 3 locations, 2 of which are shown in Figure 6-1.

Guideline 6.21   *Locate shared methods:*
*(a) In a subsection before the first **Methods** body subsection to an aim.*
*(b) In a subsection after the last **Methods** body subsection to an aim, or*
*(c) In the first **Methods** body subsection to an aim.*

Regardless of the organizational alternative that you select for the **Shared Methods Section**, in the introduction to the **Methods Section** you need to mention that there are shared methods and where these shared methods are discussed.

Guideline 6.22   *In the introduction to the **Methods Section**, identify the subsection where you describe the shared methods.*

This forecast of shared methods can help reviewers better understand the organization of the **Methods Section** and your research design. The forecasts to shared methods subsections in Examples 28 and 29 alert reviewers to shared methods and their locations.

Example 28
[1]We hypothesize that apoER2′and GPIb mediate PC-platelet binding and activation, and the conversion of PC to APC on the platelet surface. [2]*In D.1, we present methods that are common across Aims 1 to 3.* [3]We then describe the methods we will use to determine the mechanisms by which apoER2′ (Aim 1; Section D.2) and GPIb (Aim 2; Section D.3) mediate PC-platelet binding and activation. [4]Section D.4 details the proposed methods to delineate the ability of platelet receptors to recruit and catalyze the conversion of PC to APC (Aim 3).

Example 29
Methods for Aims 1–3 are described in Sections D.2, D.3, and D.4, respectively; *methods common to the experiments for Aims 1–3 are described in D.5.* (From Case 5-10, sentence 4.)

An organizational alternative to creating a shared methods subsection is to explain the shared method in the *first* relevant procedural body subsection of a proposed aim. Then in the later, relevant procedural body subsection that uses the shared method, you cross-reference back to its first description. Example 30 illustrates an entire **Methods** procedural body subsection to an aim; it consists of just one sentence that cross-references back to an earlier, relevant subsection that describes the shared method. Example 31 presents a sentence within a procedural subsection that refers back to another subsection that describes the relevant shared procedures.

Example 30
Data analyses and interpretation. These methods are the same as those described above for Aim 2 (see Section *D.2.4*).

Example 31
To achieve Aim #2, the subjects will perform 5 sets of protocols: the 2 protocols as described in Shared Methods (See C.2.3) and 3 additional protocols as described in this section.

Guideline 6.23 *Whenever you include a shared methods subsection, cross-reference to it at the relevant points in the procedural description for each aim.*

6.12 Ending the Methods Section and the narrative

The ending subsection of the **Methods Section** is equivalent to the **Ending Section** of the narrative. Funding agencies often consider the ending optional, and few funding agencies even mention it in their submission requirements. When they do mention the ending, it is usually identified as the location for your anticipated time schedule for completing the methods.

You also might consider the narrative ending optional and dispensable if you are struggling not to exceed prescribed page-length limits for the narrative. However, the ending should not be considered optional.

Guideline 6.24 *Include an ending section to explain your next steps in the line of research after your anticipated successful completion of your proposed research.*

The ending, comprising the last sentences of your narrative, is your last chance to mention any information that you want your reviewers to notice and to remember. Typical important information you can include in the ending is:

● Significance of your proposed research.

● Novelty of your proposed research.

● Expected major outcome(s) from your proposed research.

● The research team's qualifications to successfully execute your proposed research.

● Next steps in the line of research, give the successful achievement of your proposed research objective and aims.

● Any other information that you want your reviewers to notice.

You should carefully craft and polish the ending in order to leave a good impression with the reviewers. Also, the ending needs to be short in order to move reviewers quickly to their next review tasks. As a result, a **Methods** ending seldom includes all of the information listed above. However, the ending should address your next research project, assuming the successful completion of the proposed research in order to help reviewers understand your line of research and the important role that your proposed research has in propelling you forward along this line.

The ending is given a heading, such as **Future Plans**, **Future Directions**, or **Summary**. Case 6-12 gives 2 different **Ending Sections**. Although the first example is very short, it still includes enough information for reviewers to understand the direction that this line of research will take and the role of the proposed research in this line of research.

Notice that the above list of content for the **Ending** does not include the time schedule for the proposed research. Of course, if your funding agency requires the time schedule in the ending, you must deliver on that submission requirement. However, in the absence of a submission requirement, we suggest including the time schedule in the introduction to the **Methods Section**, in an overview table (Chapter 5.2.2) so that reviewers can evaluate the schedule while reading the proposed methods. However, even if the time schedule is included in the ending and not in the introduction to the **Methods Section**, an overview table is still useful in the **Methods** introduction to help reviewers understand your proposed methods.

The ending section in Case 6-13A presents a paragraph that discusses the timeline, which we have converted into a table in Case 6-13B. Alternatively and preferentially, this table can be used in the **Methods** introduction.

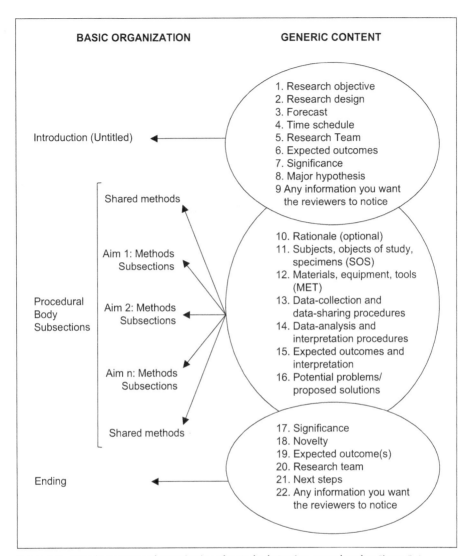

BASIC ORGANIZATION

GENERIC CONTENT

Introduction (Untitled)

1. Research objective
2. Research design
3. Forecast
4. Time schedule
5. Research Team
6. Expected outcomes
7. Significance
8. Major hypothesis
9 Any information you want
   the reviewers to notice

Shared methods

Aim 1: Methods
       Subsections

Procedural
Body
Subsections

Aim 2: Methods
       Subsections

Aim n: Methods
       Subsections

Shared methods

10. Rationale (optional)
11. Subjects, objects of study,
    specimens (SOS)
12. Materials, equipment, tools
    (MET)
13. Data-collection and
    data-sharing procedures
14. Data-analysis and
    interpretation procedures
15. Expected outcomes and
    interpretation
16. Potential problems/
    proposed solutions

Ending

17. Significance
18. Novelty
19. Expected outcome(s)
20. Research team
21. Next steps
22. Any information you want
    the reviewers to notice

Figure 6-1. Generic content and organization of a Methods Section, reproduced as Figure 5-1.
Left-hand side: Generic subsections of a **Methods Section**. Right-hand side: Generic content.

Table 6-1. Organizational alternatives for data-collection and data-analysis subsections.

A. Data-collection and data-analysis procedures are different for each experiment.	B. Data-collection procedures are different for each experiment, but data-analysis procedures are the same across experiments for an aim.	C. Data-collection procedures are different for each experiment, but the same data-analysis procedures are shared by all experiments/aims.
C. Methods Section Untitled Introduction C.1 Rationale C.2 Methods for Aim 1 • Exp. 1: Data-collection procedures • Exp. 2: Data-collection procedures • Exp. 1: Data-analysis procedures • Exp. 2: Data-analysis procedures C.3 Methods for Aim 2 • Exp. 3: Data-collection procedures • Exp. 4: Data-collection procedures • Exp. 5: Data-collection procedures • Exp. 3: Data-analysis procedures • Exp. 4: Data-analysis procedures • Exp. 5: Data-analysis procedures C.4 Methods for Aim 3 • Exp. 6: Data-collection procedures • Exp. 6: Data-analysis procedures	C. Methods Section Untitled Introduction C.1 Rationale C.2 Methods for Aim 1 • Exp. 1: Data-collection procedures • Exp. 2: Data-collection procedures • Exps. 1&2: Data-analysis procedures C.3 Methods for Aim 2 • Exp. 3: Data-collection procedures • Exp. 4: Data-collection procedures • Exp. 5: Data-collection procedures • Exps. 3–5: Data-analysis procedures C.4 Methods for Aim 3 • Exp. 6: Data-collection procedures • Exp. 6: Data-analysis procedures	C. Methods Section Untitled Introduction C.1 Rationale C.2 Methods for Aim 1 • Exp. 1: Data-collection procedures • Exp. 2: Data-collection procedures C.3 Methods for Aim 2 • Exp. 3: Data-collection procedures • Exp. 4: Data-collection procedures • Exp. 5: Data-collection procedures C.4 Methods for Aim 3 • Exp. 6: Data-collection procedures C.5 Data-analysis procedures for all experiments

Table 6-2. Organizational alternatives for data-collection and data-analysis subsections for aims, distinguished by different data-analysis procedures.

A. Organizational Alternative for Shared and Different Data-collection Procedures to a Generic Study.	B. Organizational Alternative for Shared and Different Data-collection Procedures to a Clinical Trial.
C. Methods Section Untitled Introduction C.1 Rationale C.2 Shared methods • Data-collection procedures #1 • Data-collection procedures #2 C.3 Methods for Aim 1 • Data-analysis procedures C.4 Methods for Aim 2 • Data-collection procedures #1 • Data-collection procedures #2 C.5 Methods for Aim 3 • Data-analysis procedures	C. Methods Section Untitled Introduction C.1 Rationale C.2 Shared methods • Subjects • Equipment • Data-collection procedures C.3 Methods for Aim 1 • Data-analysis procedures C.4 Methods for Aim 2 • Data-analysis procedures #1 • Data-analysis procedures #2 C.5 Methods for Aim 3 • Data-analysis procedures

Case 6-1. Heading outline of a Methods Section reproduced from Case 5-9.

4. PLAN OF RESEARCH

Introduction (untitled)

4.1 Project Overview

} Introduction

4.2 Preparing the Physical Model System (Objective 1)
  4.2.1 Design the 2D physical model (Task 1.1)
    • Selection of materials
    • Preliminary reactive-transport modeling
  4.2.2 Construct physical model system (Task 1.2)
  4.2.3 Prepare analytical methods for Objectives 2 and 3 (Task 1.3)

} Body Subsection

4.3 Studying Horizontal Effects (Objective 2)
  4.3.1 Establish and characterize horizontal steady-state (Task 2.1)
  4.3.2 Apply sequential treatments (Task 2.2)
  4.3.3 Model horizontal effects (Task 2.3)

} Body Subsection

4.4 Studying Vertical Effects (Objective 3)
  4.4.1 Establish and characterize vertical steady-state (Task 3.1)
  4.4.2 Apply Sequential Treatments (Task 3.2)
  4.4.3 Model vertical effects (Task 3.3)

} Body Subsection

4.5 Characterizing Microbiological Processes (Objective 4)
  4.5.1 Map microbial community structure (Task 4.1)
  4.5.2 Determine role of dissimilatory iron reducing bacteria (Task 4.2)

} Body Subsection

Case 6-2. Headings from a Methods Section.

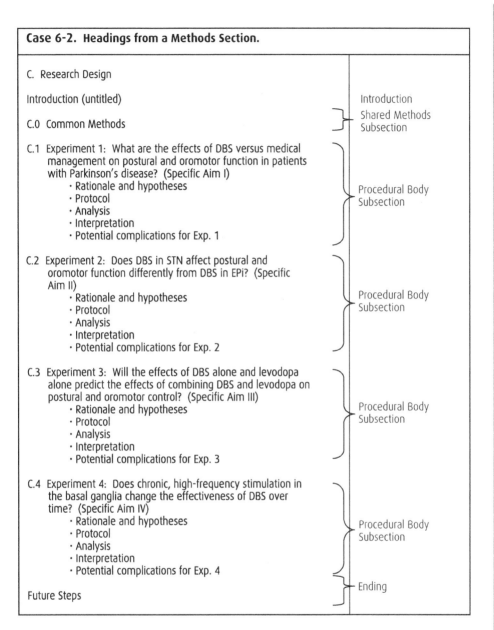

C. Research Design

Introduction (untitled)   Introduction

C.0 Common Methods   Shared Methods
                                                           Subsection

C.1 Experiment 1: What are the effects of DBS versus medical
     management on postural and oromotor function in patients
     with Parkinson's disease? (Specific Aim I)
        · Rationale and hypotheses   Procedural Body
        · Protocol   Subsection
        · Analysis
        · Interpretation
        · Potential complications for Exp. 1

C.2 Experiment 2: Does DBS in STN affect postural and
     oromotor function differently from DBS in EPi? (Specific
     Aim II)
        · Rationale and hypotheses   Procedural Body
        · Protocol   Subsection
        · Analysis
        · Interpretation
        · Potential complications for Exp. 2

C.3 Experiment 3: Will the effects of DBS alone and levodopa
     alone predict the effects of combining DBS and levodopa on
     postural and oromotor control? (Specific Aim III)
        · Rationale and hypotheses   Procedural Body
        · Protocol   Subsection
        · Analysis
        · Interpretation
        · Potential complications for Exp. 3

C.4 Experiment 4: Does chronic, high-frequency stimulation in
     the basal ganglia change the effectiveness of DBS over
     time? (Specific Aim IV)
        · Rationale and hypotheses   Procedural Body
        · Protocol   Subsection
        · Analysis
        · Interpretation
        · Potential complications for Exp. 4
                                                           Ending
Future Steps

Case 6-3. A data-collection subsection. Chronological transitions in *bold italics*.

Aorto-iliac graft procedure. [1]This procedure will be performed by Dr. John Smith, a qualified cardiothoracic surgeon at our university. [2]This procedure is well-established in our lab (8, 161-166) and is described in Section F. [3]In brief, ePTFE (5-7 cm long, 4 mm i.d.) will be implanted bilaterally as aorto-iliac grafts (Fig. 5). [4]*After a midline incision*, the aorta and iliac arteries will be dissected free from surrounding tissues. [5]*Following anti-coagulation with heparin*, the aorta will be ligated at the bifurcation of the iliac arteries. [6]Graft implants will be anastamosed to the aorta, proximal to the aortic ligation and to each common iliac artery just proximal to the takeoff of the internal iliac arteries. [7]End-to-side anastomoses will be constructed using proline sutures. [8]*Following graft placement*, flow will be re-established, and hemostasis will be verified at the graft anastomoses. [9]*At the time of graft placement and at graft harvesting*, blood flow rates through the grafts will be measured with a flow meter (Transonics, Inc., Model T206). [10]Flow rates have averaged 150-250 ml/min, depending on animal size. [11]*At the time of sacrifice*, the grafts will be regionally pressure-perfused with 10% formalin under arterial pressure (100 mm Hg) and will be maintained by restricting flow through the perfused arteries distally. [12]The aorta and graft segments will be removed en bloc, dissected free of surrounding tissue, blocked, sectioned, and stained. [13]Overall graft patency has averaged about 85% (163, 166). [14]This method has proven safe, with advantages that: 1) results are reproducible, so that the number of animals required is minimized, 2) the quality of information obtained per animal is high, and 3) paired comparisons are possible between contralateral grafts and between proximal and distal anastomoses, thereby improving statistical outcomes.

Case 6-4. Sequential data-collection procedural subsection.

Preparing slices

[1]Fish will be deeply anesthetized in cold anesthetic (MS-222, 1:10,000), the brain will be rapidly exposed, and the cranial nerves of the medulla will be cut. [2]Transverse cuts will be made through the brain just rostral and just caudal to the central lobes, and the block of tissue will be placed in cold artificial cerebral spinal fluid (ACSF). [3]For transverse slices, the block of tissue will be glued onto a metal plate with rostral face down. [4]Transverse slices will always include portions of both the central lobes and the valvula. [5]For sagittal slices, one side of the block will be trimmed under cold ACSF, and the trimmed surface will be glued to the metal plate. [6]The metal plate will then be rapidly mounted in the cutting chamber of a Leica vibratome, which will be filled with cold sodium-free ACSF. [7]The composition of the normal ACSF (in mM) will be as follows: NaCl-124; KCl-2.0; KH_2PO_4-1.25; $CaCl_2$-2.6; $MgSO_4$-7; H_2O-1.6; $NaHCO_3$-24; Glucose-20. [8]The sodium-free ACSF will be the same, but the NaCl will be replaced with 213 mM of sucrose. [9]The ACSF will be saturated at all stages with 95% O_2, 5% CO_2. [10]The sodium-free ACSF reduces depolarization and its associated cell damage during slice preparation.

[11]Three sagittal or five transverse slices of 350 µm thickness will be cut from each brain. [12]One or two slices will be transferred to the recording chamber immediately after slicing. [13]The rest of the slices will be stored in a mixture of sodium-free ACSF and normal ACSF (1:1) at room temperature for 40 min, and then in normal ACSF until needed.

[14]All initial experiments will be done in an interface chamber with the top surface of the slice exposed to a moistened mixture of 95% O_2 and 5% CO_2, and with the bottom surface exposed to a steady flow of normal ACSF at a rate of 1-1.5 ml per minute. [15]Both the interface and submerged methods have been used in the past. [16]The interface method has the advantages of better oxygenation of the tissue, smaller shock artifacts following electrical stimulation of fiber bundles, and greater recording stability. [17]Application of drugs into the bath solution is usually satisfactory with the interface chamber. [18]The submerged chamber will be used for some pharmacological studies, however, because drugs and ions can be washed in and out more quickly with the submerged chamber. [19]All experiments will be done at room temperature (the normal temperature for the fish).

Case 6-5. Data-collection subsection. Sentence 1 presents background information for the procedures in Sentences 2-3 and 5-7. Sentences 4 and 8-9 present expected outcomes and interpretation information. Sentences 10-11 present ancillary methods.

B. Characterization of the SarA DNA-binding site upstream of cna

Method #1: **Localization of the SarA-binding site upstream of cna.** [1]EMSA experiments employing a series of short, overlapping fragments derived from the region upstream of cna and purified Sar indicate that at least one SarA-binding site exists within approximately 200 bp upstream of the cna start codon (see Preliminary Results, Fig. 12). [2]To further localize the SarA-binding site(s) upstream of cna, we will synthesize DNA fragments corresponding to progressively smaller regions of the DNA upstream of cna and use these fragments in EMSA experiments utilizing purified SarA. [3]Appropriate fragments will be generated either by synthesizing and annealing complementary pairs of oligonucleotides or by PCR. [4]The fact that we have already sequenced the region extending 930 pb upstream of the cna transcriptional start site (19) will greatly facilitate the synthesis of the appropriate DNA targets. [5]EMSA experiments will be done using standard procedures (5). [6]Briefly, 32P-labeled DNA fragments will be mixed with purified SarA diluted to concentrations ranging from1 to 100 nM (this concentration range was chosen based on preliminary experiments demonstrating a band shift with 55 nM SarA). [7]After an appropriate incubation period, SarA-DNA complexes will be resolved by native gel electrophoresis as illustrated in Preliminary Results. [8]It is anticipated that these experiments will allow us to localize the SarA-binding site to a region spanning 50–100 bp. [9]However, we recognize that SarA-binding sites may exist across a relatively long stretch of DNA with intervening and perhaps irrelevant nucleotides in between each site. [10]To address that issue, all fragments that are bound by SarA will be used in "mix and match" competition experiments aimed at defining the relative affinity of SarA for different binding sites. [11]The experiments described below will be done starting with the DNA fragments that exhibit the highest affinity for SarA (i.e., that exhibit a gel shift with the lowest concentration of SarA and are not competitively inhibited in the presence of other, unlabeled DNA fragments).

Case 6-6. Data-collection subsection.

Balance Rehabilitation Protocols. [1]Subjects will be randomly assigned to 2 groups. [2]The control group will perform standard balance exercises with a physical therapist, and the experimental group will perform the same exercises with a physical therapist while using the ABF device. [3]The balance exercises are a modification of Frenkel's exercises (1902), widely used in rehabilitation to improve balance in patients with somatosensory ataxia (ref). [4]For each subject group, we will add standing and walking exercises, with and without vision, under challenging surface conditions. [5]Subjects from both groups (with and without ABF) will receive balance training once a day for 45 minutes per session, 5 days a week for 2 weeks, while they are inpatients in the University's Rehabilitation Hospital. [6]During each daily session, the subjects will perform the following tasks first with eyes open and then with eyes closed:

1. Standing and walking in tandem (heel-to-toe) or semi-tandem stance.
2. Standing and walking slowly on uneven support surfaces (inclines and support surfaces with irregular foam obstacles placed under a soft mat).
3. Walking sideways (crossing over feet ['braiding']) and backwards.

[7]Three physical therapists will be trained by Dr. Smith to provide these rehabilitation protocols and to teach subjects how to use the ABF device. [8]The therapists will be assigned an equal number of neuropathy subjects in the both of the ABF and control groups.

Case 6-7. Data-collection subsection from a modeling grant proposal.

D.1.2.4 OMAG imaging of directional flow using a novel digital frequency modulation technique. [1]In preliminary studies (Section C.3), we showed that the imaging of directional flow is feasible in OMAG by moving the mirror back and forth. [2]The movement of the reference mirror towards the incident beam images blood flow in one direction, away from the direction of the incidence beam. [3]When the reference mirror moves away from the incident beam, OMAG images blood flow in the opposite direction, towards the direction of the incidence beam. [4]However, the consequence of the mirror moving back and forth is that: 1) the OMAG imaging speed is reduced by half; 2) the computational load on OMAG is doubled to obtain meaningful blood flow images because OMAG needs to acquire two, 3D volumetric images. [5]This multiple imaging is not desirable for fast imaging.

[6]In our proposed research, we will determine whether two, 3D volumetric images are needed to image the direction of blood flow. [7]We anticipate that only one 3D volumetric image is sufficient when the mirror moves either forward or backward. [8]We give a brief mathematical proof to support this hypothesis.

[9]Assume the real function of a spectral interferogram is:

$$B(t_1, t_2) = \cos\left(2\pi f_0 t_1 + 2\pi(f_M - f_D)t_2 + \phi\right) \tag{7}$$

(see Section C.1.2 for details and variable definitions). [10]If we construct the analytic function of Eq. (7) by performing the Hilbert transform in terms of t_1 (note that t_2 is constant), then this analytic function is always:

$$\hat{B}(t_1, t_2) = \exp\{j[2\pi f_0 t_1 + 2\pi(f_M - f_D)t_2 + \phi]\} \tag{8}$$

because f_0 is always > 0, which is guaranteed by placing the sample surface below the zero delay line (see Figs.1 and 2). [11]Thus, given Eq. (8), it becomes a trivial matter, to transform Eq. (7) into

$$B'(t_1, t_2) = \cos\left[2\pi f_0 t_1 - 2\pi(f_M + f_D)t_2 + \phi\right] \tag{9}$$

by digitally multiplying Eq. (8) with a complex function $\exp[-j4\pi f_M t_2]$ and then taking the real part of the results. [12]It is feasible because this modulation f_M is known a priori, determined by either the forward or the backward movement of the mirror. [13]Now, we can construct the analytic function of Eq. (9) by considering t_2 a variable, while t_1 is constant. [14]Then, if $f_M + f_D < 0$, the analytic function is

$$\hat{B}'(t_1, t_2) = \cos[2\pi f_0 t_1 - 2\pi(f_M + f_D)t_2 + \phi] + j\sin[2\pi f_0 t_1 - 2\pi(f_M + f_D)t_2 + \phi] \tag{10}$$

but if $f_M + f_D > 0$, the analytic function becomes

$$\hat{B}'(t_1, t_2) = \cos\left[2\pi f_0 t_1 - 2\pi(f_M + f_D)t_2 + \phi\right] - j\sin\left[2\pi f_0 t_1 - 2\pi(f_M + f_D)t_2 + \phi\right] \tag{11}$$

[15]Consequently, Eq. (7) – (11) simulate the exact situation of the mirror moving in an opposite direction, as described in Section C.1.2. [16]This derivation of the analytic function provides a solid mathematical basis for obtaining information on the direction of blood flow from only one 3D spectrogram image captured by OMAG.

[17]We will incorporate this novel digital frequency modulation approach into our optimal OMAG system. [18]As a result, we will use the following approach: the analysis in Section C.1.2 will give one image of blood flow direction when the mirror moves *towards* the incident beam, and the analysis in this section will give an image of blood flow in the opposite direction, as if the mirror moves *away from* the incident beam.

Case 6-8. Data-sharing procedures. (A) A data-sharing subsection from a center grant proposal. **(B)** Introduction to a data-collection subsection from a grant proposal with a collaborator and the PI in different institutions.

A.

Internal communications and team building: [1]Central to our proposed STC is a highly collaborative effort among all participants. [2]Our internal and external advisory committees will contribute to this collaboration, and the management team will encourage all STC participants to invest in our interdisciplinary objectives. [3]The STC will support internal communications and team-building activities that will create a unified STC community. [4]These activities will include a dedicated, multi-campus videoconference system (see Shared Facilities); a website; a newsletter for all participants; periodic inter-campus workshops and seminars; and an annual retreat in a mountain environment, in which participants will present their research.

B.

Methods for Exp. 4. [1]Additional experimental data will not be collected for this experiment. [2]The methods will involve the simulation of postural responses using the IC model in a variety of environmental and stimulus conditions that mimic conditions from previous experiments using single-input and dual-input stimuli (refs and Exp. 3). [3]Results from the simulations will be compared with the experimental results to identify performance criteria that can predict the experimental results.

[4]We have established a collaboration with Dr. John Smith at State University, who has developed an optimal control model of postural control (114, 115). [5]We will share our experimental results with him via email, phone calls, and access to restricted data websites, and will compare our predictions of optimization methods based on the IC model with predictions from the Smith model.

Case 6-9. Interpretation subsection from a narrative.

Interpretation of *Exp. 1*. [1]If significant tilts of postural verticality are revealed in subjects with chronic UVL, this finding will be consistent with our hypothesis that incompletely compensated vestibular asymmetries affect postural verticality even years after a UVL. [2]If postural bias is revealed during any of the tasks above, but not when subjects with chronic UVL stand on an unmoving surface, this finding will support the hypothesis that chronic UVL subjects substitute somatosensory information to compensate for their UVL and that postural bias is revealed when surface somatosensory information is altered, forcing subjects to rely on vestibular information and thus revealing a central vestibular imbalance. [3]We expect subjects to orient their trunks with reference to the moving surface when tilted to the side of the vestibular lesion but to orient more to gravitational vertical on the intact side. [4]If chronic UVL subjects do not show significant postural bias, this may suggest that central vestibular compensation mechanisms have restored symmetry of postural verticality; alternatively, postural verticality may be determined primarily from somatosensory graviception sources rather than from vestibular information (ref).

[5]A clinical measure of such biased postural verticality will be useful in following subjects' compensation after UVL, since compensation of various postural and eye-movement symptoms are known to have different time courses and underlying mechanisms (ref). [6]If subjects with chronic UVL tilt in the direction of their loss in several tasks in *Exp. 1*, this will suggest that clinical, behavioral recovery of posture following UVL results from somatosensory substitution rather than from central compensation for asymmetrical vestibular inputs. [7]If the degree of postural bias (trunk tilt toward the side of their vestibular loss) during sinusoidal surface rotations is found to be related to the frequency or amplitude of surface rotations, this will suggest that vestibular information is more critical for postural orientation under surface conditions in which somatosensory information is inadequate. [8]*Exp. 5* will then be used to determine whether the sinusoidal surface conditions that best unmask asymmetrical postural verticality are those conditions that required the greatest use of vestibular cues for postural control.

[9]If subjects with chronic UVL show biased postural verticality in their final equilibrium position following lateral-ramp surface rotations, this finding will suggest that the goal of final equilibrium for postural responses can be altered by an incompletely compensated vestibular asymmetry. [10]In addition, this finding will be consistent with the postural bias we observed in the current grant cycle, when control subjects were exposed to asymmetrical GVS during surface translations (ref).

[11]If chronic UVL subjects' postural verticality or stability during quiet stance is affected by the incline of the surface, this finding will support our hypothesis that bilateral vestibular information is required, together with somatosensory information, for the CNS to calculate the surface (or foot) incline in space. [12]The vestibular system may provide a centrally represented "uprightness set-point", around which the somatosensory system controls upright postural control (ref). [13]If postural stability is

worse in chronic UVL subjects than in control subjects when the surface is rotated or inclined, this will suggest that one intact vestibular system does not completely take over the function of bilateral vestibular systems in attempting to control postural equilibrium.[14]We expect that even chronic, clinically well-compensated UVL subjects will show a variety of levels of compensation based on their postural asymmetries and their answers to the questionnaires. [15]If there is a significant correlation between our quantitative measures of postural asymmetry and subjective complaints, this will suggest that compensation is most complete in UVL subjects who achieve greater left-right balance of tonic activity in vestibular pathways, such that posture is more symmetrical when they are forced to rely on vestibular information for postural orientation. [16]If there is no significant correlation between UVL subjects' postural asymmetries and their answers to the questionnaires, we will relate the degree of postural compensation with clinical factors, such as the number of years post lesion, diagnosis, and age. [17]We will also determine whether differences in postural compensation across subjects are continuous or bimodal (well-compensated and poorly compensated).

Case 6-10. Example of anticipated problems and proposed solution integrated into an outcomes subsection.

C.1.c. Outcomes, potential problems and difficulties, and alternative approaches. [1]The goals of this first section are the characterization of any important variation, flexibility, and asymmetry in the parvovirus capsid structures, and effects on capsid permeability and the exposure of internal components important for infection. [2]Structural variation clearly occurs in the capsid, and many of the variants or mutants we will be examining are already known to show differences in flexibility and/or protease cleavage susceptibility. [3]The effects of structural changes on the capsid functions are harder to predict, but functions known to be altered include the affinity of receptor binding (in particular the host-specific binding to the glycosylated canine TfR), as well as sialic acid binding, antibody binding, and exposure of various peptides and of the viral DNA. [4]Receptor binding will also be examined in detail in the experiments in Section C2, and the relationships between antibody and TfR binding will be examined in Section C3. [5]There are some potential challenges in these analyses of the capsids. [6]First, we cannot predict the functional effects of some of the structural changes, and some effects may be relatively subtle. [7]It is clear that the capsids act as finely regulated machines and perform their functions through controlled transitions, and do not (for example) simply fall apart, even to release the DNA, but likely expose internal structure or the DNA 3'-end only under controlled circumstances during infection. [8]Asymmetry can be difficult to detect and quantify, but in this system we should be able to detect that using a variety of biochemical or other approaches. [9]We know some of the capsid changes that control protease cleavages and can use that information to predict alternative sites, and have enough preliminary information to be able to prepare informative mutants for our functional analyses. [10]The divalent ion binding sites are well defined from the X-ray structures, as are natural mutations that are known to prevent one ion binding and also to increase the flexibility of the adjacent loop that controls sialic acid binding (60). [11]Altering the binding sites of the other two ions is quite straightforward, although we do not know if capsids with mutations affecting the binding of those ions will be viable. [12]However, we can also control ion binding by chelation with EDTA or EGTA, or by incubation at low pH, so there are several alternative approaches available. [13]If mutant viruses are non-viable, we can test many functions by expression of capsids by plasmid transfection or by expression from baculoviruses (58).

Case 6-11. Shared Methods subsection. The procedures described in this **Shared Methods Subsection** will be used in experiments that cross multiple proposed aims. The procedures specific to an aim follow in Sections D.2 to D.4 (not shown).

D.1 Shared Methods

PC, APC and PC/APC mutants (Exps. 1, 3, 4): [1]Human, plasma-derived PC, APC, and Gla-less PC and APC will be purchased from Hematologic Technologies (Essex Junction, VT). [2]These PC preparations are free of detectable APC (<0.1%) and thrombin (<0.04 pM in 100 nM PC), as determined by amidolytic assays.[75] [3]Recombinant wild-type and mutant PC expression vectors will be constructed and purified from conditioned media, as described.[76,77] [4]Concentrations of recombinant wild-type APC and APC mutants will be determined by active-site titration using APC at 8 μM in HBS and p-nitrophenol-guanidino benzoate at 0.1 mM and using an extinction coefficient for p-nitrophenol of 11 400 $M^{-1}cm^{-1}$ (at pH 7.4), as described.[76,78] [5]Mutant PC and APC will be provided by Dr. John Smith (Affiliation).

Expression of soluble GPIb and apoER2' (Exps. 2, 5): [6]Glycocalicin (the extracellular portion of GPIb) will be isolated from outdated human platelets and eluted with a linear salt gradient of 0 – 0.7 M NaCl in 20 mM Tris/HCl, as previously described.[79,80] [7]Recombinant human soluble GPIbα will be cloned and expressed, as previously described.[25] [8]Soluble apoER2' will be cloned and expressed, as previously described.[44] [9]ApoER2' splice variants and domain deletion mutants will be cloned into the expression vector PTT3-Srα-GH-HISN-TEV. [10]The apoER2' expression vector is constructed from the pTT3 and pSGHV0 expression vectors, and is a kind gift from Dr. John Smith (Affiliation). [11]Recombinant protein concentrations will be measured with a BCA protein assay kit (Pierce, Rockford, IL). [12]Protein purity and molecular weight will be assessed by SDS-PAGE and then by a PageBlue protein staining solution (Fermentas) according to the manufacturer's protocol.

Platelet Purification (Exps. 6, 7, 8): [13]Human venous blood from healthy volunteers will be drawn by venipuncture into sodium citrate and ACD, as previously described.[27,43] [14]Platelet-rich plasma will be prepared by centrifugation of whole blood at 200 g for 20 min. [15]Platelets will be isolated from the plasma by centrifugation at 1000 g for 10 min in the presence of prostacyclin (0.1 αg/ml). [16]Platelet pellets will be washed and resuspended in a modified Tyrodes buffer (in mM: 129 NaCl, 0.34 Na_2HPO_4, 2.9 KCl, 12 NaHCO₃, 20 HEPES, 5 glucose, 1 $MgCl_2$; pH 7.3).

IP/Western blotting (Exps. 9, 10): [17]Following lysis with an ice-cold IP buffer, platelet samples will be pre-cleared for 1 hr at 4°C with protein A- or G-Sepharose (50% (w/v) in Tris-buffered saline plus Tween 20). [18]Antibodies will be added and samples rotated overnight at 4°C. [19]The Sepharose pellet will be washed in a lysis buffer and then in TBS-T before a Laemmli sample buffer is added. [20]Proteins will be separated by SDS-PAGE on 10% gel and electrically transferred onto polyvinylidene difluoride membranes. [21]The membranes will be blocked in 10% (w/v) BSA dissolved in TBS-T. [22]Antibodies will be diluted in TBS-T containing 2% (w/v) BSA and then will be incubated with the membranes for 1 hr at RT. [23]The membranes will be washed in TBS-T after each incubation and developed using an enhanced chemiluminescence system.[41,43]

Case 6-12. Two examples of Ending Sections (A and B).

(A)

Future Plans – [1]The results from this study will provide preliminary data to calculate sample sizes and show feasibility for a more definitive NIH study. [2]In that study, we will expand the sample sizes to determine the optimal dosage and timing of AMES pre-conditioning and to characterize the fundamental sensorimotor elements associated with laparoscopic surgery.

(B)

Future directions: screening for additional Ifitm-binding proteins
[1]The focus of this R21 is on evaluating the Bat5, Sae2 and Rnf41 proteins as the primary candidates for Ifitm binding. [2]However, we do not assume that this list is inclusive of all possible partners, especially since the action of interferon may induce the production of additional Ifitm cargo proteins. [3]In the future, we plan to initiate experiments to use all five of the epitope tagged Ifitm proteins to bind to lysates from resting and activated cells, as well as cells treated with IFN-β, to determine if additional binding proteins can be identified. [4]Using either Baculovirus or E.coli produced His-tagged Ifitm proteins, we will bind and elute cytoplasmic lysates from normal splenocytes (plus or minus IFN-β treatment for 10 hrs). [5]We will also use the analogous lysates from the Ifitm KO splenocytes that have no free Ifitm proteins that could potentially compete for binding. [6]Samples of lysates will be separated by SDS-PAGE and analyzed by silver staining to determine the complexity of binding proteins. [7]Immunoblotting will also be performed to test for the presence of predicted binding proteins. [8]Two strategies can then be employed to analyze these binding proteins. [9]The first will be to excise protein bands of interest from the stained gels and use them in single Mass Spec protein identification assays. [10]A second strategy will be to sequence the sample in bulk and derive all possible protein constituents in the preparation. [11]Our university's Mass Spectrometry and Proteomics Core can assist in both of these approaches. [12]Potential target proteins can then be evaluated using the same approaches as described elsewhere in this application.

Case 6-13. Example of an Ending Section to a narrative. (A) Ending section with a timeline paragraph. **(B)** Timeline paragraph in **(A)** converted into a table.

(A)

C4. Overall Summary and Conclusions

[1]These studies will integrate our understanding of the structures of viral capsids with a more detailed knowledge of the binding properties and functions of host cell receptors and antibodies. [2]The results will be correlated with biochemical and structural analyses of the capsid flexibility, variation, and/or asymmetry. [3]These are central questions that apply to any non-enveloped animal virus and have parallels to the structural changes and interactions seen for many enveloped virus glycoproteins, and so the results will clarify some of the underlying rules about how viruses interact with their host ligands and infect cells. [4]The work builds on a solid intellectual and methodological foundation resulting from our previous studies, and we have most of the materials and background information required. [5]For each of the projects we combine well-established methods with new approaches and have alternative approaches for each of the experiments where the technology is novel or untested. [6]Studies already underway will be continued in the first phase of the funding period, while studies requiring the development of reagents or information from previous studies will be done later in the project.

TIMELINE

[7]This project will take 5 years to complete. [8]The sequence of studies will initiate in years 1 and 2 with the preparation of the capsid mutants and their testing, along with development of the new methods for sample preparation for cryoEM and collection of the cryoEM data for analysis. [9]Analysis of the role of cleaved and stabilized capsids will initiate with the currently available mutants and continue through years 3 and 4. [10]Mutant forms of the TfR and antibodies will be prepared in the first years and tested in later years up to year 5. [11]The preparation of capsids with altered receptor binding sites (peptides or domains) and selected on mutant receptors will occur during years 3 to 5.

(B)

Year 1	Year 2	Year 3	Year 4	Year 5
Prepare capsid mutants and their testing	Prepare capsid mutants and their testing	Analyze role of cleaved and stabilized capsids	Analyze role of cleaved and stabilized capsids	
Develop new methods to prepare samples for cryoEM and to collect cryoEM data for analysis	Develop new methods to prepare samples for cryoEM and to collect cryoEM data for analysis			
Prepare mutant forms of the TfR and antibodies	Prepare mutant forms of the TfR and antibodies	Test mutant forms of the TfR and antibodies	Test mutant forms of the TfR and antibodies	
		Prepare capsids with altered receptor binding sites (peptides or domains) and selected on mutant receptors	Prepare capsids with altered receptor binding sites (peptides or domains) and selected on mutant receptors	Prepare capsids with altered receptor binding sites (peptides or domains) and selected on mutant receptors

Other prose considerations

7.1   Abstracts

An abstract is a section separate from the narrative of a grant proposal. The abstract summarizes the most important information from each major section of the narrative. The abstract typically precedes the narrative and is often made available to the public separately from the narrative. Reviewers will likely read the abstract before the narrative, so it plays an important role in helping them form favorable impressions about you and your credibility, and in attracting their interest about your research topic and its significance and novelty.

Different funding agencies name the abstract differently. For example, NIH refers to its abstract as the **Project Summary**. The Michael J. Fox Foundation refers to the abstract as the **Grant Abstract**, and the Society of Family Planning calls it a **Project Abstract**. NSF refers to a similar section as the **Project Summary** but cautions that the NSF **Project Summary** "should not be an abstract."[1] However, as discussed in Chapter 7.1.2, for all practical purposes the NSF **Project Summary** is an abstract. The Robert Wood Johnson Foundation also calls its abstract a **Project Summary**.[2] Regardless what a funding agency calls its summary to the narrative, in this book it is called an *abstract*.

---

[1] http://www.nsf.gov/pubs/policydocs/pappguide/nsf13001/gpg_2.jsp#IIC2b (accessed February 10, 2013).
[2] anr.rwjf.org/applicationPrintPreview.do?phaseId=2459 (accessed November 28, 2014).

7.1.1   Length, layout, and level of technicality

When writing an abstract, you need to follow all submission requirements from the targeted funding agency. Submission requirements usually specify the length and layout of the abstract, its level of technicality, and its required content. Table 7-1 gives submission requirements for NIH and NSF abstracts.

• **Length and layout**. In most cases, funding agencies require that an abstract be of a particular character count or word count. Commonly prescribed lengths for an abstract are one page or a maximum of 250 words.

If the funding agency's submission requirements do not specify the layout for the abstract, you can present your abstract: (a) in one or multiple paragraphs, termed an *unstructured* abstract (e.g., Case 7-1A), or (b) in multiple, short subsections, termed a *structured* abstract (e.g., Case 7-2), in which each subsection presents key content from each major section.

• **Level of technicality**. Submission requirements often specify the level of technicality for an abstract. For example, the National Niemann–Pick Disease Foundation, Inc. (NNPDF) requires that the abstract, which it calls **Summary**, be written for a lay audience because "although the Scientific Advisory Board makes a funding recommendation to the NNPDF Board of Directors, it is the family members on the NNPDF Board who make all funding decisions."[3] Other funding agencies, such as the National Multiple Sclerosis Society,[4] require 2 abstracts: one for a lay audience and one for a scientific audience.

Abstracts written for lay audiences use less technical vocabulary and fewer abbreviations than those written for technical audiences, and they use definitions for any key technical terms that cannot be avoided. In abstracts written for a technical audience, technical vocabulary for key terms are often used but are only sometimes defined. (See Chapters 7.1.2 and 8.2.)

7.1.2   Content

An abstract presents key content from each major section of the narrative. It can also include any other information that a funding agency wants to include, whether or not the information is mentioned in the narrative. Including particular information to satisfy submission requirements from the funding agency does not change the fundamental nature of an abstract, which is to represent the major content of the narrative.

> **Guideline 7.1**   *Limit the abstract to key content from every major section of the narrative, unless submission requirements indicate otherwise.*
> **Exception:** *Do not include any proprietary information.*

---

[3] http://www.nnpdf.org/npresearch_15.html (accessed November 28, 2014).
[4] http://www.nationalmssociety.org/for-professionals/researchers/get-funding/research-grants/index.aspx (accessed November 28, 2014).

NSF notes that its **Project Summary** is not an abstract. One possible reason for this effort to distinguish the NSF **Project Summary** from an abstract could be to draw attention to NSF's requirement that its **Project Summary** must include particular significance information: intellectual merit and broader impacts[5] (see Table 7-1B). Including significance in the abstract is understandable and even desirable, especially if the abstract is made available to the public separately from the grant proposal and if the reviewers include lay readers, such as board members of foundations, legislators determining appropriations, and the general public, including family members of relatives with particular diseases, taxpayers wanting to find out where their tax dollars are going, and comedians looking to write jokes about government spending.

The widespread use of abstracts by lay readers makes information about significance critically important – so important that it is required content not only by NSF but also by other funding agencies, such as NIH[6] when it specifies that the **Project Summary** should "describe the relevance of this research to public health" (see Table 7-1A) and the Michael J. Fox Foundation, which specifies that abstracts to its grant proposals need to address the **Relevance to Diagnosis/Treatment of Parkinson's Disease** (e.g., see Case 7-2).

The abstract is simultaneously an independent and a dependent unit of prose. It is independent in that it should stand alone as a coherent (but brief) description of the narrative. An abstract is usually dependent on the narrative in the sense that its content is limited by: (1) the content in the narrative, (2) any special content requested by the funding agency, and (3) terminology and abbreviations used in the narrative. The following list identifies the key information that the abstract should briefly mention.

• **Research purpose: objective, aims, and long-term goal.** An abstract needs the proposed research objective clearly identified (see Chapter 2.2.4 on goals, objectives, and aims). To help reviewers identify the research purpose quickly, you should use standard terms, such as *objective*, *aim*, or *goal*, as shown in Cases 7-1A and 7-2. Case 7-1A identifies the proposed research objective in sentences 6 and 12. Especially in NIH grants and in grants that use the NIH model of the narrative, the proposed aims are also identified, illustrated in Case 7-1A, sentences 13 and 15. The abstract to a grant proposal submitted to the Michael J. Fox Foundation in Case 7-2 identifies the research objective in sentence 1 and the goal in sentence 4.

• **Research hypothesis.** An abstract for a project that involves basic research usually has its hypothesis identified in the abstract, but sometimes an abstract for clinical or translational research does not include a hypothesis. The abstract in Case 7-1A includes a hypothesis in sentence 11, but no hypothesis is given for the abstract in Case 7-2. For clarity, it is important that you use the term *hypothesis* or *hypothesize* when presenting your hypothesis.

---

[5] http://www.nsf.gov/pubs/policydocs/pappguide/nsf13001/nsf13_1.pdf (III-2 and -3; accessed November 28, 2014). Broader impact and technical merit correspond to significance from societal and scientific and/or technical perspectives (see Chapter 2.2.5).

[6] http://grants.nih.gov/grants/funding/phs398/phs398.html (I-34; accessed November 28, 2014).

• **Definitions**. You need to define key terms that are central to your research topic in order to help readers quickly understand the topic of your research. Also, if the entire meaning of the abstract hinges on a few key terms, you should define them, whether or not reviewers might be familiar with them (see Chapter 8.2). Definitions in an abstract should not be long or complex, and the explanatory terms used in the definitions should be readily understandable by the reviewers. The abstract in Case 7-1A defines *hemodynamic forces* (sentence 2) and *heart outflow tract* (sentence 8); the significance of the proposed research would not be clear unless reviewers understood these 2 key terms. Similarly, in Case 7-2, sentence 2 defines *APL-130277*.

• **Background, research issue, and significance**. An abstract presents background information that directly relates to the research issue (need or problem) that you will be investigating. Significance should also emerge from the discussion of the background and research issue. In Case 7-1A, the first half of sentence 1 gives background information; the second half of sentence 1, and sentences 2 and 3 identify the problem, and sentence 4 identifies the research need. In Case 7-2, sentences 10–15 address the background, research issue, and significance.

• **Methods information**. The abstract identifies primary features of the research design, **SOS** (subjects, objects of study, or specimen), key **MET** (materials, equipment, or tools), and key procedures to achieve the proposed aims. In Case 7-1A, methods are presented in sentences 7–10, and after each aim, in sentences 14 and 16. In Case 7-2, methods are presented in the **Project Description**, sentences 5 to 9.

• **Expected outcomes**. Expected outcomes are important in an abstract. They are the major results that you expect or predict from executing your proposed methods. Case 7-1A gives an expected outcome in sentence 11, and Case 7-2 gives expected outcomes in **Anticipated Outcome**, sentences 16–17.

With the abstract drawing heavily on the content from every major section of the narrative, it is most efficiently written *after* the entire narrative is stable, typically when the narrative is close to the final-draft stage. With this said, however, some writers prefer drafting the abstract before writing any section in the narrative, to help them clarify to themselves the conceptual framework of the proposed research. If you write the abstract before the narrative is stable, you should allow extra time to substantially revise it.

7.1.3 Types of abstracts

There are 3 types of abstracts, based on how the content is presented: (1) substantive, (2) topical, and (3) mixed. Case 7-1 gives an example of each type of abstract.

Guideline 7.2 *Use either a substantive or a mixed abstract. Avoid a topical abstract.*

• **Substantive**. A substantive abstract has high information value. It not only identifies the topics addressed in the narrative, but it also gives specific content from the narrative. Case 7-1A presents a substantive abstract, with topics and details. In this example, the topic *congenital heart disease* is identified, and key epigenetic factors that can lead to congenital heart disease (*CHD*) are mentioned.

● **Topical**. A topical abstract mentions only the major topics in the major sections of the narrative. An abstract that is purely topical is rare because its information value is low. Case 7-1B gives a topical abstract in which the research topic is identified in sentences 1 and 2. The topic *problems in quantifying shear forces* is mentioned in sentence 3, but the problems are not actually specified. Likewise, in sentences 4 and 5, the hypothesis, aims, and methods are mentioned as topics but are not specified.

● **Mixed**. A mixed abstract has features of both substantive and topical abstracts. This type of abstract commonly accompanies the narrative to a grant proposal. A mixed abstract identifies the main research topic and gives substantive content from the narrative. However, it also includes a few topics without detailing them. Case 7-1C presents a mixed abstract with substantive information in sentences 1–7 and 9. Sentences 8 and 10 identify the topics of the hypothesis and aims, but give no details. A mixed abstract is useful; it not only provides actual content from the narrative, but it also topicalizes some content, which can reduce its total length.

7.1.4   Other characteristics of abstracts

● **Coherent unit**. An abstract should be written in full sentences, with each sentence logically following from previous sentences. Taken together, all sentences need to represent the key information from each major section of the narrative while expressing sound scientific methodology. One way to establish a clear connection between scientific methodology and your proposed research as summarized in the abstract is to use key terms from your narrative and also from scientific methodology, such as *aim, hypothesis, methods, protocol,* and *research objective.*

● **Key terms and abbreviations**. You need to use the same key terms and abbreviations in the abstract that you use in the narrative, and no terms or abbreviations that you do not use in the narrative. For abstracts written for lay or technical reviewers, if you believe that they might not be readily familiar with the terms, you need to briefly define them (see Chapter 7.1.2).

Abbreviations, such as acronyms, are popular in abstracts because they can help shorten the text. However, any abbreviations that you include should also be used in the narrative. If you decide to use abbreviations in the abstract, you should use them sparingly – perhaps only 2 or 3 different ones. An abbreviation requires abstraction for readers to understand it and to associate it with its full meaning. When you use many abbreviations in an abstract, some reviewers may have difficulty quickly understanding.

> **Guideline 7.3** *Limit abbreviations in an abstract to 2 or 3. Terms that you select for abbreviations should occur multiple times in the abstract and should be terms that you also abbreviate in the same way, in the narrative.*
> *Exception: Consider using abbreviations for units of measurement, without introducing the full terms.*

With few exceptions, for each abbreviation that you use, you need to introduce it in parentheses, after the full term. For example, the abstract in Case 7-1A uses just 2 abbreviations: *CHD* for *congenital heart disease* (sentence 1) and *OFT* for *outflow tract* (sentence 8). You do not need to avoid abbreviations for common size and time units, such as *cm* for centimeter, *s* or *sec* for second, and *h* or *hr* for hour. When you use abbreviations for common units of measurement, you do not define them.

● **Pronouns**. Pronouns can reduce the length of an abstract. However, pronouns can be problematic if their antecedents are unclear (Chapter 8.6.3). Clarity is paramount, so you should consider reducing pronouns, and if you do use them, you should double-check that they are clear. Note that a few funding agencies, such as NIH and NSF, require that you avoid using first-person pronouns (*I, we, me, us, my, mine, our,* and *ours*) in the abstract or that you use only third-person pronouns (e.g., *it, he, she, they, them*).

● **Grammatical articles**. In an effort to write a concise abstract, you might be tempted to use a telegraphic style that occasionally omits an article (*a, an,* and *the*). A telegraphic style is not appropriate in an abstract. Articles contribute to clarity, providing valuable information that helps distinguish unique information, information that has been previously mentioned (i.e., **given** information), and information that has not been previously mentioned (i.e., **new** information; see Chapter 8.4). Example 1 shows an excerpt from the abstract in Case 7-1A, with its articles bracketed; these articles should not be omitted.

Example 1
Although one defect may have [a] genetic cause, altered blood-flow conditions produced by [the] defect may also detrimentally affect heart development. [A] detailed study of [the] effects of hemodynamic forces on heart development requires [the] quantification of these forces in vivo. (Case 7-1A, sentences 3–4.)

● **Grammar, spelling, and punctuation**. Reviewers – whether technical or lay – start forming judgments about your credibility and your research abilities while reading the abstract. The language in the abstract should help promote a good first impression of you personally and of your proposed research – or at the very least, its language should not breed a negative response. If the audience finds spelling errors, punctuation problems, or basic grammatical problems such as subject–verb disagreement, your credibility may suffer, and ***you may have difficulty overcoming the initial negative first impression***.

With the abstract most effectively written after the narrative is stable, you might be very tired by the time you write the abstract, so you need to be especially diligent when reviewing it. You might ask a colleague for assistance in reviewing the abstract to ensure that it is clear and polished. Notice that both abstracts in Cases 7-1A and 7-2 are well-written and reflect high standards of educated prose.

● **Font.** You need to follow font requirements for the abstract, as specified by the funding agency. If the font is not specified, you should use the same font – both in style and size – that you use in the narrative. Some funding agencies expressly prohibit the use of font highlights, such as underscoring or bold italic font, possibly because their software to transfer abstracts does not translate the highlights. However, other funding agencies may be silent on this point. If abstract requirements expressly allow font-highlighting techniques, then you can consider using them, but only sparingly so that they do not lose their emphatic impact.

7.1.5   The abstract as a scientific argument

The coherence of an abstract also involves scientific argumentation, which comprises content reflecting **TRACS** (Chapter 1.9): the *topic* that you propose to research; a brief *review* of the background literature that defines the research problem or issue; an *analysis* of the topic in relation to the background, such as the research issue (need or problem); the *connection* between the previous research context and your proposed research purpose and key methods, such as the research objective and aims; and the *significance* (relevance and importance) of the proposed research.

The abstract in Case 7-1A includes the content of a scientific argument: the *topic* of *shear stresses* is identified in sentences 2 and 5, and is discussed throughout the abstract; background is *reviewed* in sentences 1–4 and 8–9; the background is *analyzed* in sentences 1 and 5 with problem statements and in sentence 11 with a hypothesis; the *connection* between the previous research and the proposed research is identified in sentences 6–15 with information about the research objective and aims, and key methods; and *significance* emerges from sentences 1 and 2.

The abstract in Case 7-2 also includes the content of a scientific argument: the *topic* of *APL-130277* is first identified in sentence 1, and this topic runs throughout the abstract to the very last sentence; the background is *reviewed* in sentences 2, 3, 11, 12, 14, and 15 and is *analyzed* in sentence 10 with a problem statement; the *connection* between the previous research and the proposed research is identified in sentence 1 with the research objective identified, in sentence 4 with the goal identified, in sentences 5–9 with the proposed methods summarized, and in sentences 16 and 17 with the information related to future studies; and *significance* emerges from sentences 4 and 5.

A component that PIs sometimes omit from an abstract is a brief review of background information leading to the research problem or issue. By providing a brief historical context for your proposed research, brief in the sense of just one or 2 sentences, the background can help reviewers better understand the novelty and significance of your proposed research.

Even if background in the abstract is not required, consider including it in a sentence or two.

7.2 Resubmissions

If your grant proposal is not funded, you have essentially 3 options:

(1) Revise and resubmit to the same funding agency.

(2) Revise and/or submit to a different funding agency.

(3) Create a new application and submit it to the same or different funding agency.

Some funding agencies, such as NIH[7] and the National Multiple Sclerosis Society,[8] have special requirements for resubmitting a grant proposal that they have previously denied funding. For example, NIH requires an additional prose section called the **Resubmission Introduction**; its length varies – either one or 3 pages – depending on the type of NIH funding mechanism. The **Resubmission Introduction** "summarizes the substantial additions, deletions and changes to the application ... [and] a response to the issues and criticism raised in the Summary Statement."[9] The National Multiple Sclerosis Society basically follows NIH guidelines for its **Resubmission Introduction**, except that it limits the **Resubmission Introduction** to 2 pages. Other funding agencies may have no special requirements for resubmissions, in which case you can use the guidelines in this book.

7.2.1 Tone control

The **Resubmission Introduction** provides an opportunity for you to respond to reviewers' criticisms and concerns, and to explain how you changed your proposed research in response to their criticisms. It is *not* an opportunity for you to argue with reviewers, to point out deficiencies in their understanding, or to comment on their inadequate reading of your narrative.

Hostility is inappropriate in a **Resubmission Introduction** (or in any other place in the narrative). If you do not agree with your reviewers, you need to express your disagreement without being hostile or disrespectful. If you decide not to change your proposed research in response to a reviewer's concerns, you need to substantiate

---

[7] NIH provides excellent information on criteria for a resubmission. URLs to sites with explanations for NIH resubmissions include: http://grants.nih.gov/grants/policy/resubmission_q&a.htm, http://grants.nih. gov/grants/guide/notice-files/NOT-OD-12-128.html, and http://www.nimh.nih.gov/funding/grant-writ ing-and-application-process/resubmitting-an-application.shtml (all accessed November 28, 2014).

[8] http://www.nationalmssociety.org/For-Professionals/Researchers (accessed November 28, 2014).

[9] http://grants.nih.gov/grants/funding/phs398/phs398.html (I-17; accessed November 28, 2014).

why, preferably using others' published findings, and then you should integrate this substantiation into the narrative. However, if you do not follow the reviewers' suggestions, they could again not score your proposal favorably.

If you believe that a reviewer misread what you wrote, did not read your narrative closely enough, or must have been asleep not to have understood what you were explaining, it is still your responsibility to explain your proposed research in a way that reviewers will understand.

Guideline 7.5   *Avoid including anything that the reviewers could interpret as revealing your hostility; instead, focus on explaining your proposed research so that reviewers will readily understand it.*

When responding to a reviewer who did not understand something that was absolutely clear or who actually missed information that you obviously included, a tactful approach is to revise the passage slightly and mention in the **Resubmission Introduction** that you have revised it. You might feel more satisfied by telling the reviewers that the original write-up was clear, but this approach will not help your reviewers view your proposal favorably. With this said, however, you do not need to grovel.

Examples 2 and 3 show how you can avoid accusing reviewers of not understanding an already clear passage. Examples 4 and 5 illustrate how you can disagree with reviewers by substantiating your initial position.

Example 2
We have revised Sections 4.2.1–4.2.3 to explain our methods more clearly.

Example 3
Based on our explanation of X, we understand that Reviewer 3 took our rationale to mean Y. We have rewritten (revised) this passage to explain X more clearly.

Example 4
We respectfully disagree with Reviewer 1's assessment of our second inclusion criterion for subjects in Aim 1. In formulating this criterion, we followed Smith *et al.* (2012). We now include a brief description of Smith *et al.* in the rationale to our methods for Aim 1 (Section C.1.1).

Example 5
Reviewer 2 perhaps was basing the criticism of Exp. 4 on essentially insufficient preliminary studies. Section C.2 now discusses 2 additional preliminary studies, one of which has been accepted for publication since our initial submission (see revised list of references), which supports our Exp. 4.

7.2.2   Content and organization of the Resubmission Introduction

The first step in drafting a **Resubmission Introduction** is to identify the reviewers' criticisms that you will address. At the very least, the **Resubmission Introduction** should address the criticisms shared by the reviewers and the major criticisms from each reviewer.

Guideline 7.6 *In the **Resubmission Introduction**:*
(a) Address criticisms that are common to all or the majority of your reviewers.
(b) Address major criticisms that are not common across the reviewers.
(c) Consider addressing non-major criticisms that are common across the reviewers.
(d) Consider addressing non-major criticisms that are not common across the reviewers.
*(e) Respond to each criticism that you identify in the **Resubmission Introduction**, and identify its impact on the resubmitted document.*
(f) Consider including a compliment that a reviewer may have given you.

Funding agencies may provide a summary of the reviewers' criticisms, in addition to each reviewer's criticisms. The summary can help you identify criticisms that reviewers have in common and the criticisms that may be major, from the reviewers' perspectives.

After identifying the criticisms that you will address, you need to decide on the organization of the **Resubmission Introduction**. The **Resubmission Introduction** has an opening statement and a body, but it rarely has an ending, due to length constraints. Table 7-2 gives 3 basic organizational alternatives for the **Resubmission Introduction**, and the example of a **Resubmission Introduction** in Case 7-3 shows a fourth organizational alternative.

• **Opening statement.** Regardless of the length of a **Resubmission Introduction**, it needs an opening statement. Typical content of this opening statement can include any of the following:

(1) Expression of appreciation to the reviewers.

(2) A compliment from a reviewer.

(3) Forecast of the body to the **Resubmission Introduction**.

(4) Identification of the major substantial changes in the resubmitted narrative, grant application, and budget.

(5) Inclusion of completed preliminary studies since the prior submission.

(6) Identification of how the substantial changes are marked in the narrative.

In Case 7-3, the 2-sentence opening statement to the **Resubmission Introduction** includes 3 types of information from the above list: (1) an expression of appreciation to the reviewers, (2) compliments from reviewers, which are quoted, and (3) a forecast of the upcoming information. Notice the use of quotes lends authenticity to the compliments.

• **Organizational alternatives for the body.** The body usually presents a list of criticisms and the PI's response to each in a list, either numbered or bulleted. A short heading accompanies each listed item in order to identify the reviewer who gave the criticism. When choosing an organization for the **Resubmission Introduction**, you should choose one that is easy to follow, that best allows you to address many criticisms, and that allows you to identify the reviewers by number.

Table 7-2A shows an organizational alternative that can be used when each of the reviewers gives different criticisms. The organization in Table 7-2B is useful when criticisms are shared, with common criticisms listed in the first section, and with criticisms unique to individual reviewers in the subsequent sections. Table 7-2C shows an alternative in which the criticisms are organized by the major sections of the narrative. This third alternative has limited application since often one criticism will relate to multiple sections in the narrative. However, if reviewers group their criticisms by major sections of the narrative, the organization in Table 7-2C can be particularly useful. Case 7-3 uses an organization in which all of the reviewer's criticisms, except for one (see Item 4), are summarized; in this case, the headings to Items 1–3 identify both sources of criticisms: the summary and the relevant reviewers.

Another style for the **Resubmission Introduction** uses only paragraphs. The paragraph format is not advised since its readability is low in terms of helping reviewers quickly locate each criticism, including quickly finding their own previous criticisms and your responses to them.

● **Content in each listed item.** Each item listed in the **Resubmission Introduction** needs to: (a) identify by number the reviewer who gave the criticism; (b) present the reviewer's criticism or question in an emphatic font (e.g., italic or bold font) or in quotation marks if direct quotes are used; (c) briefly present the PI's response; and (d) identify the location of the consequent changes in the resubmitted narrative. Notice in Case 7-3, the only numbered section that does not include a cross-reference to the consequent change in the resubmitted narrative is Item 1, in which the PI explains why the scope of the project is not too large.

Guideline 7.7 *Consider responding to a reviewer's criticism by including:*
(a) The PI's position regarding the criticism.
(b) Support for the PI's position.
(c) A possible justification for the reviewer's criticism.
(d) The PI's action to address the reviewer's concern.

As noted in Guideline 7.7(a), you can include your position regarding a reviewer's criticism. For example, in Case 7-3, in response to the issue of the scope of the project being too large, the PI disagrees by offering: *"we are confident that we can complete the proposed research on schedule."* The support for your position can be drawn from: (a) relevant features of the proposed research, such as sentence 7 where the PI notes "*We already have available the 3 field-resistant mosquito lines . . .*" and (b) review of background literature and citations, such as in sentence 12 of Case 7-3. If you include background in the **Resubmission Introduction** that you have not yet included in the narrative, you also need to include it in the narrative and in the list of references.

For tone control, you can accept the reviewer's criticism, especially to take ownership of lack of clarity on your part, illustrated in sentence 19 of Case 7-3, where the PI notes: "*This concern may have come from a lack of clarity in our previous submission.*"

An important part of the response is indicating the action that you have taken to address the reviewer's criticism and where in the narrative the reviewers can find mention of this action. For example, in Item 3 of Case 7-3, the PI notes in sentence 20, *"In the current proposal, we specifically discuss our innovations in a separate section (Section B)."*

Because the **Resubmission Introduction** is short, you will likely have trouble including all of the information that you want to discuss and still create a layout that is easily readable. Notice that an itemization scheme and brief headings, such as in Case 7-3, can help readability. You can also create textual space for your responses by either summarizing the reviewers' criticisms or using quotations with ellipses for information that you leave out, such as the quote in Item 4 of Case 7-3. However, if you abbreviate any criticism, you still need to represent it accurately.

Another strategy to create more space in the **Resubmission Introduction** is to use an ampersand (&) in headings and to use common abbreviations, such as *Exp. 1* for *Experiment 1, U.* for *University, Dept.* for *Department*, and *R1* for *Reviewer 1*. If you use such abbreviations, you do not need to introduce them on their first mention in the **Resubmission Introduction**.

7.3 Creating an Ending Section

Funding agencies seldom require an **Ending Section** to the narrative. As a result, the narrative typically stops very abruptly after the **Methods Section**. However, a grant proposal needs an **Ending Section**. Among other information, the ending identifies the very next research project, given the successful outcome of the proposed research. By providing this information, you help the reviewers better understand your line of research and its long-term goal, and how you are organizing your research projects to achieve this goal. Chapter 6.12 further explains an **Ending Section**.

7.4 Combining sections in short narratives

Some funding agencies, especially foundations, specify short narratives, such as 5 or fewer pages. We offer a few organizational and layout strategies to help you meet this short length limit and to provide a few points about how you can include scientific argumentation even in a short narrative. Chapter 8.6 offers strategies to shorten the text.

7.4.1 Organizational and layout alternatives for short narratives

Funding agencies that require short narratives usually specify the organization for its narrative, or they offer a series of questions or identify the content that the narrative needs to address. If your funding agency does not specify an organization for a short narrative, but identifies questions or the content that it needs to cover, the alternatives in Table 7-3 will help you organize the narrative.

Both alternatives in Table 7-3 are very similar to the basic organization described in this book for longer narratives. Table 7-3A shows a background section with 4 major sections, and Table 7-3B, with 5 major sections. The main difference between them is that Table 7-3A gives an organization in which both types of background – other researchers' and the PI's previous studies – are in the same major section, and in Table 7-3B, both types are presented in 2 major sections. Whether the background is presented in one or 2 major sections, the section following the background section(s) is the **Methods Section**. The **Methods Section** in a short narrative usually runs close to one-half the total length of that narrative. Also, in a short narrative, the headings to all sections and subsections are usually in run-in style (Chapter 8.6.2) in order to save space.

Sometimes funding agencies require a short narrative divided into 3 sections. The Christopher & Dana Reeve Foundation requires the following 3 sections in its 5-page narrative: **Aims**, **Background/Significance**, and **Methods and Experimental Design**.[10] As mentioned above (Chapter 7.3), although very few funding agencies require an ending section, we suggest that you include one in order to help reviewers understand the relationship of your proposed research to your overall line of research. To save space, rather than presenting the ending in a section with a heading, you can present it as the last paragraph in the **Methods Section**.

7.4.2   Scientific argumentation in short narratives

Whether you are writing a short or a long narrative, you need to argue for your proposed research. A short narrative should have at least 2 scientific arguments (Chapter 1.9). The first occurs in the **Aims Section**, and one (or more) in the **Background Section**. Thus, in each of these 2 sections, you need content that constitutes **TRACS**: a *topic*, a *review* of background and/or preliminary research, an *analysis* of the reviewed information (e.g., a problem statement, hypothesis, or speculation), the *connection* to your proposed research in terms of your research purpose and/or methods, and the *significance* of your proposed research.

7.5   Letter of intent

Sometimes a funding agency will require a letter of intent (LOI) before allowing you to submit a grant proposal. An LOI is a business letter that identifies your plan to submit a grant proposal to the funding agency. The LOI briefly explains your proposed research and may even include a preliminary budget. The PI does not submit an LOI unless the funding agency requests one in its submission requirements.

A funding agency sometimes uses an LOI to screen out applications that obviously do not fit within the goals of the funding agency or that do not qualify for other reasons

---

[10]   http://www.christopherreeve.org/atf/cf/%7B3d83418f-b967-4c18-8ada-adc2e5355071%7D/RESAPPLICA TIONGUIDELINES2011.PDF (accessed November 28, 2014).

(e.g., the PI or the applicant's organization does not meet eligibility guidelines). A funding agency may also use an LOI for administrative purposes, such as to schedule enough staff to process the high number of applications that it expects to receive. An administrative LOI is usually no longer than one page.[11] A generic administrative LOI is illustrated in Case 7-4. In addition to standard elements of a business letter (i.e., letterhead, subject line, the salutation, and closing signature block), an administrative LOI typically has the following information, plus any other information required by the funding agency:

- Identification of the PI and any Co-I.

- Title of the research project.

- Name (and number) of the program or funding mechanism that the PI is targeting.

- PI's contact information: email address, and phone and fax numbers.

- Research objective for the proposed research.

- Research aims.

7.6   Biographical information

A grant proposal provides biographical information about the PI and any Co-Is in order to help reviewers assess whether their training and skills will likely support their successful execution of the proposed research. Some funding agencies may require biographical information on all professionals who will be working on the project, including collaborators, while others require biographical information on only PIs and Co-Is. Biographical information is usually presented in a non-narrative section, variously called a biosketch, biographical sketch, biographical profile, bioprofile, curriculum vitae (CV), or resumé. In this book, this part of the grant application is called the CV. Sometimes the funding agency will have its own CV form for each member of the research team to use.

7.6.1   Consistency

Although the CV is not part of the narrative, it needs careful crafting. Since a CV is commonly drafted and updated over time, one of the hardest features of the CV to control is its internal consistency. All bibliographic entries need to be of the same format to enhance your credibility and to reflect your attention to detail. Ideally, when multiple CVs are submitted with the same application – such as for proposed research with multiple investigators, or a center or network proposal with multiple sites and investigators – you

---

[11] The LOI is sometimes an online form that the PI fills out and submits before the submission deadline for the grant proposal (e.g., Strong Start for Mothers and Newborns, http://www.innovation.cms.gov/initiatives/Strong-Start [accessed November 28, 2014]).

should strive to have all CVs formatted similarly, including the lists of publications. However, such consistency is hard to achieve within one CV, let alone across multiple ones.

The following list identifies features of a CV that are sometimes inconsistent and that need your attention before submission:

● **Font.** To help unify all CVs, they should all use the same font, unless otherwise specified by submission requirements.

● **Abbreviations.** You need to use the same abbreviations throughout a CV. If you use an abbreviation for a proper noun that is frequently used, such as *UC Berkeley* for the *University of California, Berkeley,* you need to introduce the abbreviation in parentheses, upon first mention of the term. If you use a common abbreviation, such as *Dept.* for *Department,* you do not need to first introduce it. Thus, for example, you would use *Dept. of Biomedical Engineering* on first mention in a CV, not *Department (Dept.) of Biomedical Engineering.*

You need to use the same punctuation for abbreviations of a particular type. For example, you need to use periods in all abbreviations of degrees or in none of them. Example 6a gives a series of degrees with inconsistent punctuation, which should be avoided. Examples 6b and 6c present all abbreviated degrees with consistent punctuation.

Example 6
a. MS, MD, M.P.H.
b. MS, MD, MPH
c. M.S., M.D., M.P.H.

● **Sequencing.** The most common sequence used for biographical information is chronological order, from the most recent to the furthest in time. However, it is common practice to sequence biographical information in all fields of a CV from the furthest in time to the most recent, perhaps due to an online example of an **NIH Biographical sketch** that has all of the entries beginning with those furthest in time and ending with the most recent.[12] Whether you sequence biographical information from the furthest in time to the most recent, or vice versa, you need to be consistent, as noted in Guideline 7.8.

> **Guideline 7.8** *Sequence all information in all fields of a CV in the same organizational pattern.*

7.6.2 Publications

If you use a software program for your list of publications in a CV, the entries will all be formatted consistently. However, you may have compiled the list at different times, some manually and some even with different software. As a result, you may need to edit your entries so that they are all consistent.

---

[12] http://www.niaid.nih.gov/researchfunding/grant/pages/samples.aspx (accessed November 28, 2014).

> **Guideline 7.9**   *Create consistent entries for the list of publications in a CV, in terms of: (1) the type of information, (2) the arrangement of information, (3) the punctuation that separates each type of information, (4) abbreviations, (5) capitalization, (6) font, and (7) font-highlighting features for particular information.*

Note that if the grant proposal application includes multiple CVs, you may likely have entries that are formatted differently. In this case, you should at least check for consistency across entries in the list of publications within each CV. The following example shows 2 consistent entries from a publications list in a CV. Each entry presents the same types of information, in the same order, in the same font, and in the same punctuation separating each type of information. Also notice the use of bold font for the PI's name, which can help reviewers quickly locate the PI's name.

Example 7

P Smith, K Jones, **W Smythe**. Increased dynamic regulation of postural tone through Alexander Technique training. *Hum Mov Sci* 2011, 30:74–89.

S Roberts, **W Smythe**, J Smith. Contributions of skin and muscle afferent input to movement sense in the human hand. *J Neurophysiol* 2011, 105:1879–1888.

Ideally, if you have the time, you should revise the capitalization of all titles to journal articles so that they match across entries, regardless of the capitalization scheme used when first published. Two alternative capitalization schemes are: (1) initial capitalization only (sometimes called *sentence capitalization*), except for proper nouns (see Example 7), and (2) conventional capitalization, usually reserved for the titles of books and involves the capitalization of all words except for articles (*a, an, the*), conjunctions (e.g., *and, but, or*), and prepositions (e.g., *under, by, as*).

Names of journals are italicized, to distinguish them from article titles. Names of journals are also abbreviated in the list of publications. There are many different internet sites that identify standard abbreviations to journals. Here are just a few sites that were accessed April 9, 2014:

http://cassi.cas.org/search.jsp
http://www.library.illinois.edu/bix/resources/journals/abbreviations.html
http://library.caltech.edu/reference/abbreviations
http://www.abbreviations.com/jas.asp
http://library.humboldt.edu/infoservices/FTitleAbbr.htm

7.7   Discussing visuals

A visual displays information using images, physical orientation, and/or dimensions. In academic documents, such as the narrative to a grant proposal, visuals supplement the text. In contrast, in non-academic writing, such as in some advertisements, the text can supplement visuals, in which case the visuals carry the burden of communicating most of the key information.

There are many reasons to use visuals in a narrative, such as: to clarify the text, to summarize passages in the text, to emphasize information, to help reviewers remember information, to help reviewers understand part-whole relationships, to help explain logical relationships and patterns in data, and to help reviewers understand complex concepts and processes that are explained in the text. On the other hand, visuals can take more space, but if you have the space, you should use them because they help clarify the text. Visuals can help save space because they can present a lot of information – especially collected or analyzed data – in less space than the same information integrated into sentences. However, you still need to discuss the visuals.

This section gives strategies on how to discuss visuals. Since an effective discussion of a visual depends on how you prepare it for discussion, we necessarily first address what this preparation entails. We do not address how to select and design different types of visuals. The term *table* refers to a visual that displays its information in columns and rows, and the term *figure* is a visual that does not display its information in columns and rows. Thus, the term *figure* as used here is a broad term that covers, for example, flow charts, graphs, photographs, maps, charts, histograms, diagrams, sketches, and cartoons.

7.7.1   Drafting visuals in advance of writing

It is easier and faster to discuss visuals clearly and precisely while looking at them rather than to discuss them while imagining what they will look like, once you create them. As a result, it is best to draft visuals **before** writing the text about them.

Guideline 7.10   *Draft your visuals before writing about them.*

However, Guideline 7.10 is not meant to suggest that you can or should use a ready-made visual from a previous publication or grant application. These pre-existing visuals sometimes include information that is unnecessary or irrelevant to the discussion in the narrative. If you are inclined to borrow one of your pre-existing visuals, you should modify it so that it is appropriate to the passage in the narrative where you intend to locate it. The bulleted item **Using pre-existing visuals** below gives suggestions for how you can modify a pre-existing visual.

The following discussion identifies characteristics of visuals that can help you explain your proposed research.

● **Simplicity and readability**. Visuals that are not cluttered and that present their information with simplicity and elegance are easier for writers to discuss and are easier for readers to understand. If unnecessarily cluttered or complex, a visual can distract, confuse, or slow readers down, instead of helping them readily understand the point you are trying to make in the text. For example, there is no need for shadow or 3-D boxes unless you are dealing with content that needs dimensionality for clarity. There is no need to include a background grid on a histogram if you want reviewers only to notice the overall slope distribution of bar heights on the histogram.

A common problem with visuals, regardless how elegantly you may design them, is their size. Sometimes they are too small for reviewers to read easily, especially when they are printed in hard copy. Postage stamp-size visuals that cannot be easily read can slow and irritate reviewers. Sometimes, however, visuals can be too large, which results in wasted space and in giving the impression that you are trying to fill space. You should print out each page with a visual, to evaluate its size and readability on the page.

Guideline 7.11   *Avoid small visuals that are hard to read and large visuals that waste space.*

Another problem with visuals involves the size of font for their labels and the explanatory information that accompanies the visual. Submission requirements will usually indicate the minimum size of font for visuals in the narrative. With this said, for readability, you might need to use a font that is larger than the minimum required size. Also, reviewers might be reading your grant proposal at a time of day, such as late evening, when their eyes could be tired, so a larger font can help the legibility of the text. Guideline 7.12 discusses the size of fonts in visuals.

Guideline 7.12   *Unless your funding agency specifies otherwise, consider **not** using font smaller than 11 point Times Roman, or 10 point Arial or Helvetica, for all text and symbols on a visual.*

• **Headings for visuals**. As illustrated in Case 7-5, a heading to a visual includes: (1) the type of visual, (2) a reference number, and (3) a title. You can use bold or italic font in the heading to make it visually outstanding. A standard location for a heading to a table is above it, where the explanatory paragraph is also located; a standard location for a heading to a figure is below it, with its explanatory paragraph. With this said, however, it is becoming increasingly common to find headings and explanatory paragraphs to both tables and figures above them. You should also consider placing headings and the explanatory paragraph to the side of the visual in order to save space (see Chapter 8.6.2).

If you have multiple figures of one type, such as flowcharts, you can create a category for this one type, so that within a narrative, for example, you might have tables, figures, and flowcharts.

There are 2 types of numbering schemes for visuals from which to choose. In the first, all visuals of a particular type are numbered sequentially, regardless of the section in which they occur and regardless of the order in which they occur. Thus, for example, if you are including 2 tables, 3 figures, 2 flowcharts, 3 more tables, one more figure, and 2 more flowcharts in the narrative, the reference number scheme would be:

Table 1	Figure 1	Flowchart 1	Table 3	Figure 4	Flowchart 3
Table 2	Figure 2	Flowchart 2	Table 4		Flowchart 4
	Figure 3		Table 5		

An alternative numbering scheme has 2 digits: the first digit designates the major section in the narrative where the visual resides, and the second, the sequence in which the visual occurs. At the beginning of each section of the narrative, you reset the sequence number to "1." For example, for visuals residing in a **Preliminary Studies Section** that is itemized as **Section C** and that includes 2 tables, 3 figures, 4 flowcharts, and 2 more tables, the numbering scheme would be:

Table C-1	**Figure C-1**	**Flowchart C-1**	**Table C-3**
Table C-2	**Figure C-2**	**Flowchart C-2**	**Table C-4**
	Figure C-3	**Flowchart C-3**	
		Flowchart C-4	

For visuals residing in a **Preliminary Studies Section** that is itemized as **Section 3**, the numbering scheme would be:

Table 3-1	**Figure 3-1**	**Flowchart 3-1**	**Table 3-3**
Table 3-2	**Figure 3-2**	**Flowchart 3-2**	**Table 3-4**
	Figure 3-3	**Flowchart 3-3**	
		Flowchart 3-4	

If you have only one visual in your entire narrative, then you can omit the number designation in the heading. However, to help reviewers understand what the visual is about, you should identify its type and title.

● **Titles to visuals.** The title to a visual briefly identifies its topic, that is, what the visual is about. The title is usually phrased as noun. The visual should be designed so that the reviewer can look at it and identify its type, so you should have no need to name the type of visual (e.g., figure, table, or schematic) in the heading.

Conciseness is needed in a title and in its explanatory paragraph. In Example 8a, the type of visual, *a schematic representation*, is not needed in the title and is omitted in Example 8b. The title in Example 8a is also wordy and includes information that belongs in an explanatory paragraph. By reorganizing the information into a title with an explanatory paragraph, not only does readability improve, but the heading is shortened (see Chapter 8.6 on strategies to shorten the text).

Example 8

a. **Figure 4. Schematic representation showing representative EMG** for the right (R) biceps femoris (BFM) and left (L) external oblique (EOS) muscles pre- (gray solid) and post-treatment initiation (black dash) during a straight leg-raise task for 2 subjects with lower back pain. (41 words, 272 characters with spaces.)

b. **Figure 4. Representative EMG.** (Right) biceps femoris (BFM) and (Left) external oblique (EOS) muscles pre- (gray solid) and post-treatment initiation (black dash) during a straight leg-raise task for 2 subjects with lower back pain. (34 words, 230 characters with spaces.)

Throughout the narrative, titles should be capitalized and punctuated consistently. Consistent punctuation, such as a period, a colon, or a dash, should follow the number of the visual. The title ends with a period (or sometimes a dash or colon).

- **Visuals with panels.** If the visual has fields or panels, each is assigned an upper case letter in parentheses, beginning with (A):

(A)	(B)
(C)	(D)

- **Labels and vocabulary.** Labels are descriptive text on the face of the visual, to help readers understand what they are seeing. Labels include key terms, abbreviations, measurement units, and symbols used to define axes, symbols, and other graphical features. Labeling text should match the vocabulary used in the text that discusses the visual. The following guideline summarizes how to label visuals.

> **Guideline 7.13**
>
> (a) *Orient labels horizontally. The one exception is the label for vertical axes of graphs. Labels to vertical axes are oriented vertically to save space.*[13]
> (b) *If there is space, use labels in their unabbreviated forms; if space is limited, use standard abbreviations, colors, and symbols as much as possible and then explain them in a legend or at the end of an explanatory paragraph.*
> *Exception: Use abbreviated forms for common measurement units, such as cm, ft, and hr.*
> (c) *Do not use periods in any abbreviations that you place on the visual and in its explanatory paragraph.*
> (d) *Use the same key terms, abbreviations, and symbols on the visual that you use in the text.*

When selecting labels for visuals, you should look to the key terms in your text. If you draft visuals and legends before the text is written, as recommended, you need to think ahead about how you will discuss the visual, and then once the text is drafted, you need to double-check that the key terms, abbreviations, and symbols on the visual still match those in the text. Because of the need to match the key terms, abbreviations, and symbols on the visual to those in the text, it may be difficult using a pre-existing visual from a publication without first modifying it to be consistent with the narrative.

- **Explanatory paragraphs.** A title to a visual in a narrative can be followed by an explanatory paragraph, in run-in layout. An explanatory paragraph, comprising one or

---

[13] Some graphs have both right- and left-vertical axes to provide units for two superimposed plots. The left axis label is rotated 90° counter clockwise, and the right axis label is rotated 90° clockwise (from the horizontal).

more sentences or sentence fragments, identifies noteworthy information in the visual that has been mentioned within the text. If the visual has panels, the explanatory paragraph identifies what each panel is about. Each item in the list bears the alpha itemization of each panel, followed by the explanation. Also, all items in the list should be in parallel grammar (see Chapter 8.6.3). Sometimes the alpha itemization is presented in bold font, to improve readability. The explanatory paragraph can end with legend information.

● **Legend information.** Legend information lists what each abbreviation, color, and symbol on the visual represents. Legend information can be listed in run-in layout, after the explanatory paragraph. However, sometimes it is placed in a frame on the face of the visual, usually in the upper or lower right-hand corner, to save space. The frame, called the *legend*, usually takes less textual space than does the same information added to the end of the explanatory paragraph. You need to assess legends for readability.

● **Using pre-existing visuals.** If you already have a relevant visual in one of your publications and will use it in the narrative, you need to include a citation to the publication. This citation, such as *(reproduced from Table 4 in ref. [3])* can be located at the end of an explanatory paragraph, after any legend information. The citation can also be included on the face of the visual, usually in the lower-right corner.

Often a visual cannot be cut and pasted from a publication, inserted into the narrative, and still be appropriate for a particular passage in the narrative. The visual might include other information that is not relevant to your proposed research, or it might include key terms or abbreviations that are different from those in the narrative. Therefore, you might need to modify your published visual for it to fit your narrative, in which case you need to identify the visual as modified, such as by noting *(modified from Table 4 in ref. [3])*.

> **Guideline 7.14** *If you intend to use a pre-existing, published visual in the text:*
> (a) *Revise the visual so that it contains only information that is relevant to your proposed research and so that it uses only those terms and abbreviations used in the text.*
> (b) *Indicate on the visual or at the end of its explanatory paragraph that it is modified, and then cite to the publication.*

7.7.2 Where to locate visuals

Visuals can be used in any section of the narrative,[14] unless submission requirements indicate otherwise. However, there are certain sections where visuals are less commonly used than others.

Visuals are not often used in an **Aims Section**. Visuals are at times included in a **Background Section**, to explain some point that ultimately clarifies or reinforces a

---

[14] The abstract is not considered part of the narrative; the abstract does not include visuals.

feature in your proposed research design or methods. However, compared with all other major sections of the narrative, visuals are the most common in a **Preliminary Studies/Progress Report Section**, where they present your preliminary data and analyses, and sometimes clarify features of your preliminary methods that are relevant to your proposed research. Visuals are also commonly used in a **Methods Section**, for example, to help explain your proposed research design and methods (e.g., see Chapter 5.2.2 on overview tables).

Visuals can be located within the text at relevant points in your discussion, or you can stack them at the end of the (sub)section where you first refer to them. Although stacking visuals can save you time in terms of positioning visuals (and their headings and explanatory paragraphs), this strategy might create readability problems. Upon being referred to a stacked visual, reviewers might not immediately skip to the end of the section to look at it, or they might skip to the end and examine all of the stacked visuals instead of focusing on the one that you want them to notice at that particular time.

7.7.3 Discussing visuals

Discussing visuals in the narrative usually involves: referring to the visual, explaining particular information on the visual that you want reviewers to notice (you cannot assume they see what you want them to see), and possibly interpreting what the information on the visual means.

• **Referring to visuals**. Guideline 7.15 explains where to refer to a visual first, relative to where it resides in the document.

> **Guideline 7.15** *Consider consistently locating the first reference to a visual:*
> *(a) In the text, shortly before the visual.*
> *(b) In the text, before or next to the visual if text is wrapped around the visual.*

By referring reviewers to the visual *before* they encounter it on the page, you have a better chance of influencing when they will first look at it, without distraction from other visuals. For a visual with text wrapped around it, you can refer to the visual before or within the wrapped text.

For readability and clarity, visuals are not often identified with the term *above* or *below* since it is unclear how much before (or after) the reference point that reviewers need to look. Instead, you need to refer to the visual by its number, such as *See Fig. 4.*

In addition, an impersonal reference to a figure, such as *It is shown in Fig. 4 that . . .* is not preferred phrasing since it is unnecessarily wordy and passive. The phrase *Fig. 4 shows that . . .* suffices. When you refer to a figure, the term *figure* is not abbreviated to *Fig.* when it leads a sentence. However, it may be consistently abbreviated to *Fig.* throughout the narrative when it occurs within a sentence.

- **Reviewing visuals.** You cannot assume that reviewers will notice the features of a visual that you want them to notice, even if the features are obvious. You need to point out anything on a visual that you want them to notice. When you explain the visual in the text, the terms that you use and the labels on the visual need to match for clarity.

> **Guideline 7.16** *Explain what the visual is about and point out what you want reviewers to notice.*

If you find that you cannot describe the visual in terms of the vocabulary that you use in the narrative, you have a writing problem with: (a) the labels you have selected for the visual; (b) the terms you use in the text to describe the visual; and/or (c) the location where you first refer to the visual. You will likely need to revise the visual so that its labels match the terms in the text. It is not likely that you will need to change the terms in the text since you should already be using the key terms from the **Aims Section** (Chapter 2.2.1).

Further, when discussing a visual in the text, especially a table, you should not just recite the information in its columns and rows. The discussion needs to identify data patterns, tendencies, trends, and statistical analyses in the visual or should direct the reviewers' attention to particular data on the visual.

> **Guideline 7.17** *In the text, identify patterns of data in the visual that you want to bring to the attention of the reviewers.*

A problem with visuals is that they sometimes include unnecessary information – information that is not discussed in the text and/or not relevant to the proposed research. This problem sometimes arises when you use a ready-made visual from a previous publication or a grant application, and it includes extra information than is needed. You should modify any pre-existing visual so that its scope matches its use in the narrative.

- **Analyzing information on a visual.** After reviewing the information on a visual that you want reviewers to notice, you can analyze the information in the text through the use of conclusions, speculations, problem statements, hypotheses, and predictions, which is how you analyze a review in a scientific argument (see Chapter 2.2.3). However, only an analysis that is a conclusion is usually also mentioned in the explanatory paragraph of a visual. If your analysis involves other types of analyses, such as speculations or hypotheses, those analyses are usually addressed only in the text, not in the explanatory paragraph.

> **Guideline 7.18** *Consider including conclusions in explanatory paragraphs, but not speculations and hypotheses.*

7.8   Working with collaborators

Many senior PIs have stories about getting collaborators to send their contributions to them *in a timely manner*. When collaborators send in their contributions, they sometimes do so the day or night before the submission deadline. Receiving a contribution at the last minute creates problems for the PI since the contribution should be reviewed for content and revised in order for its style to be consistent with the rest of the narrative. This section has suggestions for editing contributions so that they reflect the style of the rest of the narrative, but if time is short, you might need to choose which features to edit (see Chapter 7.8.2).

7.8.1   Clarifying the contribution that you want

To receive a contribution to the narrative that will require light editing, you likely will need to provide information to the collaborator well in advance of any deadline that you give the collaborator, and then again a couple of weeks before the deadline. This information should include: the targeted funding agency and type of grant proposal, the collaborator's role in the proposed research, and the collaborator's role in writing the narrative and providing information for other parts of the grant application.

Guideline 7.19   *To clarify the collaborator's writing task, identify:*
(a) *The topic of the contribution for the narrative.*
(b) *The location where you will insert the contribution in the narrative.*
(c) *The maximum length of the contribution.*
(d) *Whether the contribution should include citations and visuals, and your numbering system.*
(e) *Whether the collaborator needs to provide biographical information, and any biographical format.*
(f) *Information that you need for other parts of the grant application.*
(g) *Your absolute deadline to receive the contribution and any other information.*
(h) *Contact phone numbers and email addresses.*

Case 7-6 gives examples from emails about an anticipated collaboration.

When explaining the contribution, you should identify: (a) the topic of this contribution, such as *to describe your alternative method for X*, (b) the subsection where the contribution will be used in the narrative, such as *in the methods for Aim 2*, (c) the maximum length of the collaborator's contribution: the contribution that you receive may still run long, in which case you will need to revise it for length or to shorten another part of the narrative to accommodate the long contribution (see Chapter 8.6), (d) whether the contribution may have citations and visuals, and if so, the citation style that you prefer and the number and format of the visuals; (e) whether the collaborator needs to provide a Biosketch or an academic resumé, as indicated in submission requirements, (f) other information for other parts of the grant application, such as the budget, letter of support, or laboratory resources, (g) the date by which you need the contribution, which should be at least a week in advance of any institutional deadline; you should also ask whether the collaborator anticipates any problem meeting your schedule, and (h) your email address and phone numbers where the contributor can contact you personally at work (office and lab numbers) and at home (land and cell lines). You also need to ask for the contributor's same contact information.

7.8.2   Editing and integrating the contribution into the narrative

Integrating a collaborator's contribution into the narrative involves more than cutting and pasting. For each contribution, you need to edit it so that it reflects stylistic characteristics used in the rest of the narrative and fits within the space that you have allotted for it. Below are a few features that you need to evaluate in the contribution (see the check-off list in Table 7-4).

● **Key terms and abbreviations**. Check that the contribution uses the same key terms as used in the rest of the narrative. For example, if a contribution uses the term *circulation* and the rest of the narrative uses the term *vessels*, you need to edit the contribution, changing *circulation* to *vessels*. You also need to check that the contribution uses the same abbreviations as used in other parts of the narrative, and no others. Even common abbreviations for time and weight need to be consistent. If the contribution uses *sec* for *second* and *hour* for *hour*, but the narrative uses *s* for *second* and *hr* for *hour*, you need to change the abbreviations in the contribution to match those in the narrative.

● **Spelling and capitalization**. Revise the contribution so that terms are spelled and capitalized consistently in the contribution and the rest of the narrative. For example, if a contribution uses the British spelling *colour*, change it to *color* if American spelling is used in the rest of the narrative. If the contribution uses *data base*, but *database* is used in the narrative, change *data base* to *database*. If the contribution uses *co-I* (for co-investigator), but the narrative uses *Co-I*, you need to edit *co-I*.

● **Tone and person**. Check that the contribution matches the tone in the rest of the narrative. If a contribution uses a more informal style with contractions, but the rest of the narrative does not, you need to revise the contractions to increase the formality. If a contribution uses unnecessarily long, multi-syllabic words that you do not use in the rest of the narrative, such as *heretofore*, you need to change them to shorter, synonymous expressions (see Chapter 8.6.3). If you use *we* to refer to the PI and the research team, but the contribution uses *I*, you need to change the contribution's first-person pronouns.

● **Numbers**. Select a style for numbers that is consistent across the contribution and the narrative. For example, if a contribution uses words for all cardinal numbers 2 and higher, but the rest of your narrative uses digits, you need to change the cardinal numbers in the contribution to digits.

• **Punctuation.** Use consistent punctuation in the narrative and the contribution. Three punctuation schemes, in particular, might be inconsistent. (1) If the style in the contribution does not include a comma before *and* in a list of 3 or more items (i.e., "*X, Y and Z*") but the rest of the narrative uses a comma (i.e., "*X, Y, and Z*"), you need to revise the contribution. (2) If the contribution uses periods in abbreviations, such as *Ph. D., Dept.* and *Exp. 3,* but the rest of the narrative does not, you need to remove the periods in the contributions' abbreviations. (3) The contribution might place citations outside final periods, and the narrative might place them within final periods, or vice versa. You need to revise the contribution to be consistent with your placement of citations in the rest of the narrative (see Chapter 8.1).

• **Cross-references and citations.** If the contribution includes information that is explained in another section of the narrative, consider adding a cross-reference to the other section and removing most of the information from the contribution. If the contribution uses numbered citations but you use citation by author's family name, you need to change the numbered citations in the contribution. If the contribution uses a trigger for a citation (see Chapter 8.1.3) but does not include a citation, you need to contact the contributor for the citation or change the sentence so that it does not use phrasing that will trigger the need for a citation.

• **Visuals.** Adjust the text in the narrative if the contribution includes a visual. To handle this situation, you will need to assign a number and possibly a title, legend, and explanatory paragraph to the visual. If you assign a number to the visual, you need to revise the narrative by renumbering all of the subsequent visuals in the narrative and changing the number of all subsequent visuals in the text. An alternative is to include the visual without a number and direct the reviewers to the numberless visual through a phrase, such as *In the following table.*

• **Length.** If the contribution takes more space than you had allotted, trim the contribution by performing any or all of the following editing maneuvers: (1) significantly reduce or omit redundancies between the contribution and the rest of the narrative, using cross-references, (2) apply text-shortening strategies to reduce the length of the contribution (see Chapter 8.6), (3) apply text shortening strategies in the rest of the narrative in order to free-up space for the contribution, (4) cut content in the contribution (consider first checking with the collaborator), and (5) time permitting, ask the collaborator to cut some information and to send it back to you.

Table 7-1. Submission requirements for abstracts. (A) National Institutes of Health (http://grants.nih.gov/grants/funding/phs398/phs398.html; I-34; accessed November 28, 2014) and **(B)** National Science Foundation (http://www.nsf.gov/pubs/policydocs/pappguide/nsf13001/nsf13_1.pdf; II-7; accessed November 28, 2014).

(A) National Institutes of Health – Submission Requirements

4.2.1 Description: Project Summary and Relevance
The first and major component of the Description is a **Project Summary**. It is meant to serve as a succinct and accurate description of the proposed work when separated from the application. State the application's broad, long-term objectives and specific aims, making reference to the health relatedness of the project (i.e., relevance to the mission of the agency). Describe concisely the research design and methods for achieving the stated goals. This section should be informative to other persons working in the same or related fields and insofar as possible understandable to a scientifically or technically literate reader. Avoid describing past accomplishments and the use of the first person.

The second component of the Description is **Relevance**. Using no more than two or three sentences, describe the relevance of this research to public health. In this section, be succinct and use plain language that can be understood by a general, lay audience. **DO NOT EXCEED THE SPACE PROVIDED.**

> **Use text only (no figures or other information not in standard text). Do not include proprietary, confidential information or trade secrets in the description section. If the application is funded, the project description will be entered into an NIH database and will become public information.**

(B) National Science Foundation – Submission Requirements

b. Project Summary
Each proposal must contain a summary of the proposed project not more than one page in length. The Project Summary consists of an overview, a statement on the intellectual merit of the proposed activity, and a statement on the broader impacts of the proposed activity. The overview includes a description of the activity that would result if the proposal were funded and a statement of objectives and methods to be employed. The statement on intellectual merit should describe the potential of the proposed activity to advance knowledge. The statement on broader impacts should describe the potential of the proposed activity to benefit society and contribute to the achievement of specific, desired societal outcomes.

The Project Summary should be written in the third person, informative to other persons working in the same or related fields, and, insofar as possible, understandable to a scientifically or technically literate lay reader. It should not be an abstract of the proposal.

Proposals that do not contain the Project Summary, including an overview and separate statements on intellectual merit and broader impacts will not be accepted by FastLane or will be returned without review. Additional instructions for preparation of the Project Summary are available in FastLane.

Table 7-2. Organizational alternatives for the body of a resubmission introduction.

(A)	(B)
Concerns of Summary Statement or Reviewers	**Responses to Reviewers' Common Criticisms**
1. "Quote" or Summarize the Issue (Reviewers 1 & 2) Response	1. "Quote" or Summarize the Issue Response
2. "Quote" or Summarize the Issue (Reviewer 1) Response	2. "Quote" or Summarize the Issue Response
3. "Quote" or Summarize the Issue (Reviewer 2) Response	3. "Quote" or Summarize the Issue Response
4. "Quote" or Summarize the Issue (Reviewers 2 & 3) Response	**Responses to Reviewer 1's Criticisms**
(C)	1. "Quote" or Summarize the Issue Response
Aims	2. "Quote" or Summarize the Issue Response
1. "Quote" or Summarize the Issue (Reviewer 1) Response	3. "Quote" or Summarize the Issue Response
Background	**Responses to Reviewer 2's Criticisms**
2. "Quote" or Summarize the Issue (Reviewers 1 & 3) Response	4. "Quote" or Summarize the Issue Response
3. "Quote" or Summarize the Issue (Reviewer 2) Response	5. "Quote" or Summarize the Issue Response
Preliminary Studies	**Responses to Reviewer 3's Criticisms**
4. "Quote" or Summarize the Issue (Reviewer 1) Response	6. "Quote" or Summarize the Issue Response
Research Design and Methods	7. "Quote" or Summarize the Issue Response
5. "Quote" or Summarize the Issue (Reviewer 1) Response	
Budget	
6. "Quote" or Summarize the Issue (Reviewer 2) Response	

Table 7-3. Organizational alternatives for a short narrative.

A. Four Major Sections	B. Five Major Sections
1. **Aims**	1. **Aims**
2. **Background** a. Others' Research b. PI's Preliminary Studies	2. **Background**
	3. **Preliminary Studies**
3. **Methods** a. Methods for Aim 1 b. Methods for Aim 2	4. **Methods** a. Methods for Aim 1 b. Methods for Aim 2
4. **Future Research**	5. **Future research**

Table 7-4. Check-off list for reviewing contributions.

1. Key Terms and Abbreviations
- Does the contribution use the same key terms as used in the rest of the narrative?
- Does the contribution use the same abbreviations as used in the rest of the narrative?
- Does the contribution use too many abbreviations?

2. Spelling and Capitalization
- Is the spelling of the contribution consistent with that in the rest of the narrative?
- Do the key terms in the contribution match those in the rest of the narrative?
- Check the spelling and capitalization of particular terms, such as *colour* and *data base*.

3. Tone and Person
- Is the use of contractions consistent with the rest of the narrative?
- Does the contribution use unnecessarily long, multi-syllabic words?

4. Numbers
- Does the numbering scheme in the contribution match the numbering scheme in the rest of the narrative (e.g., digits for all cardinal numbers >1)?
- Are all numbers that begin a sentence written as words?
- Is the numerical precision consistent with the rest of the text in terms of real numbers versus integers and number of decimal places, for example?

5. Punctuation
- Does the contribution include a comma before *and* in a list of 3 or more items? Is this usage consistent with the rest of the narrative?
- Does the contribution use periods in abbreviations? Is this punctuation consistent with the abbreviations in the rest of the narrative?
- Does the contribution place citations outside final periods? Is this usage consistent with the placement of citations in the rest of the narrative?
- Is there similar punctuation at the ends of lines in vertical lists?

6. Cross-references and Citations
- Does the contribution use the correct cross-reference to another passage in the narrative?
- Does the contribution use the same style of citation as does the rest of the narrative – either by author's family name or by number?
- Does the contribution use a citation trigger (see Chapter 8.1) without a citation?
- If 2 authors are cited, does the contribution use "and" or "&"?

7. Visuals
- Does the contribution include a visual with a correct number, given the numbering scheme for visuals in the rest of the narrative?
- Do the visuals in the narrative have the correct number?
- Are tables formatted and colored similarly?

8. Length
- Is the contribution too long? If so, edit the contribution to shorten it, and/or edit the rest of the text to shorten it. (See Chapter 8.6)
- Does the contribution omit important detail that is needed to support other parts of the narrative?

Case 7-1. Three versions of an unstructured abstract. (A) Substantive abstract: original, **(B)** Topical Abstract: Synthesized from original, and **(C)** Mixed abstract: synthesized from original.

A. Substantive abstract

[1]About 1% of newborn babies have congenital heart disease (CHD), the cause of which is not well understood. [2]Hemodynamic forces, which are the shear stresses and pressures generated by the flow of blood inside the heart, are key epigenetic factors that regulate heart development and, if abnormal, can lead to CHD. [3]In addition, although one defect may have a genetic cause, altered blood-flow conditions produced by the defect may also detrimentally affect heart development. [4]A detailed study of the effects of hemodynamic forces on heart development requires the quantification of these forces in vivo. [5]Quantification of shear forces, however, is extremely difficult using current experimental techniques. [6]In this project, we propose a combination of experimental data and finite element modeling to quantify hemodynamic forces during the cardiac cycle.

[7]We will use chick embryos early in development (2.8 days and 3.5 days, of a 21-day incubation period) as models of heart development. [8]We will focus on the heart outflow tract (OFT), which at this age is a curved tube that connects the ventricle to the aortic sac and acts as a primitive valve by expanding and contracting. [9]Both the OFT wall motion and the difference in pressure between the ventricle and the aortic sac determine blood flow through the OFT. [10]To alter normal blood-flow dynamics, we will band the OFT by placing a suture around the mid-section of the OFT and constraining the maximum cross-sectional area at the band. [11]Our hypothesis is that even though the heart rapidly adapts to constriction of blood flow, hemodynamic forces remain significantly different from normal. [12]Our objective is to determine the dynamic changes in hemodynamic forces and OFT wall motions during the cardiac cycle in normal and banded chick embryos.

[13]Aim 1: Characterize physiological parameters of normal and banded hearts of HH18 and HH21 chick embryos. [14]We will measure the movement of the OFT wall during the cardiac cycle and the difference in pressure between the ventricle and aortic sac.

[15]Aim 2: Determine shear stresses on the inside of the OFT walls of normal and banded hearts of HH18 and HH21 chick embryos. [16]We will use a dynamic finite element model of the chick OFT, which will incorporate the movements of the OFT wall and the pressures (from Aim 1), to calculate shear stresses.

B. Topical abstract, synthesized from the substantive abstract in Case 7-1A

[1]The effects of hemodynamic forces on heart development, which requires the quantification of these forces in vivo, will be investigated. [2]The focus of investigation will be on the heart outflow tract (OFT). [3]Problems in quantifying shear forces using current experimental techniques will be explained, and modeling techniques will be proposed to quantify hemodynamic forces during the cardiac cycle. [4]A hypothesis related to heart adaptation to blood flow constriction will be identified and investigated. [5]Two aims are proposed to investigate physiological parameters of chick embryos and to determine shear stresses from hemodynamic forces on the inside of the OFT walls.

C. Mixed Abstract, synthesized from the Substantive Abstract in Case 7-1A

[1]About 1% of newborn babies have congenital heart disease, the cause of which is not well understood. [2]Hemodynamic forces, the shear stresses and pressures generated by blood flow inside the heart, are key epigenetic factors that regulate heart development and, if abnormal, can lead to congenital heart disease. [3]Although one defect may have a genetic cause, altered blood-flow conditions produced by the defect may also detrimentally affect heart development. [4]A study of the effects of hemodynamic forces on heart development requires the quantification of these forces in vivo. [5]Quantification of shear forces, however, is difficult using current experimental techniques. [6]In this project, we propose modeling techniques to quantify hemodynamic forces during the cardiac cycle. [7]Our focus will be on the heart outflow tract (OFT), a curved tube at 2.8 days and 3.5 days of development that connects the primitive ventricle to the aortic sac and acts as a primitive valve by expanding and contracting. [8]We will investigate a hypothesis related to heart adaptation to blood flow constriction. [9]Our objective is to determine the dynamic changes in hemodynamic forces and OFT wall motions during the cardiac cycle in chick embryos. [10]Two aims are proposed to investigate physiological parameters of chick embryos and to determine shear stresses on the inside of the OFT walls.

Case 7-2. Structured abstract from a grant proposal to the Michael J. Fox Foundation. Abstract from https://www.michaeljfox.org/foundation/grant-detail.php?grant_id=1019 (accessed September 23, 2012).

Objective/Rationale:
[1]The objective is to compare the pharmacokinetics of a single 11-mg dose and 22-mg dose of APL-130277 compared to 2 mg and 4 mg of Apokyn subcutaneous (SQ) injection in a study of 15 subjects per cohort. [2]APL-130277 is a sublingually administered, thin-film product for use as rescue medication for intermittent OFF episodes in Parkinson's disease (PD). [3]The product is easy to self-administer under the tongue. [4]The goal is to achieve pharmacokinetics that mimic Apokyn injection, without the needle.

Project Description:
[5]A comparative biostudy will measure the pharmacokinetics of APL-130277 as compared to the injectable form of apomorphine, Apokyn. [6]The study is a 2-period, placebo-controlled, cross-over design using healthy volunteers (N=30). [7]The first cohort (n=15) or subjects will receive a dose of APL-130277 followed by a 24-hour washout and then dosed with 2 mg of Apokyn, [8]Our second cohort (n=15) will receive a higher dose of APL-130277 followed by a washout and then 4 mg of Apokyn. [9]This study will determine dose proportionality and will be used to make final adjustments before starting the pivotal bioequivalence trial in 2013.

Relevance to Diagnosis/Treatment of Parkinson's Disease:
[10]Apomorphine is an under-utilized medication in PD. [11]Despite its strong efficacy and rapid onset of action, patients find injections painful and resist use until the latest stages of PD. [12]Physicians find the dose initiation cumbersome. [13]Eliminating some of these barriers are key objectives of Cynapsus' APL-130277. [14]APL-130277 is an innovative, fast-dissolving, sublingually administered, thin-film product for use as rescue medication for OFF episodes in PD. [15]APL-130277 is easy to self-administer under the tongue.

Anticipated Outcome:
[16]This comparative biostudy will lead directly into the registration study examining the bioequivalence (BEQ) of the thin-strip formulation of APL-130277 with subcutaneous apomorphine. [17]Completing this comparative biostudy will allow Cynapsus to make an accurate assessment of sample size for the BEQ study, to further examine the PK profile of the final (clinical/commercial grade) formulation of APL-130277, and to further confirm, in a controlled study, the appropriate Tmax of APL-130277 as directly compared to SQ apomorphine.

> **Case 7-3. NIH Resubmission Introduction.** In its original format, this Resubmission Introduction was 1 page long.

Introduction to Our Revised R21 Application

[1]We thank the reviewers for recognizing (a) our research is a study that "may generate knowledge on pyrethroid resistance that will have important implications on vector control programs," (b) our research team is "outstanding," (c) our proposed methods are "appropriate," and (d) our preliminary data are "strong." [2]We respond here to the issues raised in the "Summary Statement" and by specific reviewers.

[3]**1. The Summary and Reviewer 1:** "*the research scope is too large...*" [4]Three aims are proposed, the scopes of which, at first glance, may seem large for a 2-year R21 grant. [5]However, we are confident that we can complete the proposed research on schedule. [6]One critical issue for completing the research on schedule is the availability of multiple field-resistant strains of mosquitoes, of permethrin-selected offspring of field-resistant mosquitoes, and of genetic lines of mosquitoes that have different levels of resistance; and the availability of full-length sodium channel cDNAs from resistant and susceptible mosquitoes. [7]We already have available the 3 field-resistant mosquito lines, the 3 permethrin-selected offspring of the field-resistant, and the laboratory-susceptible mosquito strains, and also the most susceptible and most resistant mosquito lines to insecticide resistance that we generated from a single-pair crossing of the field mosquito strains. [8]We also already have full-length sodium channel cDNAs from susceptible and resistant mosquitoes. [9]Further, our research team (a PI and 2 Co-Is) has extensive experience in conducting all of the experiments proposed. In preliminary studies, we successfully executed our proposed procedures, which include insecticide bioassays, RNA and DNA isolation, SNP determination, site-directed mutagenesis, sodium channel expression, and electrophysiology. [10]Thus, we do not anticipate any major technical problems and we expect to complete the research on schedule.

[11]**2. The Summary and Reviewer 2:** "*...to some reviewers it is not clear that this application will clarify the mechanism of insecticide resistance development*" [12]A large number of previous studies clearly demonstrated that mutation-mediated sodium channel insensitivity is one of the important mechanisms in pyrethroid resistance (ref). [13]However, those studies focused on only **partial** sodium channel sequences in insects, in which previously confirmed *kdr* mutations had been detected (ref) and **did not globally and comprehensively** analyze all mutations in the entire sodium channel that could contribute to insecticide resistance in an individual resistant insect. [14]Several questions still need to be answered in order to better understand how sodium channel insensitivity-mediated resistance develops (see Specific Aims, paragraph 2). [15]Our R21 research aims to address these questions by characterizing all mutations and mutation combinations in the entire mosquito sodium channel, in insecticide resistance by defining the relationship between the level of insecticide resistance, and the prevalence of sodium channel mutations and mutation combinations in mosquitoes (Aim 1); and by determining the roles of these mutations and mutation combinations in sodium channel insensitivity to pyrethroids (Aim 2). [16]This application will, for the first time, clarify (a) how many

mutations, both synonymous and nonsynonymous, are involved in insecticide resistance, in an entire sodium channel of the mosquito, (b) how frequently certain mutations co-occur in insecticide-resistant mosquitoes, and (c) specific thresholds of insecticide concentrations, at which particular mutations or mutation combinations occur in a mosquito population. [17]Ultimately, this study will shed new light on molecular mechanisms underlying the development of insecticide resistance.

[18]**3. The Summary, and Reviewers 1 and 3:** *"the application lacks innovation."* [19]This concern may have come from a lack of clarity in our previous submission. [20]In the current proposal, we specifically discuss our innovations in a separate section (Section B), addressing our novel shifting of the current paradigm in insecticide-resistance research by refocusing the approach on all mutations and mutation combinations that naturally occur in an entire mosquito sodium channel in resistance; emphasizing our novel theoretical concepts (i.e., the importance of synonymous mutations and their interactions with nonsynonymous mutations in resistance); highlighting our novel methodological approaches (i.e., our focus on multiple mosquito strains, genetic lines, and groups of mosquitoes with different levels of insecticide resistance), which will facilitate not only our pinpointing key synonymous and nonsynonymous mutations and mutation combinations involved in insecticide resistance, but also, for the first time, our analyzing specific thresholds of insecticide concentrations at which particular mutations or mutation combinations occur in a resistant mosquito population.

[21]**4. Reviewer 1:** (a) *"The work of several groups... is not cited and discussed properly..."* [22]We agree that there are many excellent studies on mosquito resistance that should be cited and discussed (e.g., 22, 34, 35–9). [23]We did not cite all in our previous submission due to space limitations. [24]In the current submission, we added citations to and a discussion of recent work on *kdr*-mediated mosquito resistance (Section A.2). (b) *"The Investigator does not describe the present and previous support..."* [25]This information is now included in the PI's biosketch.

[26]**5. Reviewer 2:** *"Approval date"* [27]Approval dates for protocols of Tech U and State U are different.

Case 7-4. Administrative letter of intent.

ORGANIZATIONAL LETTERHEAD

Date

Name of Designated Contact Person, PhD, MD
Name of Organization
Street Address, Room #, Building #
City, ST Zip Code
By Email Only: xxxxxxxx@mail.gov

Subject: Letter of Intent to Apply for an Award under RFA #

Dear Dr. xxxxxxxxxxxxxxx:

I am interested in applying for an award under RFA#. Details about my application follow:

Title of grant application: "xxxxxxxxxxxxxxxxxxxxxxxxxxxxxxxxxxxxxxxxxxxxx."

Research objective: To develop high throughput assays for xxxxxxxxxxxxxxxxxxxxx.

Names and addresses of investigators and participating institutions:
PI: xxxxxxxxxxxxxxxxxx, PhD, Associate Scientist, Dept. XXX, Affiliation

Co-I: xxxxxxxxxxxxxxxx, MD, Associate Scientist, Dept. XXX, Affiliation

Number and title of this funding opportunity: RFA #, xxxxxxxxxxxxxxxxxxxxxxxxxxxx

I intend to submit this application on or before February 21, 20XX. Thank you for your consideration.

Sincerely yours,

xxxxxxxxxxxxxxxxxxxxxxxxx, PhD

Case 7-5. Heading information for visuals. (A) Heading, consisting of the type of visual, the number of the visual, a period, the title of the visual, and a period. **(B)** Optional paragraph explaining what the visual shows.

(A) Heading   (B) Explanatory Paragraph

Type   Number   Title

Figure 1. S6K1 is activated upstream of Rac1. (A) DIC images of purified human platelets (2×10^7/ml) treated with vehicle (DMSO), the Src inhibitor PP2 (20 μM), the Syk inhibitor BAY 61-3606 (1 μM), or the Rac1 inhibitor EHT 1864 (50 μM) on a surface of fibrinogen (FG). Scale bar = 10 μm. **(B)** Lysates were incubated with GST-PAK-CRIB to capture activated GTP-bound Rac1. **(C)** Platelets were treated with inhibitors as in (A) and analyzed for S6K1 activation by blotting for S6K1-pThr389.

Case 7-6. Variations on an email seeking agreements with collaborators.

(A)

Just to follow up our meeting yesterday, I want to thank you for joining our research team. There's a July 1 submission deadline from the funding agency, but I need to submit on 6/28 since my son's getting married on 7/1 and I need to get done with it in advance. So I'll need your section on statistical analyses by **June 20th**, to give me time to review it and integrate it into the Methods Section, and to meet an internal deadline with the Foundation Office. I'm reserving 3/4ths of a page for your write-up.

If you include any citations, could you use author's name style and give full bibliographic info for the list of references? I'll need your biosketch by June 20th also (see attached NIH biosketch form).

Thanks, and if you have any questions, please let me know, and if any problems come up, where meeting the June 20th deadline is at risk, could you let me know as far in advance as possible?

I'll be working mainly from home until the proposal is finished; that number is 531/222-3344. My other numbers: office #: 531/771-4412 and cell: 220/717-2155. Or you can email me. Could you also give me a couple phone numbers where I can reach you? Thanks.

(B)

About your statistical analysis section for the grant proposal, just a reminder I'll need it by **June 20th** for the 7/1 submission deadline. My son is getting married on 7/1 (he lives 1500 miles away) so I'll be submitting early. I'm saving about 3/4ths of a page for you.

I prefer any citations that you include be in author style, but I'll work with any style you give me. And could I get info on the citations for the references? Also, could you send me your CV for the application? I've attached the NIH biosketch form.

Thanks, and let me know if I can help in any way. (office: 531/771-4412; cell: 220/717-2155; home: 531/222-3344). What numbers can I reach you at, if need be? And please keep me posted if you're having problems meeting the deadline . . . Thanks again.

(C)

Thank you for agreeing to collaborate on my grant proposal. Here is information that summarizes our discussion:

1. You will write your section on statistical analysis, keeping it to about $^3/_4$ of a page (single-spaced text).

2. The submission deadline is 7/1, but I'll need it by **6/20** at the latest (I need to meet an internal university deadline that precedes the funding agency's published submission

date by about a week, plus my son's getting married 7/3 and so everything needs to be done early).

3. If you include any citation, please use author style and include info for an entry for the list of references.

4. I'll need a CV from you, on the NIH Biographical Sketch form (I've attached it).

5. What phone numbers can I reach you at, if something comes up? My phone numbers are:

Office: 531/771-4412   Home: 220/222-3344   Cell: 531/717-2155

I'm looking forward to our first collaboration together – Thanks.

Technical features of sentences

This chapter addresses technical issues with sentences that PIs typically encounter when drafting the narrative of a grant proposal. These issues involve citations, definitions, active grammar and passive grammar, and guidelines for shortening the text while keeping content changes to a minimum.

8.1 Citations

A *citation* is a reference that identifies the source of information, whether that source is from a journal, a book, the internet, a committee, or a person. Citations are important in academic documents in general, and they are particularly important in narratives to scientific grant proposals. Citations can enhance your credibility by demonstrating to reviewers the quality, breadth, and depth of your understanding of published research that relates to your research topic.

Some writers include citations while drafting the narrative, others begin including citations when they are close to the final draft, and others are somewhere in between: early in the drafting process, they fill in citations that they readily know and leave the others to later stages of drafting. If you do not include citations in early drafts of your grant proposal, you should at least include placeholders for them. *Citation placeholders* are symbols, such as XX, that do not ordinarily show up in a search of text and that you insert into the text while drafting to remind you where you need citations. Inserting citation placeholders while drafting can help you write faster and keep track of the review and analysis components of scientific arguments (see Chapters 2.2.2 and 2.2.3).

8.1.1 What to cite and not to cite

Throughout most sections of the narrative, you need citations to indicate the source of any information that has been published. An exception is the **Aims Section**, where citations are typically optional, and another exception is the abstract, which rarely includes citations, depending on submission requirements from funding agencies.

> **Guideline 8.1** *When deciding what to cite and what not to cite in the narrative:*
> *(a) Include citations in all sections of the narrative, with the possible exception of the* **Aims Section.**
> *(b) Include citations to information that has been published.*
> *(c) Consider not including citations for: (i) common knowledge and (ii) non-published information.*

- **Published information.** In the context of the narrative to a grant proposal, published information is information that has been made public or shared among researchers in hard copy format or on the internet.

When including citations, you should consider citing to information published in a refereed publication as much as possible in order to enhance your credibility. A refereed publication has undergone peer scrutiny and evaluation. Other possible authoritative sources are conference abstracts, conference proceedings, and book chapters. However, the currency of book chapters might be problematic since information in a book can become outdated by the time it is published.

Citations are not needed for common knowledge that has been published. *Common knowledge* is information that has widespread acceptance by professionals in a field and is considered correct, undisputed, or widely uncontroverted. Common knowledge includes standard definitions of known terms. Sometimes you might be unsure whether particular information is common knowledge or not, in which case you should cite to a review article or to one or 2 well-received, published research papers on the topic.

- **Non-published information.** Non-published information is not cited. Examples of non-published information are: (a) forecast statements, which identify where certain information is located in a document (see Chapter 5.4.3), are themselves typically located in introductions to sections, (b) your unpublished analysis of reviewed information; such analyses can include problem statements, hypotheses, predictions, speculations, and conclusions (see Chapter 2.2.3), (c) connection statements, such as your proposed research objective and methods, and (d) personal communication (Chapter 2.2.4).

An unauthoritative source of published information is a person who passes information to you by word of mouth. This information is known as *personal communication*, and its citation is illustrated in Example 1. Personal communication (oral communication) that results from a speaker–listener relationship is analogous to a document (written communication) that results from a writer–reader relationship. Hence, the information passed via word of mouth is considered a type of published information. When citing to personal communication, you should identify the speaker's family name, at least the initial letter in the first name, and the date this person shared the information with you. You can also include the person's affiliation.

Example 1
To put this response into perspective, protecting wild-type mice from carotid occlusion at 5% $FeCl_3$ requires a super-physiological dose of heparin (one that leaves the animal prone to fatal hemorrhage), while aspirin does not protect mice at all at 5% $FeCl_3$ (personal communication, John Smith, PhD, U Oregon, 2012).

A citation to personal communication indicates that you are aware of the need to support a statement, such as a tentative conclusion (i.e., a claim). However, such a citation does not necessarily help your credibility. You may have misunderstood the communication, and there is no way for reviewers to evaluate its accuracy and whether it can be substantiated short of contacting the person who gave the information to you.

8.1.2 Styles of citations

Three styles of citations are: (1) citation by number, (2) citation by author's family name (i.e., surname), and (3) citation by URL.

(1) Citation by number. Citation by number uses digits to indicate sources of information. The first citation is numbered "1," which corresponds to the first numbered entry in the list of references.

The numbers can be presented in parentheses or brackets, or can be superscripted. A benefit of the citation-by-number style is that it takes much less space than does citation-by-family name style. However, citation by number is far less informative, for reviewers need to check the reference list in order to decode the number, unless you integrate the author's name into the sentence (see below). In addition, if you add or remove a reference from the narrative after your citations are nearly completely stable, you will need to renumber some or most of the pre-existing references. In the right column of Case 8-1, example (a) illustrates the citation-by-number style.

(2) Citation by name. This style of citation involves identifying the family name(s) of the authors to a publication, such as the author to a research article, an editor(s) of a book, or a colleague. Citation by name is much more informative than is the citation-by-number style since the source of information is identified within the text. A drawback of this style is that it uses much more textual space than does the citation-by-number style. The following table presents a simplified citation for single and multiple authors:

Type of citation	Example
Citation to one publication with a single author	(Smith 2012)
Citation to one publication with 2 authors	(Smith and Smythe 2011)
Citation to one publication with more than 2 authors	(Smith *et al.* 1998)
Citation to a first-listed author who is being cited for multiple articles published in the same year: lower case letter typically follows the year in the citation to distinguish that author's individual articles	(Smith *et al.* 2010a,b)
Multiple citations within one set of parentheses, arranged in alphabetical order and separated by semicolons	(Jones *et al.* 2011; Kelly 2012; Smith *et al.* 2010b; Smythe *et al.* 1998)

Notice that a comma is not needed between the parenthetical citation to the author and the date of publication since the citation is clear without the comma. Also, a semi-colon, not a comma, is used to separate multiple sources listed within one set of parentheses. The Latin term *et al.* requires a period since it is the abbreviation for *et alia*.

(3) Citation by URL. The URL citation is used for information that you retrieve from the Internet. Because information is added to and is removed from the Internet so quickly, you should not cite to information retrieved from the Internet unless: (a) you identify the source, (b) the source is from a journal or a site that your technical reviewers will recognize as reputable, and (c) the information has not been published in print format. If the information is available in print, you should cite to the printed source, unless instructed otherwise by the journal.

If you cite to information that is retrievable only on the internet, such as from an e-journal, you should use the bibliographic information provided on the e-journal's website. A URL citation can be cited either by number or by the URL address. For any URL that you include, you should identify when you accessed the information. Example 2 shows a citation to a published article and a sentence with a URL:

Example 2
False Discovery Rate (FDR) program software (http://www.math.tau.ac.il/~ybenja/software.html; accessed February 2, 2013) will adjust p values using an FDR of 5% ($q = 0.05$, Jones and Smith 2009) for multiple comparisons, to reveal significant differences in gene expression and enzyme activity among samples (Smith *et al.* 2009). (Modified from original.)

8.1.3   Where to cite

Where you locate citations, relative to the information and punctuation in a sentence, is important, for the location of a citation indicates its scope. Guideline 8.2 identifies where to place a citation in order to indicate the appropriate scope, and Case 8-1 provides examples.

Guideline 8.2

(a) *To indicate that all of the information in one sentence is from the same source, place the citation at the end of the sentence, either consistently before the period or after the period.*

(b) *To indicate that all of the information in a series of sentences is from the same source, place the citation at the end of the **first** sentence in the series, either consistently before or after the period.*

(c) *To indicate that information in different parts of one sentence comes from different sources, place a citation immediately after the information and before any punctuation.*

(d) *If a sentence contains both published and non-published information, sequence published information and its citation before the non-published information that does not receive a citation.*

(e) *Do not share citations across paragraph (or subsection) boundaries. Instead, repeat the same citation in the first sentence of the paragraph that you use in the last sentence of the previous paragraph.*

Authorities on scientific writing sometimes do not agree on where to place a citation, relative to punctuation. The main issue about citation placement is to be consistent. If you place a citation before punctuation in one sentence, you should do so in all sentences that include citations. Similarly, if you place a citation after punctuation in one sentence, you should do so in all sentences with citations.

Example (a) in Case 8-1 shows 2 sentences, each of which ends with different citations, indicating that all of the information in the first sentence comes from one set of sources (citations *13–18*), and all of the information in the second sentence comes from a different set of sources (citations *1, 4,* and *5*). Example (b) shows a passage with 5 sentences and 2 sets of citations. The first 4 sentences are from the same source, yet only the first sentence receives the citation. The scope of this citation is assumed to extend through sentence 4 because sentence 5 has a different citation. Example (c) is one sentence with 4 different sets of citations, indicating that the information in different parts of the sentence comes from different sources. Example (d) also consists of just one sentence. Notice that the first half of the sentence receives the citation (*144*); the second half of the sentence, beginning with *suggesting that*, consists of unpublished information, in this case the PI's analysis (a speculation) about the information in the first half of the sentence. As an analytical statement, the speculation does not receive a citation unless it has already been published (see Guideline 8.2(c)).

Guideline 8.2(e) notes that the scope of citations ends at the end of a paragraph, subsection, or section. In other words, citations do not cross paragraph, subsection, or section boundaries. If you end any of these units with a citation, you should repeat these citations in the upcoming paragraph, subsection, or section if the citation is still applicable.

- **Integration of an author's name into a sentence.** If you are using a citation-by-number style and you want to identify the source of information, you can integrate the author's name into the sentence, as illustrated in Examples 3 and 4. These examples also use common phrases – *following* and *as described in* – for this integration.

Example 3
Following Thrane *et al.* [12] to model the OCT signal, we will: (1) mathematically describe our imaging geometry in the sample arm; (2) formulate equations that describe light propagation from the sample arm lens to a point in the tissue and back, and include mathematical descriptions of the effects of interest (i.e., spectral effects) in these expressions; and (3) integrate the preliminary analytical expressions over all points at depth z and the lens plane using Mathematica (Wolfram Research) to obtain the final equations that describe the OCT signal.

Example 4
We will perform a frame-based spectral analysis of the source and target speech, and will compute LPC coefficients using an autocorrelation method *as described in* Smith and Jones (1978).

- **Citation triggers.** Some phrases trigger the use of citations, so you need to be aware of them so you can include citations. In Examples 3 and 4 above and in Examples 5–9 below, bold italics identify such phrases, which often indicate conclusions. Notice how each sentence includes citations to substantiate the conclusions. The second sentence in Example 9 does not have a citation because its source of information is identical to the source of information in the first sentence (see Guideline 8.2(b)).

Example 5
The algorithm is *well-analyzed* (e.g., 12, 14, 20) and comes with *theoretically proven* statistical properties (21, 32).

Example 6
Those studies focused on only partial sodium channel sequences in insects, in which *previously confirmed kdr* mutations (Smith 2007) had been detected.

Example 7
[3]*In our previous psychophysical studies*, the pattern of vibration (i.e., constant frequency) bore little similarity to natural patterns of muscle spindle firing during movement (e.g., Jones *et al.* 1981; Smith *et al.* 2002; Smythe 1972).

Example 8

The pharmacological targeting of Rac1 GTPase *has been proposed* as a strategy to combat thrombosis,[6–8] stroke,[9] atherosclerosis,[10] and cancer,[11,12] despite an incomplete understanding of the mechanistic role that Rac1 plays in these diseases.

Example 9

Many connections between PDGF-mediated signaling and PKA *have been established* in fibroblasts and smooth muscle cells.[22,23] PKA is activated and translocated from the cell membrane to the cytosol following stimulation of cells with PDGF and *has been reported* to antagonize and to promote cell migration and proliferation.

8.2 Definitions of key terms

A key term is a word or phrase that represents a concept that is central to your proposed research and is frequently used throughout the narrative, beginning in the **Aims** or **Background Section**. You need to define key terms in the narrative, where you first mention them. Thus, you will likely need definitions in the **Aims** and **Background Sections** more frequently than in later sections of the narrative. Case 8-2 shows the first paragraph in an **Aims Section** opening with a definition in the very first sentence.

> **Guideline 8.3**   *Define the key terms in the narrative where you first mention them.*

You might not want to define key terms that are technical, especially if you expect that none of your reviewers will be lay or if you know your technical reviewers already know the terms. Also, you might not want to use valuable textual space for definitions. Such reasoning misses the point of definitions. You may be correct; your technical reviewers might already share the same definitions of key terms that you are providing; and you might be very tight on space. However, there are many reasons to include definitions.

● **Credibility.** Definitions help technical reviewers evaluate *your* understanding of key terms.

● **Cache of terms.** Definitions provide valuable terms from which you can readily select to write faster due to the information structure of sentences. Definitions provide terms that help you sequence information within sentences from given to new information. (See Chapter 8.4.)

● **Clarity.** Less technical reviewers will need definitions to better understand what you are discussing, and technical reviewers will need definitions if professionals in the field define the key terms slightly differently.

8.2.1   The classic definition

The *classic definition*, sometimes called an Aristotelian definition, consists of 3 types of information: (1) *the key term*, (2) *the class*, and (3) *the distinguishing feature(s)*. This information is traditionally represented by the following 3-part structure:

A/an/the/xx _____ is a/an/the/XX _____ that _____
      Key Term   **Class**   **Distinguishing Feature(s):**
                                                  Function, Type, Structure

Examples 10–12 below illustrate classic definitions.

(1) Key term. In a definition, the key term is the word or phrase that is being defined. It is usually defined in the singular. As a result, a key term that is a *count noun* is commonly preceded by *a* or *an* (depending on whether the term begins with a vowel sound when spoken) unless it is unique or singular by its nature, in which case the count noun is preceded by *the*, such as *the central nervous system*. A count noun is capable of having its individual items counted, such as *systems*. A key term that is a non-count noun, such as *oxygen* and *sand*, receives no article (represented by XX in the 3-part structure of a classic definition). A non-count noun, sometimes referred to as a *mass noun*, is not countable; instead, its measurement unit is countable, such as *2 ml of water*, or its composite materials are countable, such as *2 grains of sand*. Also, as shown in Example 10 below, if an abbreviation has not yet been introduced for a key term, it is introduced in parentheses in the definition, after the key term.

Example 10
Staphylococcus aureus (*S. aureus*) is ***a well-armed opportunistic pathogen that produces a diverse array of virulence factors and causes a correspondingly diverse array of infections.***

Example 11
MS is ***a demyelinating disease of the central nervous system that manifests in a range of neurologic symptoms, progressive disability, and physical and psychosocial decline over 30–40 years.***

Example 12
The caudomedial neustriatum (NCM) is ***a major avian auditory brain center that is possibly equivalent to supragranular layers of the mammalian auditory cortex.***

(2) Class. The class is the group, category, or classification of which the key term is a member. Similar to the key term, the class is also presented in the singular if it is a count noun. The class in Example 10 is *a well-armed opportunistic pathogen* (italicized in Example 10); in Example 11, the class is *a demyelinating disease of the central nervous system*; and in Example 12, the class is *a major avian auditory brain center*.

(3) **Distinguishing feature(s)**. The third part of a classic definition identifies an important feature(s) that distinguishes the key term from other members of its class, usually in terms of one of 3 features: its function or purpose, a characteristic relating to the item's type, or a structural characteristic. When you define a key term, you need to decide which type of information (function, type, or structure) best characterizes and distinguishes your key term from other members of its class, given the key term's use in the narrative. Examples 10, 11, and 12 present a functional distinguishing feature, a typological distinguishing feature, and a structural distinguishing feature, respectively (bold italics). As illustrated in these examples, the distinguishing feature is phrased in a *that* adjective clause (technically, a **restrictive adjective clause**) because the distinguishing feature limits the class to the key term. The *that* adjective clause is iconic to the classic definition.

Because the classic definition typically comprises a sentence, the classic definition is sometimes called a **sentence definition**. Chapter 8.6.3 gives strategies for shortening the classic definition and for integrating it into another sentence in order to save space.

8.2.2 Definitional and interpretive synonyms

A synonym is a word or phrase that is similar in meaning to another term. In scientific writing, a synonym is not frequently substituted for a key term because of the need to be precise and clear, and to lessen the chances that reviewers will mistake a synonym for another key term, not a substitution for the key term.

There are 2 types of synonyms that are useful in scientific writing: a definitional synonym and an interpretive synonym.[1] A definitional synonym is a word or phrase that defines a key term and is usually placed in parentheses after the key term. A definitional synonym gives reviewers who do not know the term a quick understanding so they can follow the discussion. After a definitional synonym is presented, the text continues to use the key term, not the synonym. A definitional synonym can be problematic: if a reader does not know the synonym, this type of definition is useless. Therefore, a definitional synonym needs to be less technically phrased than the key term it is defining. In Example 13, the synonym *light touch* defines haptic. After this sentence, you would continue to use the term *haptic*.

Example 13
We hypothesize that vision or haptic (light touch) information is used to recalibrate central somatosensory references for both postural and locomotor orientation.

Another type of synonym, the **interpretive synonym**, is frequently used in scientific writing. This type of definition consists of the demonstrative adjective *this* and a term that defines or explains a preceding passage, not just a key term. The passage that an

[1] Tyma, Deborah. Anaphoric functions of some demonstrative noun phrases in EST. In Larry Selinker, Elaine Tarone, and Victor Hanzeli, Eds., *English for Academic and Technical Purposes, Studies in Honor of Louis Trimble*. New York, NY: Newbury House Publishers; 1981, pp. 65–75.

interpretive synonym defines or explains is typically information in a clause, a sentence, or a group of sentences. Example 14 uses the interpretive synonym *this approach* in sentence 4, which not only refers to the *Smith model* in sentence 1 but also its description in sentences 1–3.

Example 14

[1]In the Smith model (Smith 1987), a set of rules is used to predict the duration of a phoneme. [2]The rules were derived from recordings in small-scale studies where phonemes were pronounced in sentences occurring in different phonetic and prosodic contexts that were systematically varied. [3]The model assumes that each phoneme has an inherent duration and that each rule in the duration model results in a percentage of increase or decrease in phoneme duration. [4]A problem with **this approach** is that it does not predict durations well when certain contexts interact, because the small-scale studies keep a contextual factor constant where in reality the levels of the factor (i.e., stressed versus unstressed) interact with the levels of another factor (i.e., accented versus unaccented) (Smith 1998).

As shown in Example 14, an interpretive synonym is particularly useful for clarifying information in scientific methodological terms.

Guideline 8.4 *Use known synonyms for definitions and use interpretive synonyms that reviewers will recognize in terms of scientific methodology.*

8.3 Explanations through examples and restatements

In addition to the use of definitions, a writing technique to help reviewers understand the text is to provide examples and to restate information. To restate information is to express a concept in different terms from those previously used.

Examples and restatements can be integrated within the grammar of sentences via commas or dashes, or they can be included within parentheses (see the variations in Example 15 below). A dash serves to emphasize the examples and restatements, so this punctuation should be used sparingly in order not to dilute its impact. The use of parentheses for examples and restatements suggests that the information is less important than the information integrated within the grammar of the sentence.

When an example or restatement is integrated within the grammar of a sentence, a phrase is used, such as *for example*. When an example or restatement is included within parentheses, it is usually abbreviated with periods. The following list shows common abbreviations to Latin terms that signal example or restatement.

Abbreviation	Latin equivalent	English equivalent
e.g.	*exempli gratia*	for example, for instance
i.e.	*id est*	that is, in other words
viz.	*videlicet*	namely, as in the following
cf.	*confer*	compare with cited reference

It is becoming increasingly common for these abbreviations not to be punctuated with periods in the narrative. However, the periods are still a feature in conservative scientific writing. When *e.g.* or *i.e.* is used in parentheses, it is usually followed by a comma, whether or not it is punctuated with periods.

Examples 15a–15d illustrate different ways to present examples and restatements in a sentence (bold italics), and Example 15d also shows a dash to emphasize a restatement. If you are emphasizing an example or restatement that is within a sentence instead of at the end of a sentence, then a dash is used on both sides of the example or restatement. Only one pair of dashes is usually used per sentence.

Example 15

a. Songbirds are among the very few animal groups that learn to produce their vocalizations (*e.g., songs and some calls*) by imitating the sounds produced by a model (*i.e., a singing father or tutor*).

b. Songbirds are among the very few animal groups that learn to produce their vocalizations (*e.g., songs and some calls*) by imitating the sounds produced by a model, *that is, a singing father or tutor.*

c. Songbirds are among the very few animal groups that learn to produce their vocalizations, *such as songs and some calls*, by imitating the sounds produced by a model, *that is, a singing father or tutor.*

d. Songbirds are among the very few animal groups that learn to produce their vocalizations (*e.g., songs and some calls*) by imitating the sounds produced by a model – *that is, a singing father or tutor.*

8.4   Information structure in sentences

Information can be packaged within sentences in particular ways to promote readability. One such packaging is termed *given-and-new* information structure.

Within a sentence, given information usually precedes new information. Given information is information that is already mentioned in the preceding text or title, or is assumed from the context of scientific inquiry. For emphasis, new information can lead a sentence, but you should not make this reversal often in order to retain the emphatic impact of leading a sentence with new information.

> **Guideline 8.5** *Sequence given information before new information in a sentence.* **Exception:** *Consider beginning a sentence with new information to emphasize it.*

In Example 16, *APA* is mentioned in sentence 1. Sentence 2 leads with *APA*, so sentence 2 begins with given information, and the new information (which happens to be about *APA* in this example) follows *APA*. Thus, sentence 2 is organized from given-to-new information.

Example 16
[1]Another potential neuromuscular measure for classifying patients with lower-back pain is EMG of the trunk muscles during tasks that elicit an anticipatory postural adjustment (*APA*). [2]An *APA* is the activation of postural muscles (e.g., in the trunk) before activation of extremity muscles that create limb movement.

Notice that in sentence 2 of Example 16, *APA* is the subject of the sentence, but you do not always have to package the given information as the subject of the sentence. In Example 17, the second sentence begins with *Based on this hypothesis*. The phrase *Based on this hypothesis* is considered given information because *hypothesized* is used in the previous sentence. Example 17 also illustrates that the given information (i.e. the noun *hypothesis*) can be of a different grammatical form (e.g., past participle of the verb *hypothesize*).

Example 17
[1]Ligand-receptor "signaling endosomes" have long been **hypothesized** to be an essential mechanism for relaying neurotrophic signaling from the plasma membrane surface to the perinuclear area in neurons.[13–16] [2]Based on **this hypothesis**, BDNF-activated TrkB receptors (BDNF-TrkB-Rs) are believed to form an active signaling complex that is internalized and transported within cells to induce phosphorylation activation of downstream signaling pathway proteins.

Example 18 shows the first 3 sentences from an **Aims Section** in case 8-2A. All 3 sentences begin with given information (bold italics).

Example 18
[1]***Staphylococcus aureus*** is a well-armed opportunistic pathogen that produces a diverse array of virulence factors and causes a correspondingly diverse array of infections. [2]The ***pathogenesis of S. aureus*** infections depends on the coordinately-regulated expression of two groups of virulence factors, one of which (surface proteins) allows the bacterium to evade phagocytes and colonize host tissues while the other (extracellular toxins and enzymes) promotes survival and multiplication at a localized site of infection. [3]***Our long-term goal*** is to elucidate the regulatory mechanisms controlling expression of these virulence factors as a prerequisite to the development of therapeutic protocols that can be used to attenuate the disease process.

The first sentence in Example 18 begins with given information since the term *Staphylococcus aureus* is first mentioned in the title of the project *sar-mediated regulation in Staphylococcus aureus* (title not given). The second sentence in Example 18 begins with given information since the term *pathogen* occurs in the previous sentence, in its related form *pathogenesis*, and *S. aureus* also occurs in the previous sentence. In addition, the third sentence also begins with the given information *Our long-term goal*, which is given information based on the context of scientific inquiry; it can be safely assumed that reviewers will already know that scientific research projects typically have long- and short-term goals.

Sometimes you might create a passage that has all the requisite information in it, but it is hard to understand. It is in these passages, in particular, that you should evaluate whether the content has been appropriately sequenced from given to new information. Example 19 is a modification of Example 18, with its third sentence arranged from new-to-given information, with new information (*A prerequisite*) preceding given information (*disease process* and *regulatory mechanisms*). This third sentence in Example 19 is harder to understand, due to the new information preceding the given information.

Example 19
[1]*Staphylococcus aureus* is a well-armed opportunistic pathogen that produces a diverse array of virulence factors and causes a correspondingly diverse array of infections. [2]The *pathogenesis of S. aureus* infections depends on the coordinately-regulated expression of two groups of virulence factors, one of which (surface proteins) allows the bacterium to evade phagocytes and colonize host tissues while the other (extracellular toxins and enzymes) promotes survival and multiplication at a localized site of infection. [3]A prerequisite to the development of therapeutic protocols that can be used to attenuate the **disease process** is the elucidation of **regulatory mechanisms** controlling expression of the virulence factors, our long-term goal.

8.5 Active grammar and passive grammar

Active grammar and passive grammar can take on particular patterns in the narrative, especially in sentences that describe methods, procedures, and processes. Also, the effective use of active grammar and passive grammar is related to the information structure of sentences (Chapter 8.4).

8.5.1 Active and passive sentences: the grammar

Active sentences have the person, thing, or process that accomplishes the action as the subject of the sentence (or clause). This type of person, thing, or process is called the **agent**. Examples 20, 21, and 22 illustrate active sentences that have the agent as the grammatical subject (bold italics) of the verb (italics).

Example 20
We *hypothesize* that Rac1 activation is regulated by the mTOR/S6K1 signaling pathway.

Example 21
The TBC system *uses* clusters of clinical measures from an LBP patient's medical history and physical examination, to categorize the patient into one of 4 types of treatments.

Example 22
This project *will*, for the first time, *clarify* how many mutations, both synonymous and non-synonymous, are involved in insecticide resistance, in an entire sodium channel of the mosquito.

In contrast to sentences with active phrasing, sentences with passive phrasing do not have the agent as the grammatical subject. Also, sentences with passive phrasing are distinguished by the grammar of the verb, which consists of 2 elements: (1) the verb *be* (*be, am, are, is, was, were, been,* or *being*), and (2) the past participle of an action verb.

Example 23a is an active sentence since the agent (*the PI, we*) is the grammatical subject, and the verb does not have both the verb *be* and the past participle of an action verb. In contrast, Examples 23b and 23c are passive sentences because in each, the agent is not the grammatical subject and the verb bears the passive form: the verb *be* and past participle (*isolated*). Also notice that in Example 23b, the agent is not even mentioned in the sentence. In contrast, the agent is mentioned in Example 23c, in the by-prepositional phrase *by the PI*.

Example 23
a. The PI (we) *will* acoustically *isolate* adult zebra finches 12–14 hours before the experiments begin, in a sound-proof chamber.
b. Adult zebra finches *will be* acoustically *isolated* 12–14 hours before the experiments begin, in a sound-proof chamber.
c. Adult zebra finches *will be* acoustically *isolated by the PI* 12–14 hours before the experiments begin, in a sound-proof chamber.
d. ‡Adult zebra finches *will be* acoustically *isolated by us* 12–14 hours before the experiments begin, in a sound-proof chamber.

When the agent is not mentioned in a passive sentence, such as in Example 23b, the sentence is called an ***agentless passive***. Agentless passives are frequently used in descriptions of methods, procedures, and processes. In agentless passive sentences, reviewers assume that the PI (or someone directly supervised by the PI) is the agent. To counter this assumption, another agent needs to be identified in a prepositional phrase (see Chapter 6.3.4). Also, as indicated by ‡ in Example 23d, when the agent is a first-person pronoun, the agent should not be identified in a *by* prepositional phrase (see Chapter 6.3.4 and Guideline 6.8).

8.5.2 Use of active and passive verbs in descriptions of procedures

There are a variety of factors that need to be evaluated when deciding whether to use active or passive phrasing in descriptions of procedures.

Guideline 8.6

(*a*) *Use agentless passive phrasing for standard, routine, or typical procedures that the PI will perform.*
(*b*) *Use active phrasing for procedures that are variations from the standard, routine, or typical, whether the PI is the agent or not.*
(*c*) *Use active phrasing for mental and communication activities.*
(*d*) *Consider sentence and paragraph focus when evaluating whether to use passive or active phrasing.*

> (e) *Consider given-new information structure when evaluating whether to use active*
> *or passive phrasing.*

● **Standard methods: agentless passive**. Passive grammar – especially agentless passive grammar – is typically used for standard methodological procedures when: (a) the identity of the agent is irrelevant to the discussion or (b) the PI (or someone immediately supervised by the PI) is assumed to be the agent and is responsible for the routine execution of the standard procedures.

The use of agentless passive grammar is prevalent in descriptions of standard procedures, protocols, and tasks in scientific writing. Example 24 gives a passage written in agentless passive grammar to describe the standard procedures of outflow tract banding and vitelline-vein ligation, where the identity of the agent is irrelevant to the description. The passive verbs in Example 24 are indicated in bold italics.

Example 24
[1]In outflow tract banding, a suture *is placed* around the outflow tract (OFT) and *tightened* to reduce the cross-sectional area of the OFT lumen, resulting in a heart with: (a) an increased after-load (a large resistance to flow at the heart outlet) and (b) larger-than-normal ventricular pressure. [2]In vitelline-vein ligation, either a left or a right vitelline vein *is clipped*, which reduces the volume of blood flowing through the heart per unit time (the volume flow rate, Q) and slightly decreases ventricular pressure.

● **Variations from standard procedures: active grammar.** Just as passive grammar is used for standard procedures, active phrasing is used for procedures that are not standard, not routine, or not typical in some way (Guideline 8.6). Notice that in sentences 1–4 in Example 25, the standard gating procedure is phrased in passive grammar (bold italics), and in sentences 5–9, active phrasing (italics) is used since these sentences describe variations to the procedures explained in sentences 1–4.

Example 25
... [1]Because of the periodic motion of the heart, however, 4D images (3D structure + time) *can be obtained* even at low image-acquisition rates, using a gating (or triggering) procedure.[59, 60] [2]In this procedure, acquisition of 2D images *is triggered* by a cardiac signal, obtained for instance from an electrocardiogram (ECG), at a precise phase (time) in the cardiac cycle. [3]In this way, image acquisitions at different locations in the heart *are synchronized* to the triggering signal, allowing reconstruction of 4D images. [4]However, for the tiny developing chick heart, the ECG signal is so small that it *can be used* to trigger 2D image acquisition only with great difficulty.

[5]In a recent study, to reconstruct in vivo 4D images of the quail embryonic heart, researchers successfully *implemented* a gating procedure that used velocity data acquired with a laser Doppler velocimeter from a vitelline vein to trigger OCT image acquisition in vivo.[55] [6]In another study, researchers *reconstructed* 4D images of a developing zebra fish heart, from non-synchronized 2D confocal microscopy images using a "gating algorithm."[34] [7]The study *showed* the feasibility of 4D

reconstruction even without a triggering signal. [8]The gating algorithm *used* the periodicity of heart motion and correlations between adjacent images in space to synchronize 2D image sequences and to reconstruct 4D images. [9]We *propose* using a similar gating algorithm to reconstruct the motion of the chick heart OFT from OCT images (Aim 1).

By using active phrasing for procedures, you can also imply that the agent is responsible for the critical decision to use particular procedures. Thus, the use of active phrasing allows you to imply the PI's responsibility for decisions about procedures.

- **Mental and communication activities.** Active grammar is commonly used to describe mental and communication activities, such as *hypothesize, determine, assume, recognize, note,* and *explain.* Example 26a illustrates the use of active phrasing for the mental-activity verb *recognize,* and Example 26b presents an inappropriate use of passive phrasing for the same mental-activity verb because the sentence does not indicate the agent, implying that the agent is unknown or unknowable. If the agent is knowable, you can use active phrasing to refer to generic researchers, as shown in Example 26c.

Example 26

a. However, we *recognize* that SarA-binding sites may exist across a relatively long stretch of DNA with intervening and perhaps irrelevant nucleotides in between each site. (Case 6-5, sentence 9.)

b. ‡However, it *is recognized* that SarA-binding sites may exist across a relatively long stretch of DNA with intervening and perhaps irrelevant nucleotides in between each site.

c. However, *researchers recognize* that SarA-binding sites may exist across a relatively long stretch of DNA with intervening and perhaps irrelevant nucleotides in between each site (refs).

In a narrative to a grant proposal, passive phrasing (e.g., Example 26b) is seldom appropriate for mental or communication activities since passive implies that, for some reason, you are deliberately not identifying the agent. Thus, in a sense, you are hiding the human who is responsible for the mental or communication activity. The following passive phrases are therefore inappropriate for mental activity verbs if you know the agent or if the agent is knowable:

‡*It will be*
‡*It is/was*
‡*It can be*   *hypothesized/determined/assumed/recognized/noted/explained*
‡*It should be*

- **Sentence focus.** If you are not describing standard procedures, variations from standard procedures, or mental or communication activities, the focus of a sentence or paragraph can help you determine whether to phrase verbs in active or passive grammar.

Guideline 8.7 *Use active grammar or passive grammar for positioning information in order to help reviewers notice the information.*

The information in the grammatical subject of a sentence is usually the focal point of the sentence.[2] As a result, if you want reviewers to notice particular information, you can make it the subject of a sentence. If you are a native or fluent speaker of English, you might already be using active or passive grammar appropriately in order to locate noteworthy information as the grammatical subject. In the first sentence of Example 27, the verb *will be assigned* results in *subjects* being positioned as the grammatical subject of the sentence. To maintain focus on the subject groups in the second sentence, the verb *will perform* is in active grammar (in contrast to the passive grammar *will be performed*), which serves to position *The control group* and *the experimental group* as the grammatical subjects of their respective clauses.

Example 27

[1]Subjects **will be** randomly **assigned** to 2 groups. [2]The control group *will perform* standard balance exercises with a physical therapist, and the experimental group will perform the same exercises with a physical therapist while using the ABF device. [3]The balance exercises are a modification of Frenkel's exercises (1902), widely used in rehabilitation to improve balance in patients with somatosensory ataxia (ref). [4]For each subject group, we will add standing and walking exercises, with and without vision, under challenging surface conditions. (Case 6-6, sentences 1–4.)

Of course, in the heat of first-drafting, you will not likely go through such detailed consideration to decide whether to use active or passive phrasing. It is far easier to remember Guideline 8.8.

Guideline 8.8 *Decide on the information that you want readers to focus on, and make that information the grammatical subject.*

● **Given-and-new information structure**. Another use of active grammar and passive grammar is to position given information before new information in a sentence in order to enhance readability (Chapter 8.8.4).

In Example 27 above, both independent clauses in the second sentence begin with given information. The first clause begins with *The control group*, which includes the term *group*, and has already been mentioned in sentence 1 (and *control* was used in a previous passage [not shown]), and the second clause begins with *the experimental group* (also previously used [not shown]). If the first clause were phrased in active grammar, new information (*the performance of standard balance exercises*) would be positioned before given information.

---

[2] *Usually* is used here because the sentence focus can be included before the grammatical subject, in grammar indicating the topic of the sentence, such as *regarding* or *as for* when located at the beginning of a sentence, such as: **Regarding the birds**, *they will be assessed* . . .

In Example 28a, the second sentence is structured from given-to-new information. It opens with *The study*, which is mentioned in the previous sentence (i.e., *a recent study*). In this second sentence, *the study* is followed by the *active* verb *showed*. If the second sentence were phrased passively, as shown in Example 28b with the passive verb *was shown*, the information in the second sentence would be organized from new-to-given since *feasibility* has not yet been mentioned, and *feasibility* is not an interpretive synonym (see Chapter 8.2.2). The sequencing of new information before given information in the second sentence of Example 28b would likely negatively impact the readability of the passage.

Example 28

a. [1]In a recent study, researchers reconstructed 4D images of a developing zebra fish heart, from non-synchronized 2D confocal microscopy images using a "gating algorithm".[34] [2]The study ***showed*** the feasibility of 4D reconstruction even without a triggering signal.

b. ‡[1]In a recent study, researchers reconstructed 4D images of a developing zebra fish heart, from non-synchronized 2D confocal microscopy images using a "gating algorithm".[34] [7]The feasibility of 4D reconstruction, even without a triggering signal, ***was shown*** by this study.

8.6   Shortening text[3]

If you are like most writers of narratives, sooner or later you will need to shorten the text in order: (1) to create space for additional information or (2) not to run beyond a prescribed maximum page or character length for a section, the entire narrative, the abstract, or the title. The process of shortening the text is difficult if you do not want to delete any existing content.

Shortening the text needs to be accomplished in many small steps, going through the whole document methodically (and painstakingly!). Chapter 8.6.1 presents suggestions on preparing to shorten the text as effectively and efficiently as possible. Chapter 8.6.2 gives strategies to shorten the text by manipulating layout, fonts, and visuals. Chapter 8.6.3 presents strategies to rephrase sentences, phrases, and words; and Chapter 8.6.4 focuses on last-resort strategies for reducing content. When shortening the text, you need to maintain clarity of content and organization.

In the upcoming discussion, you will notice that the text-shortening strategies, when applied to isolated passages, such as in the examples presented below, do not add up to substantial savings of textual space. However, the cumulative effect of these strategies, when applied to multiple sentences and passages throughout the narrative, can

---

[3] This section includes text-shortening strategies and approaches from *Writing Shorter Legal Documents: Strategies for Faster and Better Editing* by Sandra Oster (American Bar Association Press, 2011). We acknowledge and thank the American Bar Association, which has granted Cambridge University Press and the authors permission to use information in this book.

substantially shorten the text. Also, some of the strategies will be useful when trying to shorten the total number of pages.

8.6.1   Preparing to shorten text

● **Become familiar with submission requirements**. You need to determine whether your grant proposal is under any length requirements from your funding agency, for example in terms of the total number of pages for the narrative, and the total words or characters for the title and the abstract. Also, you need to double-check that any text-shortening strategies that you intend to use do not violate submission requirements.

● **Work from a stable draft of the narrative**. When shortening text, you should use a stable draft of the entire narrative – the prose text of the narrative, the explanatory paragraphs and titles of visuals, the title to the grant proposal, and the abstract. A stable draft is one that: (a) has no unclear passages, (b) does not need many content or organizational changes, and (c) those changes that it does need are minor and do not significantly impact the length of the narrative.

● **Begin early**. Trying to shorten the narrative while writing an early draft is difficult and usually unsuccessful. Shortening the narrative takes time. You need to begin writing the narrative sooner rather than later in order to give yourself time to stabilize the text and then to apply text-shortening strategies before the submission deadline.

● **Determine which text-shortening strategies to apply consistently throughout the entire narrative or only in selected text**. Some text-shortening strategies affect the style of your prose and the appearance of a page. These strategies should be applied throughout the entire narrative for consistency. Other text-shortening strategies have less impact on the style of your prose and the appearance of the page, and can be applied to selected text. Table 8-1 identifies the text-shortening strategies from Chapters 8.6.2–8.6.4 that should be consistently applied throughout the narrative and those that can be applied to selected text.

8.6.2   Manipulating layout, font, and visuals

As used here, the term *layout* refers to the placement of text and visuals on the page, relative to white space. Layout-manipulating strategies are presented not only to shorten the text but also to aid in readability through judicious use of white space. However, if misapplied, layout-manipulating strategies can obscure the organization of text, can violate submission requirements, and can negatively impact readability. As a result, such strategies need to be applied particularly cautiously.

The strategies presented here do not include adjusting margin width or changing font size since both are usually prescribed by the submission requirements. Also, if any of the following layout strategies run counter to submission requirements, you should not apply them but move on to other text-shortening strategies.

(1) **Removing blank lines between paragraphs and indenting paragraphs**. Removing blank lines between paragraphs throughout the narrative is a common strategy to reduce the total number of pages. However, this strategy can be problematic because inter-paragraph white space enhances readability. White space is critical in helping the readers' eyes move quickly across lines and down the page, and in helping them understand the internal organization of content in (sub)sections. Therefore, if you remove blank lines between paragraphs, you also need to indent paragraphs by 4–5 characters in order to add white space.

> **Guideline 8.9** *To shorten the total number of pages in the narrative:*
> (a) *Consistently remove blank lines between paragraphs in all sections (but retain lines between (sub)sections).*
> (b) *For readability, indent each paragraph 4–5 characters.*

Case 8-3A shows a readable layout with no blank lines between paragraphs but with indented paragraphs. The layout in Case 8-3B does not support the readability of the text as well as the layout in Case 8-3A. Case 8-3B has no blank lines between paragraphs and no paragraph indentations; as a result, this layout is not preferred even though the passage is shorter by one line. Case 8-3C shows a layout with blank lines between paragraphs and no indentations.

(2) **Combining paragraphs**. One way to reduce a few lines of the narrative is to occasionally combine 2 short paragraphs (e.g., 8 or fewer lines) into one long paragraph. However, the longer, *resultant paragraph* usually has lower readability than the 2 shorter paragraphs, in part because the reduction of paragraphs involves the elimination of white space. If you combine paragraphs, you need to compensate for the longer, resultant paragraph in order to improve its readability and to help readers focus on its key information.

You can enhance the font of key information in the middle of a long, resultant paragraph to compensate for its lower readability. *Compensatory font* (also termed *emphatic font*) usually involves the use of **bold** or *italic bold* font, or underscoring, to highlight key information and organizational cues.

The long paragraph in Case 8-2A is the first paragraph from an **Aims Section**. It uses compensatory font for key information in its middle: the hypothesis (sentence 4) and the objective of study (sentence 12). It also uses highlighting font for the organizational cues *first, second* and *third*. The long paragraph in Case 8-2B introduces and then describes the methods for 2 different assays. As shown in this example, each assay is underscored, which helps reviewers identify the different types of content in the passage.

Any compensatory font that you use needs to be constrained in 2 major ways: (1) its highlighting feature, such as bold italics, should match the style of highlighted font that you use in other sections of the narrative, and (2) compensatory font should be used sparingly since it can lose its emphatic impact if overused.

Some writers do not use any compensatory font in resultant paragraphs, instead opting to reposition key information into one of the 2 focus positions of a paragraph: at the beginning or at the end, where bordering white space helps direct readers to information in these positions (see Chapter 1.8.1). However, if you are up against a submission deadline, you might not have the time to reorganize a resultant paragraph, in which case you should consider using emphatic font.

Guideline 8.10   *If you combine paragraphs, compensate for the long, resultant paragraph by using emphatic font and/or possibly revising its organization.*

(3) Using run-in headings. A run-in style for subsection and sub-subsection headings (in contrast to a stand-alone style) can save a substantial number of lines. In Case 8-4A, the 2 headings to the subsections are in a stand-alone style. In Case 8-4B, the 2 lower-level, underscored headings are run-in, which shortens the text by 2 lines.

Guideline 8.11   *Consider using a run-in style for headings to subsections throughout the narrative in order to save space.*

(4) Changing vertical lists into horizontal lists. A *vertical list* is one that has each item in the list beginning directly under the previous item. A vertical list uses more white space than does a *horizontal list*, which has each item beginning on the same line as the previous item in the list, in paragraph style. A vertical list, through its white space, attracts the reader's attention, but the attraction comes with a price. A vertical list usually takes more textual space than does the same information presented in a horizontal orientation.

By convention, certain information in the narrative, such as a set of aims in the **Aims Section**, is presented in a vertical list for emphasis, regardless of the space consumed. However, almost all lists in the narrative, except for the proposed aims in the **Aims Section**, are usually presented horizontally to save space.

Guideline 8.12   *Consider orienting a list horizontally instead of vertically.*

Case 8-5A gives an **Innovation Section** from an NIH narrative, which presents the novelties of the proposed research in a vertical list. This same information is presented in a horizontal list in Case 8-5B, resulting in a saving of 3 lines.

(5) Presenting footnotes and endnotes in a horizontal list. Narratives to grant proposals usually have very few or no footnotes or endnotes. However, if you use them, word processing programs give you the choice of locating them at the bottom of a page (above any bottom margin) or at the end of a section. They do not allow you to select their format; the default style for footnotes and endnotes is a vertical list.

To save space, you can force a change in the vertical orientation of footnotes and endnotes to create a horizontal orientation by: (1) placing the cursor on the number to the second (and subsequent) note and (2) backspacing. The end result is footnotes formatted into a paragraph. In paragraphs of foot- or endnotes, however, the numbers might be hard for readers to find, so you should use compensatory font by placing the numbers in parentheses or using bold font to make them more noticeable.

Guideline 8.13 *Consider changing the format of multiple footnotes or endnotes from vertical to horizontal lists.*

(6) **Allowing paragraph widows and orphans**. A *paragraph widow* is the last line of a paragraph that is alone at the top of a page, with the rest of the paragraph on the previous page. In contrast, a *paragraph orphan* is the first line of a paragraph that is alone at the bottom of a page, with the rest of the paragraph continuing on the next page. Case 8-6A shows the last line of a paragraph at the top of page 2 (i.e., a paragraph widow).

You might have a word-processing program that, by default, does not allow orphans and widows. To prevent orphans and widows, some word processing programs automatically add a blank line at the bottom of a page to position 2 lines of text at the top of the next page. This process unnecessarily increases the total length of the narrative. Case 8-6B shows the 2 last lines of a paragraph at the top of the second page, the result of a word-processing program that has pushed the second-to-last sentence of the paragraph to the top of page 2.

To avoid your word processing program automatically adding blank lines, you need to adjust it to allow paragraph widows and orphans.

Guideline 8.14 *Consider adjusting your word-processing program to allow paragraph widows and orphans.*

(7) **Revising visuals**. Visuals can be redesigned and manipulated in different ways to save space.

● **Editing explanatory paragraphs**. Edit the text of the explanatory paragraph to a visual with strategies in Chapter 8.6.

● **Wrapping text around a visual**. If a visual does not extend from the left to the right margin, you might have space to wrap text to the side of a visual. Text-wrapping should result in at least 4–5 words, on average, fitting on a line next to the visual. If you can fit only one to 3 words on the line next to the visual, you should consider not using this strategy since such short lines of text could lower readability.

The default in word-processing programs is not to wrap text around a visual. The user's manual of your word-processing program will explain how to wrap text.

Guideline 8.15 *Consider wrapping text around a visual that does not extend from the left- to right-hand margins.*

Sometimes, the most challenging aspect of wrapping text is to reduce the width of the visual in order to create space for wrapped text. Case 8-7A presents a visual that is unnecessarily wide. The columnar space can be revised to create space for text wrapping (Case 8-7B).

• **Redesigning a visual**. Redesigning a visual can be time-consuming, and it can also involve revising textual passages that discuss the visual.

Guideline 8.16   *After redesigning a visual to save space, double-check all discussions of the visual in the text to assess whether they need to be revised.*

You need to evaluate the content and the space that each of your visuals occupies in order to determine: (a) whether the visual is *space-worthy* – that is, whether the content warrants the amount of space that it occupies; (b) whether the visual can be redesigned to take less space; and (c) whether you have time before the submission date to redesign the visual and the passages that refer to it. Case 5-4 in Chapter 5 presents an overview table (Case 5-4A) that has been redesigned (Case 5-4B) because the vertical space for the time line was deemed not to be space-worthy.

Sometimes you can quickly redesign and shrink a visual. For example, if you present a visual with a flow chart, by shortening the arrow shafts you can save lines of text. Or you may have a table that legitimately extends beyond one page. You can reduce the text by using column headers on the first page only and eliminating their repetition on subsequent pages (if submission requirements allow).

Sometimes you might need to make multiple changes to a visual in order to reduce it. For example, in order to reduce the line length of the text in Case 8-7A, the visual can be redesigned to reduce its width, which will facilitate text wrapping. Case 8-7B shows a narrower and shorter table. Space was created on the right-hand side of the visual for text-wrapping by: (1) moving the table to the left-hand margin by 3 characters, and (2) reducing the width of columns. Column 1 is narrowed by changing the line-length of the header and reducing the row header; column 2 is narrowed by omitting *Number of* in the header and changing the dates to numbers (whether or not the text uses the full dates); and column 3 is narrowed also by changing the line-length of its header. The length of the table has been shortened by reducing the spacing before and after each line of text. In Case 8-7A, the spacing was 0 points before a line and 6 points after a line of text; in Case 8-7B, the spacing has been changed to 0 points both before and after a line of text. (Check the instructions to your word-processing program for how to change space between lines.) With these manipulations, Case 8-7A was shortened by 5 lines, as shown in Case 8-7B.

(8) **Reducing emphatic font**. Text that is in emphatic font, such as bold or italic bold, takes slightly more space than does standard font and so might extend the text. As a result, you should be highly selective when deciding which words to place in emphatic font. Emphatic font used for readability to compensate for long paragraphs is useful, but some writers use emphatic font excessively, to make sure that reviewers notice particular information. Reviewers read the entire narrative closely, so a lot of emphatic font is usually not needed.

Guideline 8.17   *Consider reducing the amount of emphatic font used in the text.*
Exception: Allow emphatic font to compensate for text-shortening strategies that
decrease readability.

8.6.3   Revising the text

In addition to manipulating the layout in order to shorten the narrative, you can also
revise paragraphs, sentences, phrases, and words that comprise the text. When focusing
on the text, you need to approach the task methodically, rather than just trying to
shorten a sentence that happens to meet your eye.

The paragraph should be the focus of your attention when revising the text to
shorten the narrative. By reducing the total number of lines in paragraphs, you can
ultimately reduce the total number of pages, words, and characters in the narrative.
Most of the strategies in Chapter 8.6.3 serve to eliminate the last lines of paragraphs, in
what is called the ***paragraph-backward revision approach***.

You may have noticed that you can shorten the text in the first 2–3 lines of a
paragraph, which will reduce the total number of words or characters but will not
always reduce the total number of lines in the paragraph. To avoid this problem, you
should use the paragraph-backward revision approach if you are trying to shorten the
total number of lines. This approach involves: (1) identifying a paragraph with a short
last line, which is a line that ends closer to the left margin than to the right margin, and
(2) then editing the last 2–3 lines of the paragraph, using text-shortening strategies such
as those identified in Chapter 8.6.3, until the short last line disappears. (3) If you cannot
find anything to revise in the last 2–3 lines that will shorten the text, then you move
backward in the paragraph to edit its middle lines, still aiming to eliminate the short
last line of the paragraph. (4) If your editing still does not eliminate the short last line of
the paragraph, you move further backward to the beginning paragraph and try to revise
the first 2–3 lines of the paragraph, still aiming to eliminate the short last line.

Guideline 8.18   *When revising the text to shorten the narrative, use the paragraph-*
backward revision approach: (a) choose paragraphs with short last lines to edit, then (b)
edit from the bottom of each paragraph with a short last line, towards its first line, and
finally, (c) edit paragraphs with long last lines, moving from the bottom of the paragraphs
up to the first lines

Time permitting, after eliminating the short last lines of paragraphs, you can then
apply the paragraph-backward editing approach to paragraphs with long last lines,
which are paragraphs with their last lines ending closer to the right margin than to the
left margin.

The paragraph-backward revision approach involves paraphrasing a word or
phrase in fewer characters, without substantially changing the content. For
example, in sentence 10 of Case 8-8A, the phrases *will undergo amplification*

and *will undergo analysis* can be rephrased without nominalizations: *will be amplified* and *will be analyzed* (see Strategy #5 below). Both of these changes result in the removal of characters that ultimately eliminates the last line of the paragraph, as shown in Case 8-8B.

Application of the following text-shortening strategies will help you shorten paragraphs without substantially affecting content. These strategies will also help you develop a more concise writing style.

(1) Filling blank space at the ends of lines. Paragraphs can have *flush-right* or *ragged-right justification.* In flush-right justification, all lines of text end at the same distance from the right-hand edge of the page; in a *ragged-right justification,* all lines of text end at various distances from the right-hand margin, depending on the total character count of each line. The text in this book uses flush-right justification. The type of justification is sometimes specified in submission guidelines. If you have the choice, you should use ragged-right justification in order to apply this text-shortening strategy.

In a paragraph with ragged-right justification, you can reduce white space at the ends of lines by hyphenating words to fill white space. Case 8-9A shows a 16-line paragraph formatted with a ragged-right justification. There is substantial space at the end of line 12 (after sentence 5) that can be filled in with text. As shown in Case 8-9B, by hyphenating both *fluorescent* and *conformational* in sentence 6, the space at the end of lines 12 is filled in, resulting in the elimination of the last short line of the paragraph.

Words of caution are necessary with hyphenation. You should not hyphenate words globally in the narrative because such extensive hyphenation could lower readability, and global hyphenation might not affect the length of paragraphs. Also, this text-shortening strategy should be applied only to paragraphs that are *completely stable*. If you hyphenate words in a non-stable paragraph and then further revise its text, the hyphenated word might move to another position in the sentence, which could create a misspelling, such as *conforma-tional.*

> **Guideline 8.19** *Hyphenate selected words of completely stable paragraphs to use white space at ends of lines and to eliminate the short last line of a paragraph. Exception: Do not hyphenate URLs.*

URLs are not hyphenated since a hyphen might be mistaken for a meaningful character. Thus, URLs within the text might waste space at the end of the line, such as in Example 29a. To avoid such wasted space, you can rephrase the sentence to locate the URL closer to the beginning of the sentence in order to fit the URL onto a single line of text. As shown in Example 29b, the revision of the sentence creates space for the URL, which ultimately saves a line of text.

Example 29
a. The data set that we will use to train the algorithm is located at
   http://xxxx.yyyy.experiment.jones.4.html (ref). This data set was collected in 2010
   prior to run-off in the estuary. (3 lines, 29 words, 181 characters with spaces)

b. The data set located at http://xxxx.yyyy.experiment.jones.4.html (ref) will be used to train the algorithm. This data set was collected in 2010 prior to run-off in the estuary. (3 lines, 27 words, 174 characters with spaces)

(2) Using digits for numbers. Authorities in writing promote the convention of using words for numbers that begin sentences – possibly to clarify that a number following a period is not part of a number with a decimal. They also promote the convention of spelling out numbers 1 to 9 when these numbers occur within sentences.[4] However, the reason for this latter convention is not readily apparent. Some authorities in scientific writing are now advocating the use of numerals "to most single-digit whole numbers that were previously expressed as words."[5] To lower the character count and possibly to eliminate the last short line of a paragraph, this latter convention can be used – if done so consistently throughout the entire narrative. Throughout this book, numbers are used according to Guideline 8.20.

Guideline 8.20 *To save space, consistently use digits for all numbers in the narrative.*
Exceptions:
(a) The number one.
(b) Any number that begins a sentence.
(c) Numbers accompanied by measurement units.
(d) The first of 2 consecutive numbers (e.g., four, 2-channel units)

The number one is spelled out in order to make it more noticeable since *1*, just one character long, might not receive much notice, or it might be confused with the lower-case L. Example 30 presents a sentence with 3 numbers presented as words instead of digits. By changing these numbers from words to digits, you can save character spaces, which could result in the elimination of the last short line of a paragraph.

Example 30
a. Stimulation procedures will be the same as in Exp. 1. We will conduct *five* step trials for each of the *four* visual conditions, totaling *twenty* trials. (26 words, 149 characters with spaces)
b. Stimulation procedures will be the same as in Exp. 1. We will conduct *5* step trials for each of the *4* visual conditions, totaling *20* trials. (26 words, 139 characters with spaces)

To shorten a sentence that begins with a number, you can revise the sentence so that the number occurs within the sentence where you can use a digit. In Example 31a, all

---

[4] See, for example, *Publication Manual of the American Psychological Association*, 6th Ed. Washington, DC: APA; 2009, p. 111.
[5] See Council of Science Editors. *Scientific Style and Format: The CSE Manual for Authors, Editors, and Publishers*, 7th Ed. New York, NY: Cambridge University Press; 2006, p. 142.

numbers, including the number that begins the sentence, are in words. Example 31b shows the numbers within the sentence revised into digits. This example is further revised in Example 31c, so that the number 45 does not lead the sentence and thus can be presented as a digit. With each revision, the sentence becomes shorter in terms of the number of characters.

Example 31
a. Forty-five subjects in total were administered the three different doses of LA in our two-week, placebo-controlled trial. (17 words, 120 characters with spaces)
b. Forty-five subjects in total were administered the 3 different doses of LA in our 2-week, placebo-controlled trial. (17 words, 114 characters with spaces)
c. In our 2-week, placebo-controlled trial, 45 subjects in total were administered the 3 different doses of LA. (17 words, 107 characters with spaces)

(3) **Using abbreviations and acronyms.** An abbreviation is the shortened form of a word or phrase. For example, the standard abbreviation *B. subtilis* is the shortened form of *Bacillus subtilis*. An acronym, a type of abbreviation, is comprised of first letters in the term. Depending on how the acronym *AD* is defined in a narrative, *AD* could be the shortened form of *Alzheimer's disease* or *accidental death.*

Abbreviations and acronyms are abstract substitutions for the longer terms they represent. As abstractions, they can harm the readability of the text, especially when different abbreviations and acronyms are used frequently in the narrative. As a result, when you use abbreviations and acronyms to save space, you should use them only for selected key terms. Selected key terms are those used in high frequency, especially in the **Aims Section** and the **Methods Section**, and they usually correspond to the topic of the research and/or significant features of the methods.

In this entire book, for readability only 3 new acronyms are introduced: **SOS** (*s*ubjects, *o*bjects of study, and *s*pecimen), **MET** (*m*aterials, *e*quipment, and *t*ools), and **TRACS** (for *t*opic, *r*eview, *a*nalysis, *c*onnection, and *s*ignificance; i.e., the content of a scientific argument).[6]

While drafting the narrative, you might not consistently use an abbreviation or acronym. Although understandable, such inconsistency, if it survives to the final draft, can harm your credibility and suggest that you do not adequately pay attention to detail. After you introduce an abbreviation, you need to use it consistently not only to reduce the total character or word count, but also to promote credibility. With the Find function, you can locate the first mention of a key term in its non-abbreviated form and then include its abbreviation after the term. You can use the Find function to locate each subsequent, non-abbreviated form and can replace it with the abbreviation. Guideline 8.21 explains when and how to use abbreviations and acronyms, followed by a list of exceptions:

---

[6] Other abbreviations are used, but they are standard in science, medicine, and technology, such as **NIH** for the *N*ational *I*nstitutes of *H*ealth.

Guideline 8.21

(a) *Use abbreviations and acronyms for only a few selected key terms that occur frequently in the text.*

(b) *As much as possible, use standard abbreviations and acronyms, if they exist, over novel abbreviations and acronyms.*

(c) *On first mention of a selected key term, use the full key term and introduce the abbreviation or acronym after it, in parentheses.*

(d) *Once an abbreviation or acronym is introduced, consistently use it throughout the narrative.*

(e) *At the beginning of a (sub)section, do not re-introduce the abbreviation or acronym with the full key term.[7]*

(f) *Use abbreviations and acronyms in legends to figures if you use them in the text.*

(g) *Consider **not** using abbreviations and acronyms in headings to (sub)sections and in titles to visuals; instead, use the full key term.*

(h) *Consider **not** using abbreviations and acronyms in an abstract unless: (a) they are standard abbreviations that are likely well-known to lay readers and (b) they are also used in the narratives.*

In the following discussion, we note some abbreviations that are exceptions to Guideline 8.21.

● **Unabbreviated, common measurement units**. You do not need to introduce abbreviations for common measurement units if you are certain that all reviewers will know them. Abbreviations for common measurement units include *cm*, *sec*, and *mm* (see Chapter 6.3.4). Because lay reviewers might not be familiar with some technical abbreviations, for example the abbreviation for angstrom, Å, you need to introduce the abbreviations after their key terms.

● **Standard abbreviated terms**. The abbreviations to some non-key terms are commonly used in academia, so you can use them without first introducing them with their key terms. Such common abbreviations and their longer terms include:

Abbreviation	Term	Example
Dept.	department	Dept. of Neurology
Exp.	experiment	Exp. 3
Fig.	figure	Fig. 2B
lab	laboratory	the Smith lab
U. or Univ.	university	Brown U.

Note that the abbreviation for the term *figure*, such as in *Fig. 2B*, is used in order to save space, but when it begins a sentence, it is usually not abbreviated.

(4) Replacing non-key phrases with shorter synonyms. As mentioned earlier (e.g., Chapter 1.5.2), you need to repeat key terms (or their abbreviations); you

---

[7] In contrast, in this relatively long book, the abbreviations **SOS**, **MET**, and **TRACS** are periodically explained since readers will likely not be reading this book from front to back, unlike reviewers of narratives to grant proposals.

should not replace them with synonyms. However, to reduce the text, you can replace *non-key* terms with shorter synonyms.

> **Guideline 8.22**   *To save space, consider using a short synonym for a long non-key term.*

Table 8-2 presents some non-key verbs and phrases, and their shorter synonyms, and Examples 32 and 33 present sentences with non-key terms replaced by short synonyms. Example 32a uses *because of*, which is replaced in Example 32b with *due to*, a saving of 4 characters. Similarly, in Example 33b, *about* replaces *approximately* twice, resulting in a saving of 16 characters, which might be enough characters to eliminate the short last line of paragraph.

Example 32
a. This finding suggests that biofeedback benefits only longer-latency postural responses, perhaps *because of* delays inherent in sensing, interpreting, and voluntarily using sensory information. (23 words, 191 characters with spaces)
b. This finding suggests that biofeedback benefits only longer-latency postural responses, perhaps *due to* delays inherent in sensing, interpreting, and voluntarily using sensory information. (23 words, 185 characters with spaces)

Example 33
a. SMA stimulation will likely occur *approximately* 2 cm rostral to the vertex, and dPMC stimulation will be located *approximately* 2 cm rostral to the primary motor cortex hand region at the middle frontal gyrus. (3 lines, 34 words, 206 characters with spaces; from Case 8-8A, sentence 9)
b. SMA stimulation will likely occur *about* 2 cm rostral to the vertex, and dPMC stimulation will be located *about* 2 cm rostral to the primary motor cortex hand region at the middle frontal gyrus. (3 lines, 34 words, 190 characters with spaces)

The term *in order to* needs special discussion. When it occurs at the beginning of a sentence, as shown in Example 34a, the *in order* can be eliminated, as shown in Example 34b. However, if *in order to* occurs within a sentence and another infinitive verb (*to + verb*) is already in the sentence, then *in order* is not eliminated for clarity since *in order* indicates the higher-level purpose of the infinitive verb. In Example 34c, *in order* is not omitted since *develop* is the higher-level purpose of *to determine* and *to use*.

Example 34
a. *In order to develop* novel, effective therapies that are capable of preventing insecticide resistance and reducing mosquito-borne diseases and their impact, our long-term goals are *to determine* molecular mechanisms in mosquito response to insecticides and *to use* this information. (39 words, 277 characters with spaces)
b. *To develop* novel, effective therapies that are capable of preventing insecticide resistance and, ultimately, of reducing mosquito-borne diseases and their impact, our long-term goals are *to determine* molecular mechanisms in mosquito response to insecticides and *to use* this information. (39 words, 283 characters with spaces)

c. Our long-term goals are **to determine** molecular mechanisms in mosquito response to insecticides and **to use** this information **in order to develop** novel, effective therapies that are capable of preventing insecticide resistance and reducing mosquito-borne diseases and their impact. (39 words, 275 characters with spaces)

> **Guideline 8.23**   *Shorten the term* **in order to** + **verb** *to the term* **to** + **verb** *when:*
> (a) *There is no other* **to** + **verb** *in the sentence or*
> (b) **In order to** + **verb** *begins a sentence.*

(5) **Choosing verbs over nouns.** Some writers of narratives tend to use nouns instead of verbs when the nouns have verb equivalents. Using nouns over verbs is often problematic because noun forms are usually longer, due to prepositional phrases that typically accompany the nouns, and they are more abstract than verb forms. For example, in sentence 10 of Case 8-8A, the nouns *amplification* and *analysis* are used, instead of their verb forms *amplify* and *analyze*. In most instances, you should phrase sentences with verbs instead of their related nouns, especially if these nouns are homographs, gerunds, or nominalizations.

● **Homographs.** Some verbs and nouns are spelled identically and have the same meaning. It is only through the word's use in a sentence that the word's grammar can be determined. Examples of such homographic verb–noun pairs are:

to approach – the approach	to form – the form
to cluster – the cluster	to image – the image
to decrease – the decrease	to increase – the increase
to design – the design	to interview – the interview
to experiment – the experiment	to phrase – the phrase
to plan – the plan	to schedule – the schedule
to reply – the reply	to string – the string
to scan – the scan	to test – the test

> **Guideline 8.24**   *Consider using the verb form of a homograpic pair in order to shorten a sentence.*

Example 35a shows a sentence with the homograph *interview* used as a noun, and in Example 35b, the sentence uses the homograph in its verb form, which shortens the sentence.

Example 35
a. Our methods will be to conduct **interviews** with women in rural towns, who have delivered babies in the 2 years immediately prior to the interview. (25 words, 144 characters with spaces)
b. Our methods will be to **interview** women in rural towns, who have delivered babies in the 2 years immediately prior to the interview. (23 words, 130 characters with spaces)

● **Gerunds**. A gerund is a noun, created by adding + **ing** to an infinitive verb. In Example 36a, the gerund *treating* in *for treating* is derived from the verb *treat*. As shown in Example 36b, the verb *to treat* can replace the longer *for treating*, which results in a saving of 4 characters.

Example 36

a. Different classes of antimicrobials could be used *for treating* chronic infection as opposed to acute infection. (16 words, 110 spaces with characters)

b. Different classes of antimicrobials could be used *to treat* chronic infection as opposed to acute infection. (16 words, 106 spaces with characters)

> **Guideline 8.25**   *Consider using the verb form of gerunds to shorten sentences, to reduce prepositional phrases, and to increase readability.*

● **Nominalizations**. A nominalization is a type of noun that comes from a verb or an adjective. A nominalization has a distinctive ending, such as + **ence**, + **icity**, + **ment**, and + **tion**. Table 8-3 gives examples of nominalizations and their associated verbs and adjectives.

A nominalization is the idea of the verb or adjective; as a result, a nominalization is more abstract than its verb. A nominalization is also usually wordier than its verb or adjective form since it has more characters and is often followed by a preposition. Thus, the more nominalizations in a sentence, the wordier and more abstract the sentence.

> **Guideline 8.26**   *Consider revising unnecessary nominalizations into their corresponding verb or adjective forms.*
> *Exceptions: Do not revise nominalizations that are standard scientific terms, are seldom or never used in their verb forms, serve as transitions between sentences, or help sequence a sentence from given to new information.*

Table 8-4A gives examples of wordy phrases with nominalizations, most of which include prepositions, and Table 8-4B shows the shorter verb forms of the nominalizations. Example 37a illustrates the nominalizations *recruitment* and *assessment*, which are revised in Example 37b to shorten the sentence. In Example 38a, the nominalizations and their associated prepositions are *the development of, intervention in,* and *application of.* To make the sentence more concise and less abstract, 2 of the nominalizations (and their associated prepositions) can be rephrased into verbs, as shown in Example 38b. However, the nominalization *intervention* has a particular meaning that would be lost if it were revised into its verb form *intervene*; therefore, *intervention* is not revised.

Example 37

a. *The recruitment of* patients who have already undergone *assessment* through our Central clinic will be conducted. (16 words, 110 characters with spaces)

b. Patients who *have already been assessed* through our Central clinic will be *recruited*. (13 words, 84 characters with spaces)

Example 38

a. We **will conduct the development of the intervention in application of** our pilot study. (14 words, 85 characters with spaces)
b. We **will develop** the **intervention to apply** it to our pilot study. (12 words, 64 characters with spaces)

A nominalization is not always a problem; it can even be useful: (1) to provide a transition between sentences by sequencing given information before new information (see Chapter 8.4), (2) to establish a topic or theme across sentences by repeating a key term, and (3) to show the PI's familiarity with a discipline, if the standard term is a nominalization. In Example 39, the first sentence uses the verb *determine*. In sentence 2, the early repetition of the verb through its noun form *determination* serves to sequence content from given to new information and to provide a transition between the sentences. In Example 40, the verb *invade* is used twice before its nominalization *invasion* in the last half of sentence 2 and twice again in sentence 3. This use of *invasion* is not a problem since it follows its verb form, thus helping the sentences relate to each other.

Example 39

[1]We will **determine** whether the two subpopulations are targeted by different mechanisms. [2]This **determination** will allow us to investigate the relative proportion of each receptor and its associated function.

Example 40

[1]We expect that the more pathogenic Bp isolates **will invade** IE cells and macrophages more efficiently than less pathogenic strains. [2]These results would be important because they would indicate that the ability of the isolates **to invade** potential target cells in the intestine correlates with virulence, and this **invasion** would in turn provide an efficient assay for follow-up studies to identify specific virulence factors. [3]If, however, **invasion** is not associated with virulence, this would suggest that intestinal infection may depend on factors other than direct **invasion**, such as uptake by M cells.

For nominalizations that are standard technical terms and not typically used in their verb forms, the use of those nominalizations can imply you are familiar with your discipline, which can enhance your credibility. Such nominalizations are *evidence*[8], *up-regulation, down-regulation, Western blot analysis*, and *intervention* (see discussion of *intervention* in Example 38 above). However, there are many nominalizations that are not standard scientific terms or that are acceptable in either their nominalization or verb form. It is these nominalizations that are problematic. Example 41a shows a sentence with 4 nominalizations that are also common in their verb forms, in scientific English (*hypothesis, discrimination, direction,* and *emphasis*). Therefore, some of these nominalizations can be rephrased in their verb or adjective forms, resulting in a substantial shortening of the sentence.

---

[8] This nominalization is related to the adjective *evident*.

Example 41

a. *It is our hypothesis* that adults with and without hearing impairments *will exhibit a greater discrimination* of speech in single- and multi-voice conditions when the speech is subjected to preprocessing through algorithms *under the direction of an emphasis of* more discriminating acoustic features. (42 words, 293 characters with spaces)

b. *We hypothesize* that adults with and without hearing impairments *will better discriminate* speech in single- and multi-voice conditions when the speech is subjected to preprocessing through algorithms *that emphasize* more discriminating acoustic features. (33 words, 249 characters with spaces)

More specifically, Example 41a consists of 4 unnecessary nominalizations[9] that can be rephrased using their more concrete and concise verbs:

Passages with unnecessary nominalizations	Concise counterparts
it is our **hypothesis** that	we **hypothesize** that
will exhibit a greater **discrimination**	will better **discriminate**
under the **direction** of an **emphasis** of	that **emphasize**

Notice that in changing the nominalizations into their verb forms, the number of prepositional phrases is reduced from 9 in Example 41a to 5 in Example 41b, and the nominalization *direction* can be entirely omitted.

Reducing the number of nominalizations can result in substantial textual savings. However, before blindly revising a sentence to eliminate a nominalization you need to evaluate the entire sentence and its surrounding sentences, for the following issues:

● Does the nominalization follow its verb or adjective form? The nominalization could be functioning as a transition between sentences or as a means to position given information before new information. *If so, consider not revising the nominalization.*

● Does changing the nominalization into its verb equivalent change the meaning of the sentence in unwanted ways? *If so, consider not revising the nominalization.*

● Is the nominalization a standard technical term? *If so, consider not revising the nominalization.*

(6) **Using *pro* substitution forms.** Many writers use *pro*-verbs, personal *pro*nouns, and demonstrative *pro*nouns to shorten the text since *pro* forms are shorter than the words for which they substitute (called **referents** or **antecedents**). The following lists present pro-verbs, personal pronouns, and demonstrative pronouns.

Pro-verbs	Personal pronouns	Demonstrative pronouns
do so, does so, did so	he, him, his, she, her, hers, it, its, they, them, theirs, you, your, I, me, my, mine, we, us, our, ours	this, these, that, those

---

[9] This sentence has other nominalizations, such as *impairment*, which are not problematic and so are not revised.

Pronouns and pro-verbs can reduce the number of characters and sometimes the number of words in a sentence, which can at times eliminate the short last line of a paragraph. However, *pro* forms should be applied only in those passages where the referents are clear.

> **Guideline 8.27**   *To shorten the number of words or characters, consider using pro-verbs, personal pronouns, and demonstrative pronouns, if their referents are absolutely clear. If the referents are not absolutely clear, repeat the terms.*
> **Exception:** *For an unclear demonstrative pronoun, add a clarifying noun after the pronoun, which changes the demonstrative pronoun into a demonstrative article (see Examples 42 and 43 below, and their discussions).*

Example 42 shows the use of *they*, which has a clear referent: the only term *they* can refer to is *muscle properties* since that term is the only plural noun preceding *they* (except for **characteristics**, but that term is in parentheses and is not part of the sentence grammar per se). In this example, since the pronoun *they* is clear, *they* can be used rather than its referent *muscle properties*.

Example 42
Muscle properties (e.g., force-length and force-velocity characteristics) are complex, and **they** vary with muscle-activation levels.

Often, however, a pronoun is ambiguous because it can refer back to more than one referent, in which case the pronoun should not be used. Example 43a illustrates an ambiguous pronoun *they*, the referent of which can possibly be: (a) the fluorescence and the current, (b) voltage steps or substrate applications, or (c) measurements of the fluorescence and the current. To avoid this ambiguity, the pronoun *they* needs to be avoided and the referent needs to be repeated, even though the sentence is not shortened. Example 43b is rephrased with the pronoun omitted and the intended referent (*these measurements*) included.

Example 43
a. [‡][1]The fluorescence and the current will be measured simultaneously in response to voltage steps or substrate applications. [2]**They** will provide information on how the different substrates affect the conformational changes of the glutamate transporters during glutamate transport. (37 words, 273 characters with spaces)
b. [1]The fluorescence and the current will be measured simultaneously in response to voltage steps or substrate applications. [2]**These measurements** will provide information on how the different substrates affect the conformational changes of the glutamate transporters during glutamate transport. (38 words, 286 characters with spaces)

Example 44a below includes an ambiguous demonstrative pronoun *this* since *this* can refer to either *decreased ERPs* or *decreased APAs*. The pronoun *this* needs clarification even though the resulting sentence is longer. To clarify a demonstrative, you repeat the referent after the demonstrative, as shown in Example 44b.

Example 44
a. ERPs recorded at the scalp over the SMA (ref) and measures of APAs (refs) were both diminished in PD. Moreover, *this* was associated with bradykinetic behavior (ref). (27 words, 163 characters with spaces)
b. ERPs recorded at the scalp over the SMA (ref) and measures of APAs (refs) were both diminished in PD. Moreover, *this decreased APA activity* is associated with bradykinetic behavior (ref). (30 words, 184 characters with spaces)

Pro-verbs help shorten the text, but they also need to have clear referents. Example 45a illustrates an ambiguous pro-verb *to do so*, since *do so* can refer to *collecting data* or *running the analyses*, or both *collecting data* and *running the analyses*. In Example 45b, for clarity, *do so* is replaced with its referent *to analyze the results*, even though the sentence is longer by one word (14 characters).

Example 45
a. The results from our 4 subjects available at the time of our first submission were included in Preliminary Studies. We have since collected data on 7 additional subjects and have run the analyses, and continue *to do so*. (38 words, 217 characters with spaces)
b. The results from our 4 subjects available at the time of our first submission were included in Preliminary Studies. We have since collected data on 7 additional subjects and have run the analyses, and continue *to analyze the results*. (39 words, 231 characters with spaces)

> **Guideline 8.28** *For clarity, use pro-verbs, personal pronouns, and demonstrative pronouns sparingly. Instead,*
> *(a) For an unclear pro-verb, repeat the verb.*
> *(b) For an unclear personal pronoun, repeat the noun.*
> *(c) For an unclear demonstrative pronoun, repeat its referent.*

(7) **Reducing grammatical parallelism**. Each item in a list needs to be in parallel grammar with all other items in the list. This sameness of grammar across items in a list is termed *grammatical parallelism*. For instance, the 3 listed items in Example 46 below are grammatically parallel to each other; each is a prepositional phrase. The first listed item in Example 46 begins with the preposition *with*, and the second and third items begin with the preposition *without*. Repetition of the grammar in each item in the list – which, in this example, is the preposition – determines whether a list is grammatically parallel. As shown in this example, the terms that lead each item in the list (i.e., the prepositions *with*, *without*, and *without*) do not all have to be identical for the items to be grammatically parallel; only their grammar needs to be identical.

Example 46
Subjects will be blindfolded and instructed to keep their balance while standing on the platform *with* minimal bending of their knees, *without* taking a step, and *without* unlocking their hands.

You might have a sentence with a list that is grammatically parallel, and each item in the list begins with the same term. You can eliminate the repetitious term(s), *as long as the resulting sentence is clear*. The remaining listed items will still be in parallel grammar. Consider Example 47a, which has 2 parallel lists.

Example 47
a. Of great benefit *to* women diagnosed with uterine fibroids and *to* their physicians would be a technology *that* is well-tolerated by patients, *that* clearly quantifies the rate of blood flow supplying the fibroid, and *that* is amenable to office and clinic use. (42 words, 254 characters with spaces)
b. Of great benefit *to* women diagnosed with uterine fibroids and *to* their physicians would be a technology *that* is well-tolerated by patients, clearly quantifies the rate of blood flow supplying the fibroid, and is amenable to office and clinic use. (40 words, 244 characters with spaces)
c. ‡Of great benefit *to* women diagnosed with uterine fibroids and their physicians would be a technology *that* is well-tolerated by patients, *that* clearly quantifies the rate of blood flow supplying the fibroid, and *that* is amenable to office and clinic use. (41 words, 251 characters with spaces)
d. Of great benefit *to* women diagnosed with uterine fibroids and *to* their physicians would be a technology *that*: (a) is well-tolerated by patients, (b) clearly quantifies the rate of blood flow supplying the fibroid, and (c) is amenable to office and clinic use. (43 words, 257 characters with spaces)

In Example 47a, the first list is *to women diagnosed with uterine fibroids* and *to their physicians*, and each item begins with the preposition *to*. The second list has 3 items, and each item begins with *that*. As shown in Example 47b, the omission of *that* from the second and third items in the second list does not change the meaning, but it does shorten the text. In Example 47c, elimination of the second *to* before *physician* creates an unwanted meaning in the first part of this sentence since it is in part saying that the *women are diagnosed with their physicians*. Therefore, the second *to* before *physician* should not be eliminated. Before you reduce grammatical parallelism in a list, you should double-check that you will not be changing the meaning of the original sentence.

Guideline 8.29 *To shorten a sentence with a parallel list, consider reducing the identical terms if clarity can be maintained.*

If length is not an issue, then for emphasis, you can repeat *that* at the beginning of each item in the list, such as in Example 47a, or you can itemize the list and omit the repetition of *that*, as shown in Example 47d.

(8) Shortening the classic definition. As discussed in Chapter 8.2.1, a classic definition usually has 3 parts: the key term, the class, and the distinguishing feature that separates the key term from other members of its class. To save space or to de-emphasize a classic definition of a term that your reviewers already know, you can shorten the classic definition by integrating the definition into another sentence that includes the key term.

> **Guideline 8.30**   *Consider integrating a classical definition into another sentence if both mention the same key term.*

Example 48a has 2 sentences. The first sentence introduces *cyclic GMP* and the second sentence defines it. The definition can be integrated into the first sentence, as shown in Example 48b. Also as shown in Example 48b, when a classic definition is integrated into another sentence, the relative pronoun *which* (preceded by a comma) is used for the class and *that* is still used for the distinguishing feature. In a comparison of Examples 48b and 48c, notice that *which is* can also be omitted, further shortening the definition.

Example 48
a. The proposed research focuses on **cyclic GMP (*cGMP*)**. ***cGMP is*** an intracellular messenger ***that***, in vascular smooth muscle, mediates smooth muscle relaxation in response to vasodilators. (26 words, 180 characters with spaces)
b. The proposed research focuses on **cyclic GMP (*cGMP*)**, ***which is*** an intracellular messenger ***that***, in vascular smooth muscle, mediates smooth muscle relaxation in response to vasodilators. (26 words, 181 characters with spaces)
c. The proposed research focuses on **cyclic GMP (*cGMP*)**, an intracellular messenger ***that***, in vascular smooth muscle, mediates smooth muscle relaxation in response to vasodilators. (24 words, 172 characters with spaces)

A classic definition can also be shortened when the key term is phrased with its class. In this case, the class can be omitted from the classic definition. This type of shortened classic definition is illustrated in Example 49. In Example 49a, both the key term and the class include the key term *model*. As shown in Example 49b, the definition can be shortened to use *model* only as the key term, eliminating the repetitious class.

Example 49
a. The Sums of Product ***Model*** is ***a model that predicts*** phoneme duration on a per phoneme class basis. (18 words, 96 characters with spaces)
b. The Sums of Product ***Model predicts*** phoneme duration on a per phoneme class basis. (14 words, 80 characters with spaces)

(9) Shortening adjective clauses. Adjective clauses modify nouns. An adjective clause can be descriptive, as signaled by the pronoun *which* (which is preceded by a comma), or an adjective clause can be definitional, as signaled by the pronoun *that* (which is not preceded by a comma). An adjective clause that describes a human noun begins with *who, whom,* or *whose* (and is also preceded by a comma if the adjective clause is descriptive but is not preceded by a comma if the adjective clause is definitional). Adjective clauses can be shortened, but how you shorten them depends on their type.

● **Adjective clauses with the linking verb** *be*. Examples 50a and 51a illustrate adjective clauses with the ***linking verb*** *be*. An adjective clause with the linking verb *be* can be shortened by omitting the relative pronoun and the linking verb, as shown in Examples 50b and 51b. A linking verb essentially equates the grammatical subject with its predicate.

Example 50
a. Since the initial submission of the proposed research, we have begun collaborating with Dr. John Smith, *who is an expert* in optical coherence tomography. (24 words, 151 characters with spaces)
b. Since the initial submission of the proposed research, we have begun collaborating with Dr. John Smith, *an expert* in optical coherence tomography. (22 words, 145 characters with spaces)

Example 51
a. Nor is it known whether certain in vitro properties such as cell invasion and replication can be correlated with in vivo virulence, *which is an essential step* in developing assays to identify enteric virulence determinants. (35 words, 221 characters with spaces)
b. Nor is it known whether certain in vitro properties such as cell invasion and replication can be correlated with in vivo virulence, *an essential step* in developing assays to identify enteric virulence determinants. (33 words, 212 characters with spaces)

In Example 50a, the relative pronoun *who* (referring to *Dr. John Smith*) is equated with its predicate *an expert*, and in Example 51a, the relative pronoun *which* (referring to *correlation*) is equated with *an essential step*.

The resulting part of the adjective clause is called an ***appositive***, which is a noun that describes another noun.

Adjective clause with a linking verb	Shortened adjective clause (appositive)
who is an expert	an expert
which is an essential step	an essential step

• **Adjective clauses with non-linking verbs**. Examples 52a and 53a are sentences with adjective clauses that have transitive, non-linking verbs. The adjective clause in Example 52a uses the active verb *includes*, and the adjective clause in Example 53a uses the passive verb *are tagged*.

Example 52
a. Our progress, *which includes* a brief description of the patents since our first funding cycle, is explained here. (18 words, 112 characters with spaces)
b. Our progress, *including* a brief description of the patents since our first funding cycle, is explained here. (17 words, 107 characters with spaces)

Example 53
a. Current technologies using proteins *that are tagged* with fluorescent dyes or genetically engineered proteins (e.g., Cy3, GFP) are dim and, therefore, prevent prolonged glimpses of discrete proteins for substantial lengths of time (> 5 seconds). (35 words, 242 characters with spaces)

b. Current technologies using proteins **tagged** with fluorescent dyes or genetically engineered proteins (e.g., Cy3, GFP) are dim and, therefore, prevent prolonged glimpses of discrete proteins for substantial lengths of time (> 5 seconds). (33 words, 233 characters with spaces)

An adjective clause with a non-linking verb can be reduced: (a) by omitting the relative pronoun and using the present participial form of the verb (+ **ing**) if the adjective clause is in active grammar, as shown in Example 52b, or (b) by omitting the relative pronoun and the auxiliary verb *be* and using the past participial form of the verb if the adjective clause is in passive grammar, as shown in Example 53b. The resulting shortened adjective clause is called a participial phrase.

Adjective clause with a non-linking verb	Shortened adjective clause (participial phrase)
which includes a brief description	**including** a brief description
that are tagged with fluorescent dyes	**tagged** with fluorescent dyes

● **Consecutive adjective clauses**. If a sentence has consecutive adjective clauses, you should be able to shorten at least one of them.

Guideline 8.31

(a) *Consider shortening the first adjective clause in consecutive adjective clauses where one of the adjective clauses is a classic definition, and*
(b) *Consider shortening one or both of 2 consecutive adjective clauses where neither comprises a classic definition.*

Example 54a shows a sentence with 2 adjective clauses, both of which comprise a classic definition. For readability and clarity, you can shorten the first adjective clause by omitting *which is*, shown in Example 54b.

Example 54
a. Those women of reproductive age **who develop uterine fibroids** and wish to preserve fertility usually undergo a myomectomy, **which is an invasive surgery** that removes the fibroids but leaves the uterus intact. (32 words, 204 characters with spaces)
b. Those women of reproductive age **who develop uterine fibroids** and wish to preserve fertility usually undergo a myomectomy, **an invasive surgery** that removes the fibroids but leaves the uterus intact. (30 words, 195 characters with spaces)

Notice that neither clause in Example 55a below comprises a classic definition since there is no distinguishing feature of a key term. In such a case, you can shorten one or both of the adjective clauses, as shown in Example 55b and 55c, as long as the meaning is not altered. However, both consecutive adjective clauses are not usually shortened if both reductions end in + **ing** forms, as shown in Example 55d with *leading* and *limiting*.

Example 55

a. Conventional therapies are ineffective in this population, ***which leads*** to lifelong motor impairment ***that*** limits the use of both upper and lower limbs. (23 words, 149 characters with spaces)

b. Conventional therapies are ineffective in this population, ***leading*** to lifelong motor impairment ***that*** limits the use of both upper and lower limbs. (22 words, 145 characters with spaces)

c. Conventional therapies are ineffective in this population, ***which leads*** to lifelong motor impairment ***limiting*** the use of both upper and lower limbs. (22 words, 146 characters with spaces)

d. ‡Conventional therapies are ineffective in this population, ***leading*** to lifelong motor impairment ***limiting*** the use of both upper and lower limbs. (21 words, 142 characters with spaces)

(10) Reducing wordy, vague sentence openers. Some writers use the wordy sentence openers "It is" and "There are/is," both of which have low informational value and are sometimes considered ***not space-worthy***. In other words, the phrasing does not give enough information to justify the textual space that is used. Such vague and wordy sentence openers are illustrated here with more informative (and shorter) paraphrases.

Vague and wordy openers	**More informative and concise openers**
It is noteworthy that	The evidence shows that
It is clear that	The data clearly show that
It is our hypothesis that	We hypothesize that
It has been found that	Researchers have found that
It has been recommended that	Jones et al. (2014) recommend that
There is no known reason for	Researchers have no data to support
There is disagreement among researchers	Researchers disagree on

Some of these problematic sentence openers also use nominalizations instead of their verb equivalents, and some use agentless passive (Chapter 8.5). In addition, the pronoun *it* in such phrases is sometimes called "empty" because *it* really has no actual referent, and *there* is sometimes called a "placeholder" for an unspecified grammatical subject.

Guideline 8.32 *Substitute specific and clear grammatical subjects and verbs instead of vague and wordy sentence openers.*

(11) Gapping. Gapping is an advanced sentence structure that results in the joining of 2 related sentences (or independent clauses) and shortens the second. If the second sentence uses the identical ***verb*** as in the first sentence, then: (a) the sentences are joined with *and* or *but*, depending on whether the sentences are contrastive, and (b) the verb in the second sentence is replaced with a comma. Similar to an apostrophe that indicates eliminated letters in a contracted word, the comma indicates an eliminated verb in the gapped (contracted) part of the sentence.

Gapping is illustrated in Examples 56 and 57. In Example 56a, both sentences have the exact same verb, *will be referred*. In Example 56b, they are joined with *and*. The second verb *will be referred* is omitted and replaced with a comma. In Example 57a, both independent clauses of the sentence have the same verb, *will be examined*. In Example 57b, gapping is applied to the second independent clause, where *will be examined* is replaced with a comma. These examples use passive grammar, but you can also use gapping with active grammar.

Example 56

a. [1]Larval and adult mosquitoes injected with candidate and key genes **will be referred** to as the ds-RNA-injected mosquitoes. [2]Larval and adult mosquitoes injected with GFP **will be referred to as** the controls. (33 words, 203 characters with spaces)

b. [1]Larval and adult mosquitoes injected with candidate and key genes **will be referred** to as the ds-RNA-injected mosquitoes, and larval and adult mosquitoes injected with GFP, as the controls. (28 words, 184 characters with spaces)

Example 57

a. A binding **will be examined** in detail in the experiments in Section C2, and the relationships between antibody and TfR binding **will be examined** in Section C3. (27 words, 156 characters with spaces)

b. A binding **will be examined** in detail in the experiments in Section C2, and the relationships between antibody and TfR binding, in Section C3. (24 words, 140 characters with spaces)

> **Guideline 8.33** *To reduce the second of 2 sentences (or independent clauses) that have the same verbs, consider gapping by joining the sentences and replacing the verb in the second sentence with a comma.*

(12) Combining multiple sentences into one sentence with a list. Sometimes information is spread over a few sentences when the same information can be condensed into a list with fewer sentences, words, characters, and lines. Combining multiple sentences into one sentence with a list is useful if the sentences are closely related in content, but in a non-causal way.

> **Guideline 8.34** *To shorten the text, consider combining multiple sentences with closely related, non-causal information, into one sentence with a list.*

Example 58a gives a passage with a sequential list from an abstract to a narrative. To shorten the text, the first part of sentence 2 (bold italics) becomes the lead-in to the list, and the remainder of sentence 2, plus parts of sentences 3 and 4 comprise the items in the list, as shown in Example 58b. The numerical itemization in the list captures the chronology indicated by *First*, *Second*, and *Last* in Example 58a.

Example 58

a. [1]The studies proposed here are intended to fill a critical void in our understanding of pathogenesis of infection with this important and emerging bacterial pathogen. [2]*First, we will use the mouse infection model of Bp infection* to determine whether most or all strains of Bp can establish enteric infection and to identify virulent and avirulent isolates. [3]*Second, we will use the model* to define the role of the intestine as a reservoir for Bp infection and to identify cells in the GI tract where the organism is maintained during chronic infection. [4]*Last, we will investigate* how Bp is disseminated to other organs during chronic enteric infection. (106 words, 654 characters with spaces)

b. [1]The studies proposed here are intended to fill a critical void in our understanding of pathogenesis of infection with this important and emerging bacterial pathogen. [2]*We will use the mouse infection model of Bp infection: (1) to determine* whether most or all strains of Bp can establish enteric infection and to identify virulent and avirulent isolates, *(2) to define* the role of the intestine as a reservoir for Bp infection and to identify cells in the GI tract where the organism is maintained during chronic infection, and *(3) to determine* how Bp is disseminated to other organs during chronic enteric infection. (101 words, 610 characters with spaces)

(13) Reducing prepositional phrases. Sometimes sentences include multiple prepositional phrases, especially those sentences that include nominalizations. Sentences with multiple prepositional phrases in a series can be wordy and can lower readability. To write more concisely and to improve readability, you can reduce the number of prepositional phrases by rephrasing nominalizations into their verb and adjective forms.

> **Guideline 8.35** *Reduce prepositional phrases by:*
> *(a) Using verbs or adjectives instead of nominalizations and*
> *(b) Changing nouns within prepositional phrases into adjectives.*

Example 59a presents a sentence with 5 prepositional phrases (prepositions in bold italics). These phrases are reduced to 4 in Example 59b, by rephrasing the nominalization *regulation* into *regulating*.

Example 59

a. These data point *to* the role *of* a non-steroidal serum factor *in* the regulation *of* encephalitogenic T cells *during* pregnancy. (20 words, 123 characters with spaces)

b. These data point *to* the role *of* a non-steroidal serum factor *in* regulating encephalitogenic T cells *during* pregnancy. (18 words, 116 characters with spaces)

In Example 60a, the noun *insecticide resistance* in the prepositional phrase *in insecticide resistance* can be rephrased into the adjective *insecticide-resistant* and then located before *Culex mosquitoes*, which is the term that it modifies. As shown in Example 60b, the result is that the entire adjective clause *that are involved in insecticide resistance* can be omitted, which substantially shortens the sentence.

Example 60
a. This study will characterize the function of different gene families and regulatory factors at the whole transcriptome level of *Culex* mosquitoes ***that are involved in insecticide resistance***. (27 words, 187 characters with spaces)
b. This study will characterize the function of different gene families and regulatory factors at the whole transcriptome level of ***insecticide-resistant*** *Culex* mosquitoes. (22 words, 166 characters)

(14) Using hyphens in compound and complex adjectives. Your targeted funding agency might have an upper limit on the total number of words that you can use in the narrative, the title, or the abstract. By using conservative hyphenation of adjectives, you can reduce the word count because some word-processing programs calculate words and phrases with hyphens differently from those without hyphens.

> **Guideline 8.36**   *Consider using conservative punctuation with hyphens to reduce the word count in sentences, titles, and abstracts.*

In conservative punctuation, hyphens are used in compound and complex adjectives, to indicate that one adjective modifies another, and as a group, they modify the noun. For example, in the noun phrase *ensemble-averaging techniques*, shown below, the hyphen indicates that *ensemble* modifies *averaging*, and as one compound adjective, *ensemble-averaging* modifies the noun *techniques*. The use of a hyphen to join *ensemble* to *averaging* reduces the word count by one. Likewise, in the noun phrase *age-specific mortality rate*, the compound adjective *age-specific* modifies the compound noun *mortality rate*, and in the noun phrase *state-of-the-art microscope*, the complex adjective *state-of-the-art* modifies the noun *microscope*.

Unhyphenated noun phrases	**Hyphenated noun phrases**
ensemble averaging techniques = 3 words	ensemble-averaging techniques = 2 words
age specific mortality rates = 4 words	age-specific mortality rates = 3 words
state of the art microscope = 5 words	state-of-the-art microscope = 2 words

Example 61a below illustrates a sentence with 4 noun phrases, each of which is modified by a compound or complex adjective: (1) in *vestibular-dysfunction subjects*, the adjective *vestibular* modifies *dysfunction*, and as a compound adjective, *vestibular-dysfunction* modifies *subjects*; (2) in *self-reported dizziness, self* modifies *reported*, and as a group, *self-reported* modifies *dizziness*, (3) in *12-fold increase, 12* modifies *fold*, and as a group, *12-fold* modifies *increase*, and (4) in *6-fold increase, 6* modifies *fold*, and as a pair, *6-fold* modifies *increase*. The omission of hyphens in Example 61b increases the word count by 5.

Example 61
a. The study also documented a potential link between vestibular dysfunction and falls, reporting that ***vestibular-dysfunction*** subjects with ***self-reported*** dizziness had a ***12-fold*** increase in the odds of falling (***6-fold*** increase in ***vestibular-dysfunction***

subjects with no dizziness; both odds ratios adjusted for other risk factors).
(44 words, 326 characters with spaces)

b. The study also documented a potential link between vestibular dysfunction and falls, reporting that *vestibular dysfunction* subjects with *self reported* dizziness had a *12 fold* increase in the odds of falling (*6 fold* increase in *vestibular dysfunction* subjects with no dizziness; both odds ratios adjusted for other risk factors).
(49 words, 325 characters with spaces)

For credibility, care must be taken to use hyphens appropriately, even if it means *not* reducing the word count. Writers sometimes mistakenly add a hyphen in a complex adjective that consists of an + **ly** adverb that modifies an adjective. Conservative punctuation does not use a hyphen in this context. In Example 62, *newly* and *developed* are not joined with a hyphen even though, as a unit, they modify the noun *algorithm*.

Example 62
We will determine the effects of a *newly developed algorithm* that provides subjects with direction-, velocity- and displacement-specific, binaural, audio information about their body sway.

Example 62 also illustrates how hyphens can be used in a series of terms,[10] each of which modifies the same adjective, in this case *specific*, in order to reduce the phrasing. The use of hyphens makes it unnecessary to repeat the word *specific* with each modifier, thus avoiding the repetitious phrasing *direction-specific*, *velocity-specific*, and *displacement-specific*.

(15) Shortening explanatory paragraphs of visuals. Shortening explanatory paragraphs of visuals is another way that you can shorten the overall length of a (sub)section and the narrative. You can target visuals with explanatory paragraphs, the last line of which is short, and then apply the paragraph-backwards revision strategies described in Chapter 8.6.3 to further shorten the explanatory text. Case 8-10 presents a figure with an explanatory paragraph with a short last line. Since *prone* is, by definition, a type of *position*, the phrase *prone position* is redundant; thus, *in a prone position* can be shortened to *prone*, resulting in the elimination of the last line of the explanatory paragraph as shown in Case 8-10B.

8.6.4   Reducing content – the last resort!

You can reduce a stable draft of the narrative with careful and deliberate editing, using the types of strategies described in Chapter 8.6.2 and 8.6.3. However, you might come to a point in the text-shortening process where you must cut content in order to reduce the length of the narrative.

---

[10] In this example, the words *direction-*, *velocity-*, and *displacement* are actually nouns functioning as adjectives to modify the adjective *specific*.

Some repetitive content in a grant proposal is useful, for example to establish a theme, to emphasize a point, to clarify content, or to create a transition between sentences. Such repetition was noted in Example 40, where the verb *invade* and its noun form *invasion* were used multiple times in a passage. Sometimes, however, repetitive content does not help establish a theme, emphasize a point, clarify content, or create a transition. Such unnecessary repetition can be eliminated to shorten the text.

(1) **Eliminating unnecessarily redundant content**. Some phrases are redundant and can be shortened, such as in the following examples:

Redundant phrases	Non-redundant phrases
the fluorescent-colored red and green QDs	the fluorescent red and green QDs
in the upper-right panel of Fig. 1, labeled A	in Fig. 1A
(See Fig. 1A)	
because of the reason that	because

(2) **Shortening or eliminating the introduction to a (sub)section**. You need to assess the introduction to each section of the narrative, except for the **Aims Section**, in order to determine whether the introduction is needed. If the introduction is needed, then you should assess whether any of its information can be eliminated.

> **Guideline 8.37** *To shorten a narrative, consider reducing or eliminating the introduction to a section.*

You should strive to draft an introduction that is not more than 15% of the total length of the section since you want to send your reviewers into the body of the section fast, where they can find the most important information. Types of information that can be eliminated from an introduction are: (a) the heading **Introduction**, (b) a forecast statement, (c) information that is later given in the body of the section, (d) a review of any studies that you have already reviewed in a previous section, (e) definitions of terms that you have already previously defined, and (f) re-introductions of abbreviations that have already been mentioned.

The heading **Introduction** is not needed since reviewers typically assume that any paragraph(s) between the section heading and the first subsection heading is introductory. If you omit the heading **Introduction** from one section in the narrative, you need to remove all such headings from other sections for consistent formatting. Another feature of an introduction that can be omitted is the forecast statement. As illustrated in sentence 7 of Case 8-11, a forecast statement helps reviewers identify the topics that will be discussed in the section and the sequence or location of their discussion. A forecast statement can be important for readability, but it is expendable. In Case 8-11, omission of the forecast statement results in a reduction of 2 lines.

(3) **Reducing the Aims Section**. If the **Aims Section** needs to be shortened, there are different ways to shorten it, as suggested in Guideline 8.38:

Guideline 8.38   *To shorten the* **Aims Section,** *consider:*
(a) *Omitting forecast statements.*
(b) *Omitting summary information in the ending paragraph.*
(c) *Omitting the long-term goal of the proposed research.*
(d) *Reducing the review of literature from Passage 1.*
(e) *Reducing methodological details.*

As explained in Chapter 2.2.4, the **Aims Section** identifies 2 to 3 research purposes: an optional long-term goal, the research objective of the proposed research, and the proposed aims. You can omit the long-term goal (which is sometimes termed the *long-term objective*) since it covers what you intend to accomplish in a line of research over an extended period of time, such as 10 years or longer. Even though purpose information is useful for reviewers to understand the ultimate achievement that you intend for your proposed research and the direction that you expect your line of research to take, the long-term goal is not critical, in contrast to the research objective and aims, which cannot be eliminated.

Another way to shorten the **Aims Section** is to reduce the review of previous research in the paragraphs before the proposed aims, which corresponds to Passage 1 (see Chapter 2.1). You need to include enough previous research for reviewers to understand the logical relationship between the previous research, and your proposed research and the novelty and significance of your research. However, an involved review in the **Aims Section** is usually not needed because the narrative includes a **Background Section** where you can include the information.[11]

Similarly, because the narrative includes a **Methods Section**, you can reduce the **Methods** information in the **Aims Section**. The **Aims Section** presents an overview of your proposed research design and methods. This overview might include **SOS**, **MET**, your basic methodological approach, and your primary procedures. However, if you believe that your reviewers will need particular methodological details to understand your research design, rather than including them in the **Aims Section**, you can provide just a few key details and then include a cross-reference to the **Methods Section**. With the cross-reference, you are telling reviewers that you recognize that the methods are only briefly characterized in the **Aims Section**, and that you will provide a more extensive description in the **Methods Section** (or **Preliminary Studies Section**).

A word of caution is needed regarding cross-references, especially in the **Aims Section**. You should be able to cross-reference (connect) all of the information in the **Aims Section** to a later section in the narrative. However, in practice, you should strive

---

[11] For a narrative to a grant proposal submitted to an agency that does not explicitly specify a **Background Section**, such as NSF, you still need to assess which background information helps you clarify how your proposed research is logically related to previous research; this will help you isolate other background information that possibly can be reduced or eliminated from the **Aims Section**.

not to use many cross-references in the **Aims Section** since they interrupt the flow of information, can interfere with readability and clarity, and might lead you to omit information that is critical in making an initial, compelling scientific argument in this important first section.

(4) Reducing the Background Section. Some writers include a lot of information in the **Background Section**, claiming that reviewers need such breadth of information to fully appreciate and understand the significance of the proposed research. However, such breadth of information is usually excessive and can be reduced (for an exception, see Guideline 8.39).

Providing background information to demonstrate your knowledge of the field, independent of key elements of your proposed research, is unnecessary. Besides lengthening the text, background that does not directly relate to your proposed research can prompt your reviewers to misunderstand your ability to focus on the critical scientific issues in your proposed research. The one exception is when you know you will have a lay reviewer who will likely need a brief discussion of fundamental information relating to your proposed research topic.

Guideline 8.39   *Omit background information that does not directly relate to your research problem, research objective, aims, proposed methods, or hypothesis.*
Exception: *Include a brief discussion of fundamental background information for lay reviewers.*

(5) Reducing redundant information across Preliminary Studies and Methods Sections. Some of your proposed methods, research design, or rationale may be similar or identical to those in your preliminary research. If so, you do not repeat a full description of these commonalities in both the **Preliminary Studies** and **Methods Sections.** As noted in Guideline 8.40, you can either reduce their description in the **Preliminary Studies Section** and, instead, cross-reference to the **Methods Section** where you can describe them in more detail; or, in a less-preferred alternative, reduce their description in the **Methods Section** and include a cross-reference to the **Preliminary Studies Section** where you can describe them in more detail.

Guideline 8.40   *For methods that are similar across preliminary and proposed research, describe them briefly in the **Preliminary Studies Section** or in the **Methods Section**, and include a cross-reference to the other section where you describe them in more detail.*

(6) Reducing redundancies in the Preliminary Studies Section. You can include preliminary results in different locations in a **Preliminary Studies Section**: in the introduction to the section, in subsection headings, and within the text. If you are trying to shorten the text, however, you can eliminate the preliminary results from the introduction. You can also shorten run-in subsection headings that have been initially phrased as sentences by changing them to nouns that indicate the topic of the preliminary result that you will discuss in the subsection.

Guideline 8.41   *To shorten a **Preliminary Studies Section**, consider:*
(a) Omitting preliminary results from the introduction.
(b) Changing subsection headings from short sentences to noun phrases.

(7) Reducing the Methods Section. For a variety of reasons, many writers resist reducing the **Methods Section**. However, longer is not necessarily better. Besides lengthening the text, any unnecessarily detailed or redundant methods descriptions can interfere with reviewers understanding the structure or elegance of your proposed research. As explained in Guideline 8.42, you can omit some descriptions of methods entirely or can briefly explain them, using citations to indicate where additional details can be found.

Guideline 8.42   *To reduce descriptions of methods:*
*(a) Consider omitting details of standard methods in your field of study and instead: (i) identify the method by name, (ii) characterize the method with the term **standard** or **established**, and (iii) include citations to seminal publications that describe the methods.*
*(b) Consider omitting details of methods for **MET** that involve your following a manufacturer's instructions. Instead: (i) identify the method and manufacturer by name, and (ii) mention that you will follow the manufacturer's instructions.*
(c) If you modify standard methods or a manufacturer's instructions: (i) identify by name the method and manufacturer, (ii) briefly identify the modification and its purpose, (iii) cite to a publication that uses this same modification, and (iv) if the modification is unique, explain why it is needed.
*(d) If you use methods that are standard or established in your laboratory, but might not be standard or established in the field: (i) identify by name the methods, (ii) state that the methods are standard or established in your laboratory, (iii) cite to one of your publications that describe the methods, and (iv) briefly describe the methods, prefacing this description with **briefly** or **in brief.***

(8) Creating a shared methods subsection in the Methods Section. You might have methods that are shared (similar or identical) across your proposed aims (see Chapter 6.11). For example, you might be using the same subjects in the proposed methods for Aims 1 and 2, but you will be using different protocols. Or you might be using the same subjects and protocols in the methods for Aims 1 and 2 but different data-analysis procedures. In such instances, your text will be unnecessarily long if you describe the shared methodological feature(s) twice: once in the methods for Aim 1 and again in the methods for Aim 2. You can shorten the **Methods Section** by describing the shared methodological feature(s) in a shared methods subsection, as explained in Guideline 8.43:

> **Guideline 8.43** *To describe a methodological feature(s) that is shared across more than one proposed aim, consider one of the following alternatives to reduce the text and to eliminate unnecessary redundancies:*
> *(a) Create a shared methods subsection for the shared methodological feature(s); locate this subsection either immediately after the introduction to the **Methods Section** (see Chapter 6.11) or immediately before the ending subsection to the **Methods Section**. Then in the **Methods** subsection that describes methods to a proposed aim, include a cross-reference to the shared methods subsection.*
> *(b) Describe the shared methodological feature(s) in the **Methods** subsection of the first relevant aim, and then in the **Methods** subsection for the second relevant aim, include a cross-reference back to the subsection where you already describe it.*

It is most efficient to identify shared methods before you draft any text since creating a shared **Methods** subsection after the text is written will consume time to reorganize and to rewrite. Chapter 5.2.3 provides heading outlines that can help you identify organizational alternatives for shared methods before you start drafting the text.

(9) **Using cross-references**. In the context of a narrative, a cross-reference is a comment within the narrative that refers reviewers to related information elsewhere in the narrative or in the grant application. A cross-reference can help shorten the text, underscore the unity of your narrative, and even substantiate conclusions. A cross-reference can also help reviewers efficiently locate related information in the narrative or in the grant application. However, extensive use of cross-references can interfere with the flow of information and can prevent reviewers from moving steadily forward in the narrative. Suggestions for reducing cross-references include:

● **Eliminating extra-textual cross-references**. A cross-reference(s) to information and data outside the application can be included. However, reviewers are under no obligation to read them and usually cannot consider the extra-textual information when evaluating the proposal, unless submission requirements explicitly allow such consideration. Therefore, there is little reason to take valuable textual space for these extra-textual cross-references.

● **Using itemization for concise cross-referencing**. Chapter 1.11.3 discusses the usefulness of assigning a reference system to the organizational hierarchy. This reference system can help reviewers understand the organizational hierarchy, and the superordinate and subordinate relationships across subsections. Itemization can also help you compose concise cross-references. For example, it is more concise to use (*see Section C.2*) than (*see Preliminary Studies*), and both are more concise than (*See the methods for monitoring different steps in the glutamate transport cycle, in the Preliminary Studies Section*).

● **Cross-referencing to definitions**. There are times when you introduce a key term and define it early in the narrative, but do not use it again until a later section in the narrative. If you expect that lay reviewers, in particular, might not recall the definition, rather than re-explaining the term, you can include a cross-reference back to the earlier definition.

Guideline 8.44 *To shorten the text, consider including a few critical cross-references to information in the narrative rather than repeatedly explaining the same information; however, avoid using multiple cross-references that can interrupt the reviewers' forward movement in the narrative.*

A final note about shortening the narrative: even though there are many ways to shorten text, many of which are explained in this Chapter 8.6, it is very hard for a writer to shorten text that he or she has drafted, especially if the writer is trying to shorten the text right after drafting it. If you are such a writer, you need to work in advance of submission deadlines in order to give yourself at least a few hours break between your writing of the narrative and your trying to shorten it so you can notice more textual features. If you still are having trouble shortening the text after such a break, you should seek assistance from a collaborator or colleague, who will view your text with fresher eyes and might therefore notice text that can be shortened.

Table 8-1. Text-shortening strategies. (A) Strategy descriptions; **(B)** and **(C)** Where to apply these text-shortening strategies.

(A) Strategy Descriptions	(B) Throughout Narrative	(C) Selected Text or Visuals
From Chapter 8.6.2: *Manipulating layout, font, and visuals*		
(1) Removing blank lines between paragraphs and indenting paragraphs	X	
(2) Combining paragraphs		X
(3) Using run-in headings	X	
(4) Changing vertical lists into horizontal lists		X
(5) Presenting foot- and endnotes in a horizontal list	X	
(6) Allowing paragraph widows and orphans	X	
(7) Revising visuals		X
(8) Reducing emphatic font		X
From Chapter 8.6.3: *Revising the text*		
(1) Filling blank space at the ends of lines		X
(2) Using digits for numbers	X	
(3) Using abbreviations and acronyms	X	
(4) Replacing non-key phrases with shorter synonyms		X
(5) Choosing verbs over nouns		X
(6) Using pro-substitution forms		X
(7) Reducing grammatical parallelism		X
(8) Shortening the classic definition		X
(9) Shortening adjective clauses		X
(10) Reducing wordy, vague sentence openers		X
(11) Using gapping		X
(12) Combining multiple sentences into one sentence with a list		X
(13) Reducing prepositional phrases		X
(14) Using hyphens in compound and complex adjectives	X	
(15) Shortening explanatory paragraphs of visuals		X
From Chapter 8.6.4: *Reducing content – the last resort!*		
(1) Eliminating unnecessarily redundant content		X
(2) Shortening or eliminating the introduction to a (sub)section		X
(3) Reducing the Aims Section		X
(4) Reducing the Background Section		X

(5) Reducing redundant information across Preliminary Studies and Methods Sections		X
(6) Reducing redundancies in the Preliminary Studies Section		X
(7) Reducing the Methods Section		X
(8) Creating a shared methods subsection in the Methods Section		X
(9) Using cross-references		X

Table 8-2. Non-key terms and their more concise synonyms.

A. Wordy non-key terms	B. More concise synonyms
• approximately	• about
• at the time of, at the present time	• when, now
• because	• due to
• because of the reason that	• because, since, due to
• despite the fact that	• although
• due to the fact that	• because, since, due to
• during the time that/of, in the meantime	• during, while, when
• fewer in number	• fewer
• in order for	• for
• in order to	• to (however, see Chapter 8.6.3)
• in the event that, if the situation should arise that	• if
• in view of the fact that	• because, since, given
• it is often/frequently/seldom the case that	• often, frequently, seldom
• it is possible that, the possibility exists that	• perhaps, might, may, likely
• on account of	• due to, because, since
• on the grounds, on the basis that	• because, since
• simultaneously	• while
• the color red	• red
• the sum total	• the total
• until such time as	• until
• up to this point in time	• until, since

Table 8-3. Nominalizations and their associated verbs and adjectives.

A. Nominalizations ending in +*tion* and their associated verbs

amplification – to amplify
determination – to determine
localization – to localize
randomization – to randomize
regulation – to relate

B. Nominalizations ending in +*ment* and their associated verbs

assessment – to assess
containment – to contain
development – to develop
refinement – to refine
treatment – to treat

C. Nominalizations ending in +*sis* and their associated verbs

analysis – to analyze
diagnosis – to diagnose
emphasis – to emphasize
hypothesis – to hypothesize
synthesis – to synthesize

D. Nominalizations ending in +*ance*/+*ence* and their associated verbs and adjectives

assistance – to assist
conductance – to conduct
difference – to differ
reference – to refer
resistance – to resist

E. Nominalizations ending in +*ness* and their associated adjectives

effectiveness – effective
indissolubleness – indissoluble
preparedness – prepared
timeliness – timely
lengthiness – lengthy

F. Nominalizations ending in +*city* and their associated adjectives

acidity – acidic
mortality – mortal
sensitivity – sensitive
spasticity – spastic
specificity – specific

Table 8-4. Wordy phrases with nominalizations and their concise counterparts that can be replaced with shorter phrases on first mention of a concept.

A. Wordy phrases with nominalizations	B. Concise synonyms
• combine together, in combination of	• combine
• conclusion, the conclusion arrived at	• conclude, end, finally, after
• conduct an investigation of, conduct a study of	• investigate, study
• have the capability to	• can, is able to
• our discussion concerned	• we discussed
• do/make/perform/give an analysis of	• analyze
• give a report on	• report (verb)
• have an effect on	• affect, change, modify, alter
• have an expectation of, there is the expectation of	• we expect, we anticipate
• lead to the discoloration of, lead to the speculation that,	• discolor, speculate,
• to the consideration of	• consider
• make a decision about	• decide
• make a reference to	• refer to
• maintenance of	• maintain
• offer a suggestion about/concerning, is suggestive of	• suggest
• results in an increase/decrease in	• increase, decrease (verb)
• send a fax to	• fax
• send an email to	• email
• speculation that	• speculate
• take into consideration	• consider

Case 8-1. Location of citations under Guideline 8.2.

Guideline 8.2	Examples
*(a) To indicate that all information **in one sentence** is from the **same** source, place the citation at the end of the sentence, either consistently before the period or after the period.*	(a) The use of QDs for cell and tissue labeling has recently increased due to the unique properties of QDs, especially their bright, long-lasting, multicolor emission.[13-18] We have recently designed QD nanoparticles to detect single protein complexes and to calculate, with high accuracy (< 1 QD probe/cell is a false positive), their location within single cells (< 50 nm), with the use of a standard epifluorescent microscope.[1,4,5]
*(b) To indicate that all of the information **in a series of sentences** is from the **same** source, place the citation at the end of the **first** sentence in the series, either consistently before the period or after the period throughout the document.*	(b) The only system currently available that allows AAC users with sufficient speaking ability to record their own speech for creating a synthetic voice is the ModelTalker system (refs). The recording tool that is part of this system prompts the user to record sentences with a total of about 1650 words and phrases, the first 80 of which include words and phrases that are likely to be of need to users of AAC devices in a variety of daily contexts. The remaining words and phrases were selected to cover all diphones of American English in a broad range of phonetic and prosodic contexts that the synthesizer can use to produce an unlimited number of sentences. The recording process is said to take approximately 50 minutes. The ModelTalker system was shown to have a better word error rate in an intelligibility task than 5 out of 6 other concatenative synthesizers (which were not named) (refs).
*(c) To indicate that information **in different parts of one sentence** comes from different sources, place a citation immediately after the information and before any relevant punctuation.*	(c) Muscle spindles fire individually in relation to static joint position (refs), dynamic joint position (refs), and the speed of movement (refs), and they fire as a group, in relation to direction (refs).
(d) If a sentence contains both published and nonpublished information, sequence published information and its citation before the nonpublished information that does not receive a citation.	(d) While intimal hyperplasia was effectively abolished within the smooth-walled grafts, intimal thickening within the adjacent vessel was unaffected (144), *suggesting that*: 1) endothelialization may not be a prerequisite for graft patency, and 2) anastomotic intimal hyperplasia within the adjacent grafted artery may occur in the absence of measurable graft thrombus accumulation or graft pannus tissue ingrowth.
(e) Do not share citations across paragraph (or subsection) boundaries. Instead, repeat the same citation in the first sentence of a paragraph following the paragraph with the shared citation.	

33

Case 8-2. Long paragraph with compensatory font and underscoring.

A.

Specific Aims [1]*Staphylococcus aureus* is a well-armed opportunistic pathogen that produces a diverse array of virulence factors and causes a correspondingly diverse array of infections.[2]The pathogenesis of *S. aureus* infections depends on the coordinately-regulated expression of two groups of virulence factors, one of which (surface proteins) allows the bacterium to evade phagocytes and colonize host tissues while the other (extracellular toxins and enzymes) promotes survival and multiplication at a localized site of infection. [3]Our long-term goal is to elucidate the regulatory mechanisms controlling expression of these virulence factors as a prerequisite to the development of therapeutic protocols that can be used to attenuate the disease process. [4]**The specific hypothesis behind the proposed research is that the staphylococcal accessory regulator (*sar*) is a major regulatory switch controlling expression of *S. aureus*virulence factors.** [5]The hypothesis is based on the following observations. [6]**First**, *sar* encodes a DNA-binding protein (SarA) required for expression of the *agr*-encoded RNAIII regulatory molecule (27). [7]The SarA-dependency of RNAIII expression is important because RNAIII module expression of many *S. aureus* virulence factors (29). [8]**Second**, phenotypic comparison of *sar* and *agr* mutants indicates that *sar* also regulates expression of certain *S. aureus* genes in an *agr*-independent manner 11, 21). [9]An example of particular relevance to this proposal is the *S. aureus* collagen adhesin gene (*cna*). [10]**Third**, mutation of *sar* results in reduced virulence in animal models of staphylococcal disease (8,10, 28). [11]Moreover, as anticipated based on the preceding discussion, *sar/agr* double mutants have reduced virulence even by comparison to *agr* mutants (8,24). [12]**Based on these observations, the experimental focus of this proposal is on the *sar* regulatory locus.** [13]The specific aims are designed to provide a comprehensive assessment of the *agr*-independent regulatory functions of *sar*.

B.

Experimental Approach. [1]Previous studies with *Salmonella* and *Shigella* have shown the enteric virulence correlates with intestinal invasion and cytopathicity (47). [2]Therefore, we will use a mouse primary intestinal epithelial (IE) cell line (mIE2) derived from the Immortomouse® (Dr. John Smith, Affiliation), and a mouse macrophage cell line (RAW267.2) to evaluate invasion and cell killing by *Bp* isolates. [3]These screens will be done using the 3 most and the 3 least virulent enteric *Bp* isolates identified above. [4]For the cell invasion assay, adherent IE or RAW cells in triplicate wells will be infected for 1h with Bp at an MOI of 5, then washed and incubated for 1 hour with 10 µg/ml ceftazidime to kill extracellular bacteria. [5]The cells will then be immediately lysed and numbers of intracellular bacteria quantitated, using techniques described previously (28). [6]The ability of different *Bp* strains to invade IE cells will be compared statistically using non-parametric ANOVA, and a similar analysis will be done for macrophage invasion. [7]The cytopathicity assay will be done using the same approach as above, except that cultures will be continued for an additional 24 hours following infection, with 10 µg/ml ceftazidime in the medium to suppress extracellular replication. [8]The number of viable IE or macrophage cells will be determined by MTT assay or by trypan blue exclusion.

Case 8-3. Comparison of layout styles. (A) Indented paragraphs, **(B)** Non-indented paragraphs, and **(C)** Double-spaced between non-indented paragraphs.

A. Total Length: 17 Lines
Introduction

The applicant appreciates the reviews on the first submission of this R21 application. He has read over the reviews carefully and has modified this submission to try to take into account the concerns. It is also clear that the "Fragilis" nomenclature for these proteins should be discontinued and the term "Ifitm" be used instead.

Critique 1 was uniformly laudatory, and the applicant appreciates the support. He has reversed the order of the specific aims as requested. The revision of the application was extensive due to the contraction of materials (12 to 6 pages) that none of it has been marked as "New" material.

Critiques 2 and 3 raised a number of weaknesses including poor scores for significance, innovation and approach. We had just taken possession of the Ifitm KO mouse at the time of submission and concerns were raised about its apparent lack of phenotype. The body of the document summarizes our findings but a recently published manuscript contains very significant data that highlights the importance of these proteins (1). This paper describes the results of a human Dharmacon siRNA library screen to test the effect of protein loss on Influenza A H1N1 virus infection of human osteosarcoma cells.

B. Total length: 16 lines
Introduction

The applicant appreciates the reviews on the first submission of this R21 application. He has read over the reviews carefully and has modified this submission to try to take into account the concerns. It is also clear that the "Fragilis" nomenclature for these proteins should be discontinued and the term "Ifitm" be used instead.

Critique 1 was uniformly laudatory, and the applicant appreciates the support. He has reversed the order of the specific aims as requested. The revision of the application was so extensive due to the contraction of materials (12 to 6 pages) that none of it has been marked as "New" material.

Critiques 2 and 3 raised a number of weaknesses including poor scores for significance, innovation and approach. We had just taken possession of the Ifitm KO mouse at the time of submission and concerns were raised about its apparent lack of phenotype. The body of the document summarizes our findings but a recently published manuscript contains very significant data that highlights the importance of these proteins (1). This paper describes the results of a human Dharmacon siRNA library screen to test the effect of protein loss on Influenza A H1N1 virus infection of human osteosarcoma cells.

C. Total length: 17 lines
Introduction

The applicant appreciates the reviews on the first submission of this R21 application. He has read over the reviews carefully and has modified this submission to try to take into account the concerns. It is also clear that the "Fragilis" nomenclature for these proteins should be discontinued and the term "Ifitm" be used instead.

Critique 1 was uniformly laudatory, and the applicant appreciates the support. He has reversed the order of the specific aims as requested. The revision of the application was so extensive due to the contraction of materials (12 to 6 pages) that none of it has been marked as "New"material.

Critiques 2 and 3 raised a number of weaknesses including poor scores for significance, innovation and approach. We had just taken possession of the Ifitm KO mouse at the time of submission and concerns were raised about its apparent lack of phenotype. The body of the document summarizes our findings but a recently published manuscript contains very significant data that highlights the importance of these proteins (1). This paper describes the results of a human Dharmacon siRNA library screen to test the effect of protein loss on Influenza A H1N1 virus infection of human osteosarcoma cells.

Case 8-4. Use of run-in headings to shorten the text. (A) All sections have standalone headings. **(B)** Section heading **3.2 Intonation model** is stand-alone; the other 2 headings are run-in.

A. Total Length: 27 Lines

3.2 Intonation model

An intonation model computes a representation of the melodic patterns in speech (the intonation contour), using a limited number of parameters obtained from an acoustic analysis of intonation contours of natural speech (ref). The acoustic parameter associated with intonation is the fundamental frequency (f_0), which corresponds to the opening and closing cycle of the vocal chords and, as such, is only present in voiced sounds. We will discuss 4 different intonation models: (i) the Tone and Break Indices model (TOBI; ref); (ii) the IPO perceptual model (ref); (iii) the Fujisaki model; and (iv) the generalized linear alignment model (ref).

i. The TOBI model.
　　The TOBI model (refs) is a linear intonation model that represents changes in the intonation contour by a series of *pitch accents* and *edge tones*. Pitch accents are used for emphasized words and edge tones mark phrase breaks. These tones are marked as being either High (H) or Low (L) or a combination of both. The drawback of the TOBI model is that there is no one-to-one mapping between a series of tone labels and an actual intonation contour; the labels do not reflect the actual intonation variations and timing as they would be observed in an intonation contour (ref).

ii. The IPO perceptual model.
　　The IPO model of intonation (ref) predicts intonation contours based on two assumptions: (i) anything that is not perceived by the human ear does not need to be modeled by an intonation model; (ii) the human ear perceives tone variations (rise/fall) and not actual tone values (high/low). The intonation contour is modeled by a grammar of allowable pitch movements that are represented by straight lines. The pitch range is fixed, restricting the accent peaks to within the allowed range, which can make the intonation sound very repetitive (ref). Furthermore, the reduction of an intonation contour into a sequence of straight lines is not straightforward and different annotators may come up with different stylizations.

B. Total Length: 25 Lines

3.2 Intonation model

An intonation model computes a representation of the melodic patterns in speech (the intonation contour), using a limited number of parameters obtained from an acoustic analysis of intonation contours of natural speech (ref). The acoustic parameter associated with intonation is the fundamental frequency (f_0), which corresponds to the opening and closing cycle of the vocal chords and, as such, is only present in voiced sounds. We will discuss 4 different intonation models: (i) the Tone and Break Indices model (TOBI; ref); (ii) the IPO perceptual model (ref); (iii) the Fujisaki model; and (iv) the generalized linear alignment model (ref).

i. The TOBI Model. The TOBI model (refs) is a linear intonation model that represents changes in the intonation contour by a series of *pitch accents* and *edge tones*. Pitch accents are used for emphasized words and edge tones mark phrase breaks. These tones are marked as being either High (H) or Low (L) or a combination of both. The drawback of the TOBI model is that there is no one-to-one mapping between a series of tone labels and an actual intonation contour (ref). The labels do not reflect the actual intonation variations and timing as would be observed in an intonation contour (ref.).

ii. The IPO perceptual model. The IPO model of intonation (ref) predicts intonation contours based on two assumptions: (i) anything that is not perceived by the human ear does not need to be modeled by an intonation model; (ii) the human ear perceives tone variations (rise/fall) and not actual tone values (high/low). The intonation contour is modeled by a grammar of allowable pitch movements that are represented by straight lines. The pitch range is fixed, restricting the accent peaks to within the allowed range, which can make the intonation sound very repetitive (ref). Furthermore, the reduction of an intonation contour into a sequence of straight lines is not straightforward and different annotators may come up with different stylizations.

Case 8-5. Use of a horizontal list to shorten the text. (A) Vertical and **(B)** Horizontal lists.

A. Total Length: 21 lines

Innovation

[1]The studies proposed here will elucidate, for the first time, the effects of hyperglycemia on ECs in an embryonic model of cardiac development – the chick heart – during early development. Innovations include:

1) [2]Development and characterization of in vivo embryonic chick models of chronic "constant" hyperglycemia and chronic "pulsed" hyperglycemia in order to study, for the first time, EC gene and protein expression and heart hemodynamics under excess glucose conditions during organogenesis;

2) [3]Development of a 3D, in vitro model of the developing embryonic heart to study ECs under flow conditions, with the ability to independently alter hemodynamic stress and hyperglycemic stress;

3) [4]Synergistic use of in vivo and in vitro models of heart development to better understand pathophysiological processes that underlie changes in embryonic ECs under hyperglycemic stress;

4) [5]Determination of the gene and protein expression patterns of hyperglycemia-induced EC dysfunction.

[6]Our preliminary research results attest to the ability of two highly experienced research laboratories (Drs. Smith and Jones), in collaboration with 2 experts in cardiac development and diabetes (Drs. Brown and Kelly) to form a research synergy that will allow us to tackle a major medical problem: the adverse impact of multiple factors involved in maternal hyperglycemia on embryonic heart development.

B. Total Length: 18 lines

Innovation

[1]The studies proposed here will elucidate, for the first time, the effects of hyperglycemia on ECs in an embryonic model of cardiac development – the chick heart – during early development. Innovations include: 1) [2]Development and characterization of in vivo embryonic chick models of chronic "constant" hyperglycemia and chronic "pulsed" hyperglycemia to study, for the first time, EC gene and protein expression and heart hemodynamics under excess glucose conditions during organogenesis;
2) [3]Development of a 3D, in vitro model of the developing embryonic heart to study ECs under flow conditions, with the ability to independently alter hemodynamic stress and hyperglycemic stress; 3) [4]Synergistic use of in vivo and in vitro models of heart development to better understand pathophysiological processes that underlie changes in embryonic ECs under hyperglycemic stress; and 4) [5]Determination of the gene and protein expression patterns of hyperglycemia-induced EC dysfunction.

[6]Our preliminary research results attest to the ability of two highly experienced research laboratories (Drs. Smith and Jones), in collaboration with 2 experts in cardiac development and diabetes (Drs. Brown and Kelly) to form a research synergy that will allow us to tackle a major medical problem: the adverse impact of multiple factors involved in maternal hyperglycemia on embryonic heart development.

Case 8-6. Sequential pages with and without a paragraph widow. (A) Paragraph widow on page 2: Last line of the last paragraph on page 1 is alone at the top of page 2. **(B)** Paragraph widow avoided: The second-to-last line from the bottom of page 1 was forced onto the top of page 2, resulting in an extra blank line at the bottom of page 1, two lines of text at the top of page 2, and one line less of text fitting onto page 2.

Case 8-6. Sequential pages with and without a paragraph widow. (A) Paragraph widow on page 2: Last line of the last paragraph on page 1 is alone at the top of page 2. (B) Paragraph widow avoided: The second-to-last line from the bottom of page 1 was forced onto the top of page 2, resulting in an extra blank line at the bottom of page 1, two lines of text at the top of page 2, and one line less of text fitting onto page 2.

A. Layout with a paragraph widow at the top of page 2

D2.2 Determine whether 5-HT can be measured in endosomal compartments.
    Nanometer-sized endosomal compartments contain internalized 5-HT; these compartments are not accessible for measurement by large or cell-intrusive probes. Our preliminary data indicated that a QD attached to receptors on the membrane may be able to detect low amounts of 5-HT since 5-HT-dependent QD quenching was observed inside cells even after wash-out of 5-HT from the extracellular medium (C.3). We will make 5-HT-dependent QD fluorescence quenching calibrations in cells and will use them to detect native 5-HT in the cell extracellular surface as well as in

nanometer-scale endosomes.

Procedures. To quantify 5-HT in native endosomal compartments, we will attach QD sensors to 5-HT receptors on the surface of CHO cells that have been treated with 5-HT to induce the internalization of QD-5-HT-5-HT receptor complexes into endosomes. We will then perform single QD fluorimetry assays on QDs inside cells to measure the amount and rate of QD fluorescence quenching. We estimate the amount of 5-HT in endosomes using fluorescence quenching calibration curves obtained in D2.3. As a control, we will repeat these same experiments but will omit treating the cells with 5-HT. We expect to find no 5-HT levels since these cells will contain QDs that are…

B. Layout without a paragraph widow on page 2

D2.2 Determine whether 5-HT can be measured in endosomal compartments.
    Nanometer-sized endosomal compartments contain internalized 5-HT; these compartments are not accessible for measurement by large or cell-intrusive probes. Our preliminary data indicated that a QD attached to receptors on the membrane may be able to detect low amounts of 5-HT since 5-HT-dependent QD quenching was observed inside cells even after wash-out of 5-HT from the extracellular medium (C.3). We will make 5-HT-dependent QD fluorescence quenching calibrations in cells and will use them to detect native 5-HT in the

cell extracellular surface as well as in nanometer-scale endosomes.

Procedures. To quantify 5-HT in native endosomal compartments, we will attach QD sensors to 5-HT receptors on the surface of CHO cells that have been treated with 5-HT to induce the internalization of QD-5-HT-5-HT receptor complexes into endosomes. We will then perform single QD fluorimetry assays on QDs inside cells to measure the amount and rate of QD fluorescence quenching. We estimate the amount of 5-HT in endosomes using fluorescence quenching calibration curves obtained in D2.3. As a control, we will repeat these same experiments but will omit treating the cells with 5-HT. We expect to find no 5-HT levels since…

Case 8-7. Redesigning visuals to shorten the narrative. (A) Passage with the original table. **(B)** Passage with a redesigned table to allow text wrapping and to shorten the text.

A. Total Length: 24 lines

... Aim 3 involves the refinement of clinical prognostic factors associated with treatment success. At the 2-year time point, we will use the data from the 180 subjects to examine their treatment success. This sample size provides a power of greater than 90% to detect a significant odds ratio of 2.5 if we take RUSI or trunk EMG data as clinical predictors of treatment success and 80% power to detect a significant odds ratio of 2.1. When restricted to those 90 subjects receiving the stabilization protocol, the sample size provides sufficient power to detect and odds ratio of 3.0 with a significance level of 0.05.

The estimated availability of people with a chronic LBP diagnosis from various UVM-associated clinics is listed in Table 1. The director of each site provided the census data for October 2010 – September 2011. If we can recruit 30% of these people, we will have sufficient numbers for our study.

Table 1. Census/Yield Estimates: Physical Therapy Clinics Associated with State University

PT Clinics Associated with UVM	Number of Patients with Diagnosis of Interest (Oct 2010 – Sept 2011)	Recruitment Yield of 30%
Jones Physical Therapy	200	60
University Hospital	75	39
E-Volve Physical Therapy	165	50
Kelly Health Care (KHC: PT Clinics – WWC/RCO)	900	270
KHC Primary Care Sites	2834	850
Estimated Total for 1 Year	**4349**	**1305**

B. Total Length: 19 lines

... Aim 3 involves the refinement of clinical prognostic factors associated with treatment success. At the 2-year time point, we will use the data from the180 subjects to examine their treatment success. This sample size provides a power of greater than 90% to detect a significant odds ratio of 2.5 if we take RUSI or trunk EMG data as clinical predictors of treatment success and 80% power to detect a significant odds ratio of 2.1. When restricted to those 90 subjects receiving the stabilization protocol, the sample size provides sufficient power to detect and odds ratio of 3.0 with a significance level of 0.05.

The estimated availability of people with a chronic LBP diagnosis from various UVM-associated clinics is listed in Table 1. The director of each site provided the census data for October 2010 – September 2011. If we can recruit 30% of these people, we will have sufficient numbers for our study.

Table 1. Census and Yield Estimates from UVM Clinics

Physical Therapy Clinics	Patients (10/10 – 12/11)	Recruitment Yield of 30%
Jones Physical Therapy	200	60
University Hospital	75	39
E-Volve Physical Therapy	165	50
Kelly Health Care	900	270
KHC Primary Care Sites	2834	850
Estimated Total, 1 Year	4349	1305

Case 8-8. Example of a paragraph revised to eliminate a short last line. **(A)** Paragraph with a short last line. Bold-italic text identifies the targets of revision in sentence 8, nominalizations. **(B)** Paragraph with the short last line eliminated. Bold-italic text identifies the text involved in the revision.

A. Paragraph with a Short Last Line (Total Length: 19 lines)

5. Repetitive Transcranial Magnetic Stimulation. [1]rTMS will be applied by a MagStim rapid rate device with a 70-mm, figure-eight, cooled-coil system to temporarily inhibit the SMA or dPMC. [2]After application of EMG electrodes to the skin, subjects will sit in a chair designed to restrain movement. [3]The Brainsight-Frameless system (ref) will provide the calibration of brain location within the skull, using each subject's MRI. [4]Single-pulse TMS will be applied at the primary motor cortex in a region suited to activate the tibialis anterior (usually 0-2 cm lateral, and 0.5 cm caudal to the vertex). [5]The optimal site of stimulation will be the location eliciting the largest amplitude motor evoked potential (MEP) at the tibialis anterior. [6]Once this region is located, the motor threshold will be systematically determined and defined by the lowest stimulation intensity that elicits 50 mV MEPs in 5 out of 10 trials to a resting TA. [7]rTMS stimulation to the SMA or dPMC will be administered at 80% of this threshold for 30 min at 1 Hz. [8]Stimulation location will be determined on-line using an obtained structural MRI scan interfaced by the Brainsight-Frameless system. [9]SMA stimulation will likely occur ***approximately*** 2 cm rostral to the vertex, and dPMC stimulation will be located ***approximately*** 2 cm rostral to the primary motor cortex hand region at the middle frontal gyrus. [10]MEP signals ***will undergo amplification***, using an A-M Systems Model 3000 AC/DC differential amplifier, and ***will undergo analysis*** with LabView software (National Instruments).

B. Paragraph Edited to Eliminate the Last Line (Total Length: 18 lines)

5. Repetitive Transcranial Magnetic Stimulation. [1]rTMS will be applied by a MagStim rapid rate device with a 70-mm, figure-eight, cooled coil system to temporarily inhibit the SMA or dPMC. [2]After application of EMG electrodes to the skin, subjects will sit in a chair designed to restrain movement. [3]The Brainsight-Frameless system (ref) will provide the calibration of brain location within the skull, using each subject's MRI. [4]Single-pulse TMS will be applied at the primary motor cortex in a region suited to activate the tibialis anterior (usually 0-2 cm lateral, and 0.5 cm caudal to the vertex). [5]The optimal site of stimulation will be the location eliciting the largest amplitude motor evoked potential (MEP) at the tibialis anterior. [6]Once this region is located, the motor threshold will be systematically determined and defined by the lowest stimulation intensity that elicits 50 mV MEPs in 5 out of 10 trials to a resting TA. [7]rTMS stimulation to the SMA or dPMC will be administered at 80% of this threshold for 30 min at 1 Hz. [8]Stimulation location will be determined on-line using an obtained structural MRI scan interfaced by the Brainsight-Frameless system. [9]SMA stimulation will likely occur ***about*** 2 cm rostral to the vertex, and dPMC stimulation will be located ***about*** 2 cm rostral to the primary motor cortex hand region at the middle frontal gyrus. [10]MEP signals ***will be amplified***, using an A-M Systems Model 3000 AC/DC differential amplifier and ***will be analyzed*** with LabView software (National Instruments).

Case 8-9. **Selective hyphenation to eliminate a line of text. (A)** A paragraph with white space at the end of line 12 and a short last line. **(B)** The terms *fluorescent* (in sentence 6) and *conformational* (in sentence 6) are hyphenated, resulting in automatic line adjustments that result in the elimination of the last line of the paragraph.

A. Total Length: 16 lines

Control Experiments. [1]The electrogenic properties of the fluorescent-labeled transporter will be characterized because the introduction of the cysteine and the subsequent fluorescent labeling of this cysteine can cause changes in the affinity and turn-over rate of the glutamate transporter. [2]The apparent affinity for sodium, potassium, protons, and glutamate will be measured and changes in these affinities will be taken into account when the effects of the different substrates are measured using fluorescent probes. [3]Radioactive-labeled glutamate uptake will also be measured for all the fluorescent-labeled transporters to ensure that the transporter protein still functions as a glutamate transporter. [4]Also, reverse uptake, stimulated by high extracellular K concentrations, will be tested to ensure that the transporter is not simply working in exchange mode. [5]We will also test for any effects of the substrates on the fluorescent probes by attaching fluorescent probes to lysine-coated microscope slides.
[6]**Fluorescence** from the probes will be tested in the same way as the fluorescent-labeled transporters to ensure that the induced fluorescence changes are due to the *conformational* change in transporters, not to the effects of the substrate on the fluorescent probe.

B. Total Length: 15 lines

Control Experiments. [1]The electrogenic properties of the fluorescent-labeled transporter will be characterized because the introduction of the cysteine and the subsequent fluorescent labeling of this cysteine can cause changes in the affinity and turn-over rate of the glutamate transporter. [2]The apparent affinity for sodium, potassium, protons, and glutamate will be measured and changes in these affinities will be taken into account when the effects of the different substrates are measured using fluorescent probes. [3]Radioactive-labeled glutamate uptake will also be measured for all the fluorescent-labeled transporters to ensure that the transporter protein still functions as a glutamate transporter. [4]Also, reverse uptake, stimulated by high extracellular K concentrations, will be tested to ensure that the transporter is not simply working in exchange mode. [5]We will also test for any effects of the substrates on the fluorescent probes by attaching fluorescent probes to lysine-coated microscope slides. [6]**Fluorescence** from the probes will be tested in the same way as the fluorescent-labeled transporters to ensure that the induced fluorescence changes are due to the **conformational** change in transporters, not to the effects of the substrate on the fluorescent probe.

Case 8-10. Visual with Its explanatory paragraph revised. (A) The explanatory paragraph has a short last line. **(B)** The words *task, in, a,* and *position* are omitted, eliminating the short last line and, thereby, adding an extra line of text in **(B).**

A.

... The examiner placed the transducer over the subject's L3 spinous process and moved it laterally about 2-3 cm until the L3/L4 facet joint was visualized (Figure 5). To control for the effects of respiration while the subject performed CAL, the examiner began RUSI at end of subject expiration. Subjects then performed a total of 4 trials on the right side and on the left side, for a total of 8 trials. In the first 2 trials, the subjects held nothing in their hands while raising their arms. For trials 3 and 4, they held a small hand weight (0.68 kg or 0.90 kg based on body weight) (ref), to provide resistance.

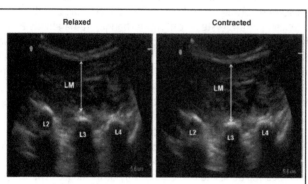

Figure 5. Ultrasound measurements of the lumbar multifidus (LM) thickness (double arrowhead) while relaxed or contracted during a contralateral arm-raising ~~task~~ with the subject ~~in a~~ prone ~~position~~. L2 = L2/L3 facet joint, L3 = L3/L4 facet joint, L4 = L4/L5 facet joint.

All images were saved for later analysis. Using programmable software, thickness of the LM at rest and during contraction was measured pre- and post-treatment (Figure 5). Thickness of LM at rest was subtracted from ...

34

B.

... The examiner placed the transducer over the subject's L3 spinous process and moved it laterally about 2-3 cm until the L3/L4 facet joint was visualized (Figure 5). To control for the effects of respiration while the subject performed CAL, the examiner began RUSI at end of subject expiration. Subjects then performed a total of 4 trials on the right side and on the left side, for a total of 8 trials. In the first 2 trials, the subjects held nothing in their hands while raising their arms. For trials 3 and 4, they held a small hand weight (0.68 kg or 0.90 kg based on body weight) (ref), to provide resistance. All images were saved for later analysis. Using programmable software, thickness of the LM at rest and during contraction was measured pre- and post-treatment (Figure 5). Thickness of LM at rest was subtracted from its thickness during contraction to obtain the change in LM thickness ...

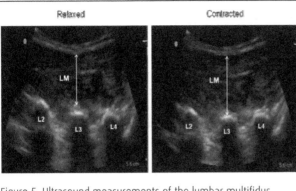

Figure 5. Ultrasound measurements of the lumbar multifidus (LM) thickness (double arrowhead) while relaxed or contracted during contralateral arm-raising with the subject prone. L2 = L2/L3 facet joint, L3 = L3/L4 facet joint, L4 = L4/L5 facet joint.

Case 8-11. Elimination of a forecast statement to save space.

C. Progress Report

[1]The main goal of the research conducted during the current grant period (April 2003–February 2007) has been to reveal the roles that female hormones, particularly estrogen, have on the eye and visual system. [2]The clinical emphasis of this current study has focused mainly on the effects of the breast cancer medications tamoxifen and anastrozole, with additional studies on related changes occurring across the menstrual cycle. [3]The 4 aims for the current grant period are: **(1)** to define the changes of visual function that occur during a standard 5-year treatment period of tamoxifen use, **(2)** to determine the extent and time course of tamoxifen-induced changes of ocular anatomy, **(3)** to distinguish the visual and ocular effects of raloxifene and anastrozole from those of tamoxifen, and **(4)** to determine the prevalence of cyclic changes of SWS-cone-mediated sensitivity across the menstrual cycle of healthy women. [4]Substantial progress has been made on all 4 aims, with some emphases changing. [5]Because of new trends in adjuvant care, and because the requested 5-year grant period was reduced to 3 years 11 months, less emphasis was placed on measuring longitudinal change than originally planned. [6]Instead, we pursued new lines of investigation that emerged from our results or that reflected evolving trends in health care, and that could be investigated productively within the shortened time period.

[7]~~This Progress Report is organized around the original 4 Aims, with an additional section to report results that relate to or are derived from several aims, collectively.~~

Glossary

Abstract – A prose section of text that precedes the narrative and summarizes key information from the narrative; the abstract is sometimes called the *project description*, *project summary*, or *summary*.

Active grammar – A clause or sentence with the agent as the subject of an action verb. See **Passive grammar**.

Action verb – A verb that identifies a task or activity that someone or something will perform in order to execute the proposed procedures if funding is received. The action verb in a procedural sentence is **will + infinitive verb** for sentences in active grammar and **will + be + past participle verb** for sentences in passive grammar.

Agent – The entity (person or thing) that executes a verb action; the agent is sometimes called the *actor*.

Agentless passive grammar – A clause or sentence that is in passive grammar and that does not mention the agent in a prepositional phrase. See **Passive grammar**.

Aim – What the PI intends to accomplish with methods that are being proposed. See also **Objective** and **Long-term goal**.

Analysis – In terms of **scientific argumentation**, the PI's educated opinion about a particular topic or research area. See **Scientific argumentation**.

Ancillary procedures – Procedures that will be executed to further investigate anticipated results.

Bolstering – A review that follows and supports an analysis in a *review-analysis-review* relationship.

Broad objective – See **Long-term goal**.

Citation – A number or an author's name, corresponding to an entry in a list of references, placed in parentheses and integrated into a sentence, to indicate the source of information within the sentence.

Citation placeholder – A symbol, such as ** or XX, that is inserted into the text to indicate where a citation is needed.

Citation trigger – A term or phrase, such as **has been established**, that indicates background information and that requires a citation for the information.

Claim – The PI's understanding of what data represent and how to interpret them.

Classic definition – A definition that explains the meaning of a term by identifying its class and features that distinguish it, in terms of function, type, and/or structure, from other members of its class.

Co-investigator (Co-I) – Researcher who helps the PI design the proposed research, write the narrative, and execute the research.

Collaborator – Researcher who performs a part of the proposed research.

Common knowledge – Information that is shared and agreed upon by a particular group.

Compensatory font – A style of font (e.g., bold italics) distinct from that used for narrative text; used to help readers attend to information in long paragraphs. See **Emphatic font**.

Competitive resubmission – A grant proposal submitted to a funding agency that had previously denied funding to a previous version of the same grant proposal.

Connection – An explanation about how the proposed research is relevant to the reviewed research, in terms of the proposed research purpose, the proposed methods, and/or the proposed outcomes. See **Scientific argumentation**.

Contractor – Researcher, engineer, or other professional who is hired on contract to perform a portion of the research and who invoices to receive payment.

Contribution – A unit of text that someone other than the PI drafts for the narrative.

Core procedural information – Required information that comprises procedural descriptions, namely the verb action comprising the procedure (*what*, *which*, and *how*), the **SOS** or **MET** that will be manipulated or affected by the verb action, and the manner in which the verb action will be executed. See **Procedural verbs, Procedures**, and **Non-core procedural information**.

Count noun – A noun that can be pluralized and that indicates its plurality through a plural ending, such as + **s** (*experiment, experiments*) or + **es** (*dish, dishes*). See **Non-count noun**.

Credibility – A characteristic of the PI in terms of how believable the PI is to reviewers.

Cross-reference – A statement that directs readers to another (sub)section in the narrative or grant application.

Customized MET – Materials, equipment, and/or tools that the PI designs and possibly manufactures, for a particular research purpose. See **Standard MET**.

Definitional synonym – A synonym that is used to define a key term. See also **Interpretive synonym** and **Classic definition**.

Descending importance – An organizational pattern in which items in a list are sequenced from the most to the least important.

Emphatic font – Font that is a variation of the font used for the main text in the narrative, the function of which is to help direct the reader's attention to particular information; also termed **Compensatory font**.

Ending – The last paragraph of a section or the very end of the narrative that briefly summarizes the key information or indicates the next step after completion of the proposed research.

Exclusion criteria – Features of SOS that prevent the SOS from being considered for participation in a study. In combination with inclusion criteria, exclusion criteria reflect the PI's attempt to control variables that can influence the collection and interpretation of data. See **Inclusion criteria** and **subjects, objects of study, and specimens**.

Explanatory paragraph – A paragraph located after the title to a visual, that describes significant features of the visual.

Extension statement – A type of analysis in scientific argumentation that identifies how proposed research builds upon strengths of previous research.

Figure – A visual that presents its information in a display other than in columns and rows. See **Table**.

Flush-left or flush-right justification – The alignment of text on a line so that the first word (flush-left) or the last word (flush-right) is always the same distance from the edge of the page as the other lines of text. See **Ragged-left or ragged-right justification**.

Focus position – In terms of a paragraph, the locations in a paragraph where white space helps draw the reader's attention – especially the first couple of sentences of a paragraph, and the last sentence or 2 of a paragraph.

Forecast statement – A phrase or sentence(s), usually located in the introduction of a (sub)section, indicating where particular information is located in the upcoming (sub)section.

Formal introduction – An introduction with a heading.

Funding agency – A public or private organization that provides a researcher's institution with money for a defined period of time in order for the researcher to conduct research.

Generalization – A statement about a phenomenon that is constant over time. See **Precise generalization**.

Generic heading – The name of a (sub)section that is phrased in standard methodological terms. See **Unique heading**.

Gerund – A type of noun that is derived from a verb and that ends in + **ing**. See **Nominalization**.

Given information – Content that has been mentioned before a particular point in the text or can be assumed from the context with certainty. See also **Information structure** and **New information**.

Goal – See **Long-term goal**.

Grammatical articles – The terms *a*, *an*, and *the*.

Grammatical parallelism – The repetition of grammar across each item in a list.

Grammatical subject – The term that controls the verb in a clause.

Heading outline – A vertical list of the intended headings for (sub)sections of a narrative, composed before drafting the narrative in order to guide the drafting of the narrative.

Hypothesis – A type of informed speculation about a specific, tentative, unproven observation; a type of analysis in scientific argumentation that typically expresses a predictor-outcome or impactor-impactee relationship. When used as a subsection heading, **Hypothesis** refers to a rationale subsection that includes the proposed hypothesis for the upcoming (sub)section. See **Rationale**.

Identifier – A short phrase or clause at the beginning of a statement that clarifies whether the statement is a hypothesis, predictor, or problem statement.

Inclusion criteria – Features of SOS that are required for them to be considered for participation in a study. In combination with exclusion criteria, inclusion criteria reflect the PI's attempt to control variables that could influence the collection and interpretation of data. See **Exclusion criteria** and **subjects, objects of study, and specimens**.

Informal introduction – An introduction without a heading.

Information structure – Different patterns of content arrangement in sentences, paragraphs, and (sub)sections. See **Given information** and **New information**.

Informed consent – A written explanation provided to a subject, which the subject or his/her legally appointed guardian needs to sign; the explanation covers the purpose of the study, the subject's type of involvement, risks to the subject, and the subject's rights as a subject, including the right to stop participating in the study whenever he/she so desires.

Interpretive synonym – A term that summarizes or explains a preceding concept or textual passage. See **Definitional synonym**.

Introduction – The first section to the narrative; sometimes called the **Aims Section** or the **Specific Aims Section**.

Introduction to a Resubmission – The first section of the narrative to a grant proposal that is being resubmitted, given a previous, unsuccessful submission. This type of introduction precedes the **Aims Section** and explains how the resubmission is significantly different from the first, unsuccessful submission, and how the resubmission is responsive to the reviewers' criticisms of the first submission.

Itemization scheme – The pattern of letters and/or numbers used to mark the sequence of (sub)sections or items in a list.

Key term – A term that represents an important, recurrent topic in the narrative; the term that a classic definition defines. See **Classic definition**.

Layout – The spatial arrangement of text and symbols on the page, in relation to white space.

Lead-in – The part of a sentence that immediately precedes a list, identifies what the list is about and, optionally, how many items are in the list.

Legend information – A list of abbreviations, colors, and symbols used on the face of a visual, that explains what each means.

Linking verb – A verb that equates a grammatical subject to the grammatical predicate. Examples of linking verbs are the verbs *be*, *appear*, and *seem*.

Long-term goal – The purpose of a line of research, for example, what the PI wants to accomplish in a specific line of research in 15 to 20 years. Some funding agencies call the long-term goal the *broad objective*. See **Aim** and **Objective**.

Maintenance procedures – Strategies that the PI will use in order to control variables in the environment from the time the SOS enter the study until data are collected from them.

Mass noun – See **Non-count noun.**

Materials, equipment, and/or tools (MET) – The tangible and non-tangible objects that investigators use to collect and to analyze data. See **MET.**

MET – An abbreviation for materials, equipment, and/or tools.

Narrative – The major prose sections of a scientific grant proposal, including the abstract.

New information – Information that has not been mentioned before a particular point in the text and cannot be assumed from the context. See **Information structure** and **Given information.**

Nominalization – A noun that is derived from a verb or an adjective, and that bears a distinctive ending, such as **+tion, +ment, +ence,** and **+ity.** See **Gerund.**

Non-core procedural information – In descriptions of procedures, non-core information includes **who** will execute the procedures, **when** they will be executed, **where** they will be executed, and **why** they will be executed. See **Procedural verbs, Procedures, Core procedural information.**

Non-count noun – A noun that is pluralized through a measurement unit or through a composite part, such as *2 molecules of water* and *grains of salt.* See **Count noun.**

Novelty – A characteristic of the proposed research design, proposed methods, and/or anticipated results, in terms of how original each is.

Objective – What the PI intends to accomplish in the methods proposed for a particular research project. See **Aim** and **Long-term goal.**

Organizational hierarchy – The relationship among sections and subsections of a document.

Outcome – In the context of a narrative to a grant proposal, the major result(s) of the proposed research and/or of the proposed methods.

Overview visuals – Tables and figures that indicate the relationship among the proposed aims and methods, and the time schedule for executing the methods.

Parallelism – The same grammar that is used in each item of a list.

Participants – Humans who are studied in a scientific, medical, and technical investigation; participants are also termed *subjects.*

Passive grammar – A clause or sentence with another element besides the agent (e.g., the undergoer or instrument) as the subject of an action verb; the action verb is formed with the verb *be* and the *past participle* of the action verb; the agent is sometimes included in a prepositional phrase. See **Active grammar**; see **Agentless passive grammar.**

Personal communication – Information, the source of which is from a person who has not published the information in a journal and has not presented it at a professional conference.

Precise generalization – A statement that describes a specific feature of a phenomenon that is constant over time. See **Generalization.**

Prediction – A type of analysis in scientific argumentation that expresses an educated guess about the results of proposed research methods.

Preliminary research – Research that the PI or a Co-I has conducted, that is related to the proposed research in some important respect.

Previous research – Research that has preceded the proposed research and has likely been published.

Primary features of methods – Characteristics defining a scientific, medical, or technological approach or experiment, as determined by experts knowledgeable in currently accepted scientific and technical methodologies, practices, and theories.

Principal investigator (PI) – The researcher who is responsible for the content of a grant proposal; for submitting it to a funding agency on time; for directing, developing, and executing the research on time; and for writing most of the narrative.

Problem statement – A type of analysis in scientific argumentation that involves identification of a shortcoming or an omission in previous research.

Procedural verb – A verb that indicates the task, activity, or action that an investigator performs in order to collect, share, or analyze data. See **Procedures**, **Core procedural information**, and **Non-core procedural information**.

Procedures – The tasks, activities, and actions that investigators perform in order to collect, share, and analyze data. Procedures are described in chronological order, in the order in which investigators intend to begin them upon funding. See **Procedural verbs**, **Core procedural information**, and **Non-core procedural information**.

Protocol – A defined set of routinized procedures to be executed in order to perform research. See **Procedures**.

Published information – Information that has been made available to the public and scientists, in professional journals, on the internet, at professional meetings, or in collections from professional meetings.

Ragged-left or ragged-right justification – The alignment of text on a line so that the first word (ragged-left) or the last word (ragged-right) is a different distance from the edge of the page relative to the other lines of text, based on the actual number of characters that the words and spaces use in each line. See **Flush-left or Flush-right justification**.

Rationale – A type of background subsection that precedes a **Methods** (sub)section and justifies particular methods in the (sub)section; its content comprises a scientific argument. See **Scientific argumentation**.

Readability – The mental and physical effort that a reader needs to exert in order to readily understand the text. A text with high readability is easy to follow and understand on a first reading; a text with low readability requires concentrated reading, mental effort, and possibly guessing in order to understand the text on the first or first few readings.

Relevance – A characteristic of the proposed research in terms of whether it is appropriately targeted to the type of research that a particular funding agency funds.

Research design – Critical features of proposed research that are described sufficiently clearly and completely so that the reviewers can theoretically replicate the experiment.

Research purpose – What the PI intends the proposed research will accomplish in terms of the long-term goal, research objective, and/or aims.

Research team – Principal investigator, co-principal investigator(s), co-investigator(s), program director, co-program director(s), collaborator(s), technician(s), contractor(s), administrator(s), and student(s) who will be involved in executing and facilitating the proposed research methods.

Restrictive adjective clause – A clause that modifies a noun and that provides information that limits or defines the term.

Resubmission – A grant proposal that a PI submits to a funding agency, after the funding agency has previously not funded it.

Resubmission Introduction – The first section of a narrative to a resubmission, in which the PI identifies major changes in the resubmission and responds to reviewers' criticisms of the initial, non-funded submission.

Resultant paragraph – A relatively long paragraph created by combining 2 or more short paragraphs.

Retention procedures – Strategies to keep subjects participating in a study until they are no longer needed.

Review-analysis-connection relationship – In scientific argumentation, a background passage that reviews preliminary or previous research and then analyzes the reviewed research; and that then connects the analysis to the proposed research in terms of the proposed methods, proposed purpose, and/or proposed outcome.

Review-analysis pair – In scientific argumentation, a passage that reviews previous or preliminary research and then analyzes the research.

Review-analysis-review relationship – In scientific argumentation, a passage that reviews research and then analyzes the reviewed research; and then presents a second review of literature in order to support the analysis and its logical relationship to the first review. See **Bolstering**.

Review of literature – Description of previous research that relates to the proposed research. See **Scientific argumentation**.

Run-in heading – A heading to a (sub)section that is followed by text on the same line. See **Stand-alone heading**.

Scientific argumentation – Particular content that describes the proposed research and clarifies the novelty and significance of the research in order to convince reviewers that the proposed research reflects features of sound scientific methodology, should contribute in novel and significant ways to scientific knowledge, and should receive a fundable score. The particular categories of content comprising a scientific argument are: topic, review, analysis, connection, and significance. See **TRACS**.

Scientific grant proposal – A request for funds to conduct original research in a discipline or in a subject area typically associated with science, technology, or medicine.

Scope – The text that a term or grammatical element influences.

Sentence capitalization – A style used for headings and titles, in which the first word is capitalized and all subsequent words (except for proper nouns) are not.

Sentence definition – See **Classic definition**.

Significance – Importance of research in terms of, for example, science, medicine, societal values, or national security; or in terms of broader impacts and technical merit. See **Scientific argumentation**.

Significance by association – Information that is impliedly important, based on other information in the immediate paragraph or subsection. See **Significance**.

SOS – An abbreviation for subjects, objects of study, and/or specimens.

Space-worthy – A feature of a unit of writing, in which the content is important enough to justify the amount of space that is used to discuss the content.

Specific aim – See **Aim**.

Speculation – A statement that expresses uncertainty or a guess about something under scientific investigation; a type of analysis in scientific argumentation.

Stand-alone heading – A heading to a (sub)section that has no text following it on the line. See **Run-in heading**.

Standard MET – Materials, equipment, and/or tools that are commonly used for a particular purpose in scientific, technical, and medical studies. See **Customized MET**.

Standard term – A vocabulary item, the meaning of which has been accepted by professionals in a discipline.

Subjects – Humans who are studied in scientific, medical, and technical investigation; subjects are also termed *participants*.

Subjects, objects of study, and specimens (SOS) – Entities that the PI uses as the target of his/her data-collection procedures.

Submission requirements – Content, organization, and layout specifications that a funding agency issues for a narrative.

Subsections – Categories of content that comprise a section and that are given headings that identify the information in the categories.

Summary analysis – An analysis statement that leads a **Background or Preliminary Studies** subsection and represents all of the review-analysis relationships in the subsection. See **Scientific argumentation**.

Table – A visual that presents its information in columns and rows. See **Figure**.

Technical question – A type of analysis in scientific argumentation, that poses a question about information that is not yet known.

Termination procedures – Methods describing what the PI will do with objects of study (primarily vertebrate animals) when the PI no longer needs them in the study.

Topic – A specific concept or question that is being investigated or discussed. See **Scientific argumentation**.

TRACS – An acronym for the content that comprises a scientific argument: topic, review, analysis, connection, and significance. See **Scientific argumentation**.

Unique heading – The name of a (sub)section that is phrased to reflect the substantive content in the (sub)section. See **Generic heading**.

Unity – A feature of a narrative, in which many different types of information complement each other, resulting in a narrative that is consistent and singular in purpose.

Up-to-date information – Information that represents – from the reviewers' perspective – the latest, reliable, and valid findings on a particular topic.

Index